Quarks, Hadrons, and Nuclei

Proceedings of the 16th and 17th Annual
Hampton University Graduate Studies (HUGS) Summer Schools on

Quarks, Hadrons, and Nuclei

16th Annual HUGS : June 11 – 29, 2001
&
17th Annual HUGS : June 3 – 21, 2002

Newport News, Virginia, USA

Editors

J. Goity

C. Keppel

G. Prezeau

Hampton University, USA

World Scientific

NEW JERSEY • LONDON • SINGAPORE • SHANGHAI • HONG KONG • TAIPEI • CHENNAI

Published by

World Scientific Publishing Co. Pte. Ltd.

5 Toh Tuck Link, Singapore 596224

USA office: Suite 202, 1060 Main Street, River Edge, NJ 07661

UK office: 57 Shelton Street, Covent Garden, London WC2H 9HE

British Library Cataloguing-in-Publication Data
A catalogue record for this book is available from the British Library.

QUARKS, HADRONS AND NUCLEI
**Proceedings of the Sixteenth and Seventeenth Annual Hampton University Graduate
Studies (HUGS)**

ISBN 981-238-804-4

Printed in Singapore by World Scientific Printers (S) Pte Ltd

2001 LECTURERS

Ricardo Alarcon	Arizona State University
Gordon Cates	University of Virginia
Robert Edwards	Thomas Jefferson National Accelerator Facility
Naomi Makins	University of Illinois at Urbana-Champaign
Michael Ramsey-Musolf	University of Connecticut, California Institute of Technology
Adam Szczepaniak	Indiana University

2001 SEMINAR SPEAKERS

Joe Mitchell	Jefferson Lab, Newport News, VA

2001 LIST OF PARTICIPANTS

Mattias Andersen	University of Lund, Lund Sweden
Fatiha Benmokhtar	Rutgers University, Camden, NJ
Robert Bradford	Carnegie Mellon University, Pittsburgh, PA
Areg Danagoulian	University of Illinois at Urbana-Champaign
Derek Glazier	University of Glasgow, Glasgow, Scotland
Mikhail Gorchetin	Institut für Kernphysik, Universität Mainz, Germany
Hassan Ibrahim	Old Dominion University, Norfolk, VA
Mihajlo Kornicer	University of Connecticut, Storrs, CT
Ludyvine Morand	CEA Saclay, France
Retief Neveling	University of Stellenbosch, South Africa
Salvador Rodriguez	George Washington University, Washington, DC
Michael Roedelbronn	University of Illinois at Urbana-Champaign
Jeffery Secrest	College of William and Mary, Williamsburg, VA
Karl Slifer	Temple University, Philadelphia, PA

2002 LECTURERS

Andrei Belitsky	University of Maryland, College Park, MD
Frank Close	University of Oxford, Oxford, UK
Gregg Franklin	Carnegie Mellon University, Pittsburgh, PA
Ron Gilman	Rutgers University, Piscataway, NJ, and Jefferson Lab, Newport News, VA
Sabine Jeschonnek	Ohio State University, Lima, OH
Patricia Rankin	University of Colorado at Boulder, Boulder, CO

2002 SEMINAR SPEAKERS

Deirdre Black	Jefferson Lab, Newport News, VA
David Gaskell	Jefferson Lab, Newport News, VA
Ioana Niculescu	Jefferson Lab, Newport News, VA
Carlos Schat	Jefferson Lab, Newport News, VA

2002 LIST OF PARTICIPANTS

Cornel Butuceanu — College of William and Mary, Williamsburg, VA

Alexandre Camsonne — University of Clement Ferrand, Aubiere, France

Benjamin Clasie — Massachusetts Institute of Technology, Cambridge, MA

Silviu Corvig — California Institute of Technology, Pasadena, CA

Martin DeWitt — North Carolina State University, Raleigh, NC

Kurniawan Foe — George Washington University, Washington, DC

David Hayes — Old Dominion University, Norfolk, VA

Lisa Kaufman — College of William and Mary, Williamsburg, VA

Jason Lenoble — Institut de Physique Nucléaire, Orsay, France

Dirk Merten — University of Bonn, Bonn, Germany

Duncan Middleton — Glasgow University, Glasgow Scotland

Yuriy Mishchenko — North Carolina State University, Raleigh, NC

Issam Qattan — Northwestern University, Evanstown, IL

Jörg Reinnarth — University of Bonn, Bonn, Germany

Gianfranco Sacco — University of Connecticut, Farmington, CT

Nikolai Savvinov — University of Maryland, College Park, MD

Steven Sheets — North Carolina State University, Raleigh, NC

Hassan Yuksel — University of Wisconsin, Madison, WI

Lingyan Zhu — Massachusetts Institute of Technology, Cambridge MA

PREFACE

"In the sciences, we are now uniquely privileged to sit side by side with the giants on whose shoulders we stand."
        – Gerald Holton, Mallinckrodt Professor of Physics, Harvard University

This volume contains contributions from lectures presented at the Sixteenth and Seventeenth Annual Hampton University Graduate Studies at the Continuous Electron Beam Accelerator Facility (HUGS at CEBAF) Summer Schools, both of which took place in Newport News, VA, USA at the Thomas Jefferson National Accelerator Facility. The Sixteenth HUGS at CEBAF was held from June 11 – 29, 2001, and the Seventeenth was held June 3 – 21, 2002. The HUGS summer school brings pedagogical lectures to graduate students who are working on doctoral theses in nuclear physics. The school has a balance of theory and experiment, and lecturers address topics of high current interest in strong interaction physics, particularly in electron scattering.

HUGS lecturers may lead major experimental efforts, or be internationally renown for contributions to the field. Regardless, they share one thing in common – their willingness to volunteer their time to educate and interest the next generation. We are extremely grateful to every HUGS lecturer, for the outstanding lectures, the engaging atmosphere that they created, and for giving so generously of their time. We are especially grateful to those who contributed as well to this volume.

The contributions included here are from Ricardo Alarcon, Gregg Franklin, David Gaskell, Ronald Gilman, Sabine Jeschonnek, Michael Ramsey-Musolf, Patricia Rankin, and Adam Szczepaniak. I periodically receive requests from HUGS students many years out for old proceedings, and have learned that these volumes are studied and utilized for years to come. Let me offer here a hearty thank you again from the HUGS organization, as well as from the students, to all the names just listed for providing this valuable resource for all of us.

The Sixteenth and Seventeenth Annual HUGS at CEBAF were made possible by others as well. The Department of Energy under grant number DE-FG05-87FR40380 provided key support. The school also received, as it has for years, invaluable support in many ways from Jefferson Lab. The suggestions of the International Advisory Committee helped shape the program, and the key efforts of the Local Organizing Committee assured its success.

Special thanks to Susan Ewing and Mary Fox for their organizational skills and unsparing effort at all stages. A special thanks to Kimberly Wilson who supported the school administratively in a variety of ways, and without whom this volume would not be possible. Finally, special recognition is deserved for Robert Williams, "Mr. HUGS", who has served as the key daily contact point for students throughout the HUGS school, morning through evening and into the weekends. The school would not be the same without him.

Cynthia E. Keppel, Ph.D.
HUGS at CEBAF Director

CONTENTS

Lectures

Lectures

ELECTRON SCATTERING FROM FEW-BODY NUCLEI

RICARDO ALARCON

Arizona State University,
Tempe, AZ 85287-1504 USA
ralarcon@asu.edu

KARL SLIFER

Temple University,
Philadelphia, PA 19122, USA
kslifer@temple.edu

Contents

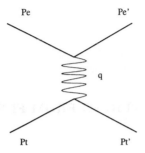

Figure 1: 1st order electron scattering

1 Introduction

Electron scattering is one of the oldest and most effective methods to examine the structure of the nucleus and constituent nucleons. The study of the few-body nuclei, in particular the A=2 and A=3 systems, has been a central theme in the field of intermediate nuclear physics for the last several decades. Important advances have been achieved in both experiment and theory. We intend here to give a comprehensive review of measurements in the field of electron scattering from few-body nuclei. The goal is to give the student as much experimental information as possible. The paper is organized as follows. A brief introduction to the formalism of electron scattering is given in Sec. 2. Section 3 is a very short introduction to the properties of the nucleon-nucleon interaction. In Sec. 4 we summarize the present formalism to calculate the nuclear electromagnetic current operator. The experimental results of the last 20 years or so, are presented in Sections 5 and 6 for the deuteron (A=2) and the A=3 systems, respectively.

2 Electron Scattering

The utility of electron scattering derives from two related facts: the weakness of the electromagnetic interaction and the availability of a precise theory, Quantum Electrodynamics (QED), with which we can calculate reaction effects. This is an advantage over strongly interacting probes since the target can be examined without greatly disturbing its structure. The relevant process is the lowest order contributing diagram shown in Fig. 1. It represents the one-photon exchange plane wave Born approximation to electron scattering.

The cross section for the scattering of electrons from nuclei, as represented in Fig. 1, is proportional to the product of two second-rank tensors [1,2,3], i.e.,

$$d\sigma \propto \eta_{\mu\nu} W_{\mu\nu} \tag{1}$$

where $\eta_{\mu\nu}$ and $W_{\mu\nu}$ are referred as the leptonic and hadronic tensors, respectively. The $\eta_{\mu\nu}$ tensor contains all the information regarding the exchange of the virtual photon and it is known accurately from QED even beyond first order. On the other hand, the hadronic tensor cannot yet be calculated from the fundamental theory of the strong interaction, Quantum Chromodynamics (QCD). The principal limitation is that nuclear forces are not yet quantitatively understood from QCD. Fortunately, many realistic models have been produced by fitting the available two-nucleon scattering data (see Section 3). These models can in principle be used to construct the hadronic tensor (see Section 4), and they are essential when we are studying few-body nuclei.

To obtain the cross section which corresponds to the actual experimental conditions (polarized electron and/or target, inclusive or exclusive scattering, etc.) one performs the appropriate average-over-initial and sum-over-final states. As a starting point we consider inclusive (e, e') scattering from nuclei for a transition from an initial nuclear state $|J_i >$ to a final nuclear state $|J_f >$. The cross section can be written in terms of projections which are longitudinal (L) or transverse (T) with respect to the momentum transfer $\vec{q}$, i.e.,

$$\frac{d\sigma}{d\Omega} = 4\pi\sigma_M f_{rec}^{-1} \left[v_L F_L^2(\vec{q}) + v_T F_T^2(\vec{q}) \right] \tag{2}$$

where σ_M is the Mott cross section, f_{rec}^{-1} is a recoil factor, and the v's are kinematic factors. The longitudinal $F_L^2(\vec{q})$ and transverse $F_T^2(\vec{q})$ form factors are expressed in terms of reduced matrix elements of Coulomb, Electric, and Magnetic multipole operators as

$$F_L^2(|\vec{q}|) = \frac{1}{2J_i + 1} \sum_{J=0} | < J_f || T_J^C(|\vec{q}|) || J_i > |^2 \tag{3}$$

$$F_T^2(|\vec{q}|) = \frac{1}{2J_i + 1} \sum_{J=1} \left[| < J_f || T_J^E(|\vec{q}|) || J_i > |^2 + | < J_f || T_J^M(|\vec{q}|) || J_i > |^2 \right] \tag{4}$$

The above reduced matrix elements are related to the matrix elements of the Fourier transforms of the nuclear charge and current densities, $\rho(\vec{q})$ and $\vec{j}(\vec{q})$. The properties of the multipole operators under parity and time-reversal invariance lead to selection rules. For example, elastic scattering can only be induced by even-J Coulomb and odd-J magnetic multipole operators. Thus, for spin-0 nuclei, we directly measure the Fourier transform of the spherically symmetric ground state charge distribution.

3 The NN Interaction

The NN interaction plays a fundamental role in the structure and dynamics of nuclei and, in particular, few-body nuclei. The effect of many-body forces (like 3-body forces) is very small and it will not be considered here. The most prominent empirical features of the nuclear force are: a finite range, an intermediate range attraction, a repulsive core, the existence of a tensor force, and a spin-orbit force. In 1935, by constructing a strict analogy to QED, Yukawa derived a force of finite range through the idea of massive-particle exchange. A comprehensive review to the NN interaction can be found in Ref. [4]. Intense theoretical and experimental work over the last few decades have yielded several models of the NN interaction. A great success of these models is the accurate reproduction of several key experimental properties of the deuteron:

- the asymptotic constants A_S (the S-wave normalization) and η (the D/S state ratio), which govern the wave function at large distances.

- the quadrupole moment Q_d, the magnetic moment μ_d, and the radius r_d.

A general feature of these models is that the long-range part of the NN interaction is due to one-pion exchange (OPE). In fact, at distances comparable to the inverse of the pion mass ($1/m_\pi \approx 1.4$ fm) OPE leads to a large tensor component in the NN interaction. This in turn gives rise to a mixture of spin and spatial degrees of freedom, a feature that makes nuclear systems, especially few-body systems, quite unique with respect to other physical systems like molecules or atoms where the dominant interaction (Coulomb) is independent of the particle's internal quantum numbers. For a modern approach to the structural effects due to the NN interaction see Ref. [5].

4 The Nuclear Electromagnetic Current Operator

The cornerstone of electron scattering is the proper construction of the nuclear electromagnetic current operator which defines the hadronic tensor. In what follows we outline the main steps and we refer the reader to the excellent review of J. Carlson and R. Schiavilla (Ref. [16]) for more details. The nuclear electromagnetic $\rho(\vec{q})$ and $\vec{j}(\vec{q})$ current operators are expanded into a sum of one- and two-body terms that operate on the nucleon degrees of freedom. The one-body operators are obtained from the covariant single-nucleon current

$$j^\mu = \overline{u}(\vec{p'}) \left[F_1(Q^2)\gamma^\mu + F_2(Q^2)\frac{i\sigma^{\mu\nu}q_\nu}{2m} \right] u(\vec{p}) \qquad (5)$$

where $\vec{p}$ and $\vec{p'}$ are the initial and final momenta, respectively, of a nucleon of mass m, and $F_1(Q^2)$ and $F_2(Q^2)$ are its Dirac and Pauli form factors taken as a function of the four-momentum transfer $Q^2 = -q_\mu q^\mu \geq 0$, with $q_\mu = p'_\mu - p_\mu$. The Bjorken and Drell convention is used for the γ matrices, and $\sigma^{\mu\nu} = (i/2)[\gamma^\mu, \gamma^\nu]$.

In order to model the composite structure of the proton and the neutron, electric G_E and Magnetic G_M form factors are introduced which are related to the Dirac and Pauli form factors in Eq. 5 via

$$G_E(Q^2) = F_1(Q^2) - \frac{Q^2}{4m^2} F_2(Q^2), \tag{6}$$

$$G_M(Q^2) = F_1(Q^2) + F_2(Q^2) \tag{7}$$

and are normalized so that

$$G_E^S(Q^2 = 0) = G_E^V(Q^2 = 0) = 1, \tag{8}$$

$$G_M^S(Q^2 = 0) = \mu_p + \mu_n = 0.880\mu_N, \tag{9}$$

$$G_M^V(Q^2 = 0) = \mu_p - \mu_n = 4.706\mu_N, \tag{10}$$

where μ_p and μ_n are the magnetic moments of the proton and neutron in terms of the nuclear magneton μ_N, and the S and V superscripts denote, respectively, isoscalar and isovector combinations of the proton and neutron electric and magnetic form factors. Further constraints on these form factors are imposed by gauge invariance.

The two-body operators fall roughly into three groups. First, two-body current operators that are required by gauge invariance and they are constructed directly from the NN interaction and contain no free parameters. Meson-exchange current operators arise as a consequence of the elimination of mesonic degrees of freedom from the nuclear state vector. Second, two-body currents which, being purely transverse, are not constrained by the continuity equation. To this class belong the currents associated with the $\rho\pi\gamma$ and $\omega\pi\gamma$ mechanisms, as well as those due to excitation of intermediate Δ-isobar resonances. Finally, there are two-body charge operators which are model dependent and may be viewed as relativistic corrections.

5   Deuteron Results

The deuteron represents the simplest system of interacting nucleons and, as such, offers a unique opportunity to examine NN interactions, meson and isobar degrees of freedom, as well as providing a method to examine the neutron in the absence of free neutron targets.

5.1   Unpolarized Elastic Scattering

Assuming single photon exchange, the electron-deuteron unpolarized elastic differential cross section can be written in the Rosenbluth form

$$\frac{d\sigma}{d\Omega} = \sigma_M \left[A(Q^2) + B(Q^2)tan^2(\frac{\theta}{2}) \right] \tag{11}$$

Here $A(Q^2)$ and $B(Q^2)$ are two structure functions that parameterize the deviation of a spin-1 nucleus from the point-like behavior of the simple Mott cross section

$$\sigma_M = \frac{\alpha^2 E' \cos^2(\theta/2)}{4E^3 \sin^4(\theta/2)} \tag{12}$$

where α is the fine-structure constant, θ is the electron scattering angle, and $E(E')$ is the electron incident (scattered) energy. $Q^2 = 4EE' \sin^2(\theta/2)$ is the four-momentum transfer of the virtual photon.

The electric and magnetic structure functions $A(Q^2)$ and $B(Q^2)$ can be decomposed in terms of the charge monopole, quadrupole, and magnetic dipole form factors $G_C(Q^2)$, $G_Q(Q^2)$, $G_M(Q^2)$

$$A(Q^2) = G_C{}^2(Q^2) + \frac{8}{9}\tau^2 G_Q{}^2(Q^2) + \frac{2}{3}\tau G_M{}^2(Q^2) \tag{13}$$

$$B(Q^2) = \frac{4}{3}\tau(1+\tau)G_M{}^2(Q^2) \tag{14}$$

where $\tau = Q^2/4M_D^2$ and M_D is the mass of the deuteron.

In the nonrelativistic limit, the three form factors are related to the distribution of the charge, the quadrupole deformation and the magnetization. In the limit $Q \to 0$

$$eG_C(0) = e \tag{15}$$

$$G_Q(0) = M_D^2 Q_D \tag{16}$$

$$G_M(0) = \frac{M_D}{M_P}\mu_D \tag{17}$$

where Q_D, μ_D are the electric quadrupole and magnetic dipole of the deuteron, while M_P is the mass of the proton.

An industrious experimental effort has been undertaken over the last 30 years at facilities such as SLAC[a], SACLAY[b], BATES[c], and JLAB[d] to measure the deuteron elastic form factors. Two experimental techniques are widely used. First, a Rosenbluth separation can be performed by measuring the cross section of Eq. 11 as a function of $\tan^2(\theta/2)$ while holding Q^2 constant. Alternatively, it can be seen that the cross section is dominated at large backward angle by the magnetic structure function $B(Q^2)$. Eq. 14 reveals that scattering at $180°$ directly measures the magnetic form factor $G_M(Q^2)$.

We will now examine examples of each method. In SLAC experiment NE4[6], electrons scattered near $180°$ were detected in coincidence with deuterons recoiling at $0°$. The cross section was dominated by the magnetic contribution, and $B(Q^2)$ was measured in this way over a Q^2 range of $1.20 \leq Q^2 \leq 2.77$ $(\text{GeV/c})^2$.

A liquid deuterium cryotarget was used as a source of deuterons. Fig. 2 details the experimental setup. To detect backscattered electrons at $180°$, a beam chicane was necessary to divert the incident beam. The quadrupole triplet of Q1-3 provided focusing for both the incident and scattered electrons, while B4 diverted the scattered electrons to the electron spectrometer. Downstream from the target, another quadrupole triplet focused incident electrons and recoil deuterons or protons on dipole B5, which deflected electrons to a beam dump and nuclei to the recoil spectrometer. Bremsstrahlung photons originating in the target were absorbed in a beam dump downstream of the target.

The momentum distributions of coincident electron-deuteron events reveals an excess of events below the elastic peak. Fig. 3 shows the missing momentum $(\delta_{mm} = \delta^e + \delta^r)$ spectrum for e-d coincidences. Here δ^e and δ^r are the percentage momentum deviations from the elastic peak settings in each spectrometer. Elastic e-d scattering results in the peak centered at $\delta_{mm} = 0$. Two possible multistep reactions were identified that could produce the large low momentum backgrounds. The first reaction is $\gamma d \rightarrow \pi^0$ at $180°$. The π^0 can subsequently decay into two photons, one of which pair produces. The resulting electron is then detected in coincidence with the recoiling deuteron.

[a] Stanford Linear Accelerator Center
[b] Centre d'Etudes Nucleaires de Saclay
[c] M.I.T. Bates Linear Accelerator Center
[d] Thomas Jefferson National Accelerator Facility

Compton scattering, $\gamma d \to \gamma d$, can also result in pair production of an electron at $180°$.

Subtraction of the background allows extraction of the magnetic structure function $B(Q^2)$ of the deuteron. Fig. 4 shows the result from this experiment (labeled SLAC-90) along with several models. The data exhibits a clear diffraction minimum and secondary maximum, and joins smoothly to the previous measurements at lower Q^2.

Next we examine experiment E91-026 [7] at Jefferson Lab. In this experiment, a Rosenbluth separation was performed by measuring coincidence electron-deuteron scattering from a liquid deuterium target. Fig. 5 shows a schematic overview of Hall A where the experiment was performed. Electrons of incident energy between 3.2 to 4.4 GeV enter from the left. They pass through a series of beam position monitors (BPM) and beam current monitors (BCM) before impinging on the target. A raster is used to widen the incident beam to distribute the heat load and reduce temperature dependent density effects. The two spectrometers contain similar magnetic sections configured in opposite polarity to detect negative or positive particles. They consist of three quadrupoles for focusing and a bending dipole. The electron detector package contains trigger scintillators, a series of wire chambers for momentum determination, a gas Čerenkov for electron-pion separation, and a Pb-glass calorimeter for further particle identification. In the hadron spectrometer, there was a similar set of wire chambers and two scintillator planes. The scintillators provided the trigger and Time-of-Flight measurements so that coincidences could be established between the two arms.

The electric structure function $A(Q^2)$ was extracted from the measured e-d cross section over the broad Q^2 range of $0.7 \leq Q^2 \leq 6.0$ $(GeV/c)^2$ under the assumption that $B(Q^2)$ does not contribute to the cross section in these kinematics [6,8].

The Hall A results are shown in figs. 6 and 7. At low Q^2 the data resolves a long standing discrepancy between earlier experiments. The JLAB results confirm previous SLAC measurements, which are in disagreement with the Bonn and Saclay data. Comparison with Impulse Approximation (IA) calculations clearly indicate the need to include Meson Exchange Currents (MEC) [e].

A similar experiment [9] that ran in JLAB Hall C published results concurrently with the HALL A results.

---

[e] For more information on IA and MEC, see discussion in sec. 5.3

5.2 Polarized Elastic Scattering

We have seen that the deuteron structure is described by three form factors; G_C, G_Q, and G_M. Measurements of $A(Q^2)$ and $B(Q^2)$ are used to separate the electric and magnetic contributions to the cross section, but that leaves the problem of completely separating the three form factors under-determined. By introducing polarization degrees of freedom, we can further disentangle the electric monopole and quadrupole terms. In elastic e-d scattering, measurements of the tensor polarization of recoil deuterons, or measurements of the asymmetries induced by a tensor polarized target are used to gain access to a third scattering observable. This chapter examines both of these experimental methods.

Recoil Polarimetry

As a spin 1 object, the deuteron possesses both a vector polarization

$$t_{10} = \sqrt{\frac{3}{2}} \left(\frac{N_+ + N_-}{N_+ + N_0 + N_-} \right) \tag{18}$$

and a tensor polarization [14].

$$t_{20} = \frac{1}{\sqrt{2}} \left(\frac{N_+ + N_- - 2N_0}{N_+ + N_0 + N_-} \right) \tag{19}$$

Here N_j is the number of deuteron atoms in a macroscopic sample that are in each of the three magnetic substates. We see that the tensor polarization is limited to the range $-\sqrt{2} \le t_{20} \le 1/\sqrt{2}$.

t_{20} along with the other tensor observables t_{21}, t_{22} can be expressed in terms of the electric and magnetic form factors of eqs. 15 - 17.

$$t_{20} = -\frac{1}{\sqrt{2}}S \left[\frac{8}{3}\tau G_C G_Q + \frac{8}{9}\tau^2 G_Q{}^2 \right.$$
$$\left. + \frac{1}{3}\tau \left(1 + 2(1+\tau) \tan^2 \frac{\theta_e}{2} \right) G_M{}^2 \right] \tag{20}$$

$$t_{21} = \frac{2}{\sqrt{3}S}\tau \left(\tau + \tau^2 \sin^2 \frac{\theta_e}{2} \right)^{1/2} G_M G_Q \sec \frac{\theta_e}{2} \tag{21}$$

$$t_{22} = -\frac{1}{2\sqrt{3}S}\tau G_M{}^2 \tag{22}$$

where

$$S = A(Q^2) + B(Q^2) \tan^2 \frac{\theta}{2} \tag{23}$$

t_{22} offers no new information, as it depends only on $G_M{}^2(Q^2)$. t_{21} gives an independent measure of $G_Q(Q^2)$ but is expected to be small because of its linear dependence on $G_M(Q^2)$. That leaves t_{20} as the most convenient tensor observable to measure experimentally.

In experiments which measure the outgoing polarization of recoil deuterons, such as those performed at Bates[11] and JLAB, t_{20} describes the likelihood of the recoil deuteron being detected in one of the magnetic substates. Experiment E94-018[15] at Jefferson Lab Hall C measured the charge form factor and tensor moments t_{20}, t_{21}, t_{22}. Unpolarized electrons were scattered off unpolarized deuterons. The scattered electrons and recoil deuterons were detected in coincidence. The scattered deuterons were then channeled to a polarimeter where they underwent secondary scattering from a liquid hydrogen target. Angular asymmetries measured in the charge exchange reaction $^1H(\vec{d},2p)n$ depend on the tensor polarization of the incident deuteron. The measured angular distribution $N(\theta, \phi)$ of the protons can be expressed in terms of t_{20}, t_{21}, t_{22}. These observables are extracted from the data by fitting the measured angular distribution to the predicted expression.

The first measurement[10] of t_{20} was performed at $70°$, so it is customary to extrapolate experimental results to this angle. The resulting value for $t_{20}(70°)$, t_{21}, and t_{22} are shown in figs. 8 - 9. The extracted values of G_C and G_Q are shown in fig. 10.

Note: The quantity $\tilde{t}_{20}$ is defined by neglecting the magnetic contribution to t_{20}.

$$\tilde{t}_{20} = -\sqrt{2}\frac{x(x+2)}{1+2x^2} \tag{24}$$

$$x = \frac{2}{3}\tau G_Q/G_C$$

$\tilde{t}_{20}$ is independent of θ and at low Q^2, $\tilde{t}_{20} \simeq t_{20}$. In all measurements to date, $\tilde{t}_{20}$, $t_{20}(70°)$, and t_{20} have not differed greatly[11], so $\tilde{t}_{20}$ is often used in comparisons with theoretical predictions.

Polarized Internal Targets

Elastic electron scattering from a polarized deuteron target is described in terms of the three analyzing powers T_{20}, T_{21}, T_{22}.

$$\sigma = \sigma_0 \left[1 + \tfrac{1}{\sqrt{2}} P_{zz} \left(\frac{3\cos^2\theta^* - 1}{2} T_{20} \right.\right.$$
$$-\sqrt{\frac{3}{2}} \sin 2\theta^* \cos\phi^* T_{21}$$
$$\left.\left. +\sqrt{\frac{3}{2}} \sin^2\theta^* \cos 2\phi^* T_{22} \right) \right] \tag{25}$$

The tensor analyzing powers are formally equivalent to the three tensor moments t_{20}, t_{21}, t_{22} already discussed. Here, σ_0 refers to the unpolarized cross section, and θ^*, ϕ^* are the spherical angles in the frame where z is aligned with the virtual photon momentum $\vec{q}$. $P_{zz} = n_+ + n_- - 2n_0$, where n_+, n_-, n_0 are the populations of each of the deuteron magnetic sublevels. In polarized target experiments, such as those performed at Novosibirsk[f], and NIKHEF[g], T_{20} describes the probability of scattering from each of the magnetic substates $m_z = -1,0,1$.

The 91-12 collaboration [12,13] measured T_{20} and T_{22} at the Amsterdam Pulse Stretcher Ring at NIKHEF using a polarized internal gas target. An atomic beam source was used to inject nuclear-polarized deuterium atoms into a windowless T-shaped target cell. A fraction of the target gas was sampled in a Breit-Rabi polarimeter. The tensor polarization of the deuterium atoms was also monitored with an ion-extraction system. The target setup is shown in fig. 11. Scattered electrons were detected using a combination of trigger scintillators, an electromagnetic calorimeter and a pair of wire chambers. Recoil deuterons were detected using multiple layers of thin plastic scintillator in conjunction with a pair of wire chambers for track reconstruction. Coincidence events were selected using the timing between the two arms and tracking information provided by the wire chambers.

From the scattering data, the following asymmetry was formed

$$A_d^T = \sqrt{2} \frac{N^+ - N^-}{P_{zz}^+ N^- - P_{zz}^- N^+} \tag{26}$$

[f] VEPP-3, Novosibirsk
[g] The National Institute for Nuclear Physics and High Energy Physics, Amsterdam

where N^+ (N^-) is the number of events detected when the target polarization is positive (negative). From eq. 25 we see that this asymmetry can be expressed in terms of T_{20}, $A(Q^2)$, and $B(Q^2)$. T_{20} and G_C were extracted from A_d^T using previous measurements of $A(Q^2)$, and $B(Q^2)$. The results are shown in fig. 12.

5.3 Threshold electrodisintegration

In the Impulse Approximation (IA), the scattering of an electron from a target nucleus is modeled as a simple interaction between incident electron and a single free nucleon. Interactions between individual nucleons are neglected. In simple systems like the deuteron, Impulse Approximation cross sections can be calculated quite reliably. Any contribution not due to IA can then be isolated unambiguously.

As a first step beyond the Impulse Approximation, interactions between nucleons can be modeled as being mediated by light mesons such as the π and ρ. The predictions using Meson Exchange Currents(MEC) prove to be quite reliable despite their apparent simplicity. When the distance between nucleons is large, interactions are well described in terms of mesonic degrees of freedom as opposed to quark degrees of freedom. It is to be expected though, that as the interaction distance is reduced, the mesonic theory will break down.

Threshold electrodisintegration of the deuteron provides a useful experimental test of the Impulse Approximation and MEC. The main component of the cross section for backward electrodisintegration of the deuteron near threshold is the magnetic dipole (M1) transition between the bound state and the (T=1) $^1S_0$ scattering state. The deuteron wave function has two components: the s-wave $^3S_1$ and d-wave $^3D_1$. The M1 transition is proportional to the sum of the transitions($^3S_1$-$^1S_0$) and ($^3D_1$-$^1D_0$). These two transitions destructively interfere in the vicinity of 12 fm^{-2}. In this region, the reaction becomes extremely sensitive to non-nucleonic degrees of freedom.

Fig. 13 shows the cross section for the inclusive reaction $D(e, e')pn$ at backward angle. Data from several experiments are compared to IA results, and to models that include two-body currents. The Impulse Approximation results, IA(v_{18}), show clear disagreement from the data even at low Q^2. On the other hand, mesonic theory performs quite well up to the large Q^2 value of 28 fm^{-2}, beyond its expected limit of validity.

5.4 Unpolarized Exclusive Scattering

The reaction D(e,e'p)n is used to examine the spectral function of the deuteron. In the plane wave impulse approximation (PWIA) the spectral function is related to the cross section by:

$$\frac{d^6\sigma}{d\omega d\Omega_e d\Omega_p dT_p} = \kappa \cdot \sigma_{ep} \cdot S(E, p_i) \tag{27}$$

where κ is a kinematic factor, σ_{ep} refers to electron scattering from a moving (bound) proton, and $S(E, p_i)$ describes the probability of finding the proton with separation energy E and initial momentum p_i.

In the one-photon exchange limit, the above cross section is written as:

$$\frac{d^6\sigma}{d\omega d\Omega_e d\Omega_p dT_p} = \sigma_{Mott}(v_L R_L + v_T R_T$$
$$+ v_{LT} R_{LT} \cos\phi + v_{TT} R_{TT} \cos 2\phi) \tag{28}$$

where the R_i, and R_{ij} are response functions containing the spectral function and the v's are kinematic factors.

There is a direct connection between the $S(E, p_i)$ and the nucleon momentum distribution inside the nucleus:

$$n(\vec{p}) = \int S(E, \vec{p}) dE \tag{29}$$

In the PWIA, the reduced cross section, defined by

$$\sigma_{red} = \frac{\sigma_{exp}}{f \cdot \kappa \cdot \sigma_{cc1}}$$

is identical to the momentum distribution. Here κ is a kinematic factor, f is a recoil factor, and σ_{cc1} is the e-p off-shell cross section of de Forest [33].

Fig. 14 shows reduced cross sections from a recent experiment [34] at Mainz. The results agree well with the PWIA prediction (using the Paris potential) up to a recoil momentum of about 350 MeV/c. For larger values of p_m the PWIA underestimates the cross section considerably.

Figs. 15 and 16 compare the cross section to predictions by Arenhövel and Schiavilla. Both models show the necessity of including meson exchange currents and isobar currents. The main difference between the two models is Arenhövel's dynamic treatment of isobar currents.

5.5   Polarized Exclusive Scattering

When both the incident electron and the target deuteron are polarized, the differential cross section [35] for electrodisintegration is given by the somewhat intimidating expression[h]:

$$\frac{d^3\sigma}{dk_2 d\Omega_e d\Omega_{np}^{c.m.}} = c \left[\rho_L f_L + \rho_T f_T + \rho_{LT} f_{LT} \cos\phi + \rho_{TT} f_{TT} \cos 2\phi + h\rho'_{LT} f'_{LT} \sin\phi \right.$$

$$+ P_1^d \left((\rho_L f_L^{11} + \rho_T f_T^{11}) d_{10}^1(\theta_d) \sin(\phi - \phi_d) \right.$$

$$\left. + \sum_{M=-1}^{1} (\rho_{LT} f_{LT}^{1M} \sin\xi_M + \rho_{TT} f_{TT}^{1M} \sin\psi_M) d_{M0}^1(\theta_d) \right)$$

$$+ P_2^d \left(\sum_{M=0}^{2} (\rho_L f_L^{2M} + \rho_T f_T^{2M}) d_{M0}^2(\theta_d) \cos[M(\phi - \phi_d)] \right.$$

$$\left. + \sum_{M=-2}^{2} (\rho_{LT} f_{LT}^{2M} \cos\xi_M + \rho_{TT} f_{TT}^{2M} \cos\psi_M) d_{M0}^2(\theta_d) \right)$$

$$+ h P_1^d \left(\rho'_T \sum_{M=0}^{1} f_T^{'1M} \cos[M(\phi - \phi_d)] d_{M0}^1(\theta_d) \right.$$

$$\left. + \rho'_{LT} \sum_{M=-1}^{1} f_{LT}^{'1M} \cos\xi_M d_{M0}^1(\theta_d) \right)$$

$$+ h P_2^d \left(\rho'_T \sum_{M=0}^{1} f_T^{'2M} \sin[M(\phi - \phi_d)] d_{M0}^2(\theta_d) \right.$$

$$\left. \left. + \rho'_{LT} \sum_{M=-2}^{2} f_{LT}^{'2M} \sin\xi_M d_{M0}^2(\theta_d) \right) \right] \tag{30}$$

$$= S(h, P_1^d, P_2^d) \tag{31}$$

with

$$\xi_M = M(\phi - \phi_d) + \phi$$
$$\psi = M(\phi - \phi_d) + 2\phi$$

---

[h]The previously examined cross section expressions of Eqs. 11, 25, and 28, are specific examples of this general expression.

Here c is proportional to the Mott cross section, and h refers to the electron helicity(± 1). The ρ are obtained from the electron vertex and are simple functions of the electron kinematic variables. The angles ϕ and θ (see fig. 17) refer to the spherical angles with respect to $\vec{q}$. The deuteron target is characterized by vector and tensor polarization parameters P_1^d and P_2^d.

The f are the response functions in the center-of-mass system, and contain all the nuclear structure information. Each of the structure functions are sensitive to different aspects of the interaction. The "fifth response function", f_{LT}', in particular, vanishes if the cross section is driven by a single amplitude, as in the case of the IA description of quasifree scattering. Hence, f_{LT}' can provide a measure of contributions to the cross section that are not included in the impulse approximation.

It is possible to rewrite the cross section in terms of the various beam, target, and beam-target asymmetries as follows:

$$S(h, P_1^d, P_2^d) = S_0[1 + P_1^d A_d^V + P_2^d A_d^T + h(A_e + P_1^d A_{ed}^V + P_2^d A_{ed}^T)] \quad (32)$$

where S_0 refers to the completely unpolarized cross section S(0,0,0). Experimental measurement of these asymmetries allows extraction of the structure functions.

Polarized Beam

In the simpler case of polarized beam incident on unpolarized target the P_j^d vanish and the cross section of Eq. 30 reduces to the more tractable form:

$$S(h, 0, 0) = c[\rho_L f_L + \rho_T f_T + \rho_{LT} f_{LT} \cos \phi$$
$$+ \rho_{TT} f_{TT} \cos 2\phi + h\rho_{LT}' f_{LT}' \sin \phi] \quad (33)$$

$$= S_0[1 + hA_e] \quad (34)$$

We see that the asymmetry

$$A_e = \frac{d\sigma_+ - d\sigma_-}{d\sigma_+ + d\sigma_-} \quad (35)$$

(where $\pm$ refers to the helicity of the incident electron beam), is proportional to the fifth structure function. The presence of the factor $\sin \phi$ in equation 33 indicates that the measurement must be made out of the scattering plane. Such measurements have only recently become possible experimentally. The first measurement [36] of f_{LT}' was performed at Bates. The results indicate that rescattering effects, which are ignored in PWIA, are important at the kinematic studied ($Q^2 = 3.3$ fm^{-2}).

A second experiment [37] at Bates measured f_{LT}, f'_{LT}, and f_{TT}.

Polarized Target

The presence of D-state components into the predominantly S-wave ground-state wave function leads to a density distribution that depends on the spin projection m_z. The deuteron exhibits a toroidal shape for $m_z = 0$ and a dumb-bell shape for $m_z = \pm1$ [38]. The spin-dependent momentum density distribution ρ_{m_z} can be written as [39]

$$\frac{\rho_0(\mathbf{p})}{4\pi} = [R_0 + \sqrt{2}R_2 d_{0,0}^2(\theta)]^2 + 3[R_2 d_{1,0}^2(\theta)]^2 \tag{36}$$

$$\frac{\rho_\pm(\mathbf{p})}{4\pi} = \left[R_0 - \frac{1}{\sqrt{2}}R_2 d_{0,0}^2(\theta)\right]^2 + \frac{9}{8}[R_2^2(1 - \cos^4\theta)]$$

with $R_0(p)$ and $R_2(p)$ the radial wave functions for $L = 0$ and $L = 2$ in momentum space. The $d_{m,m'}^j$ are the rotation functions and θ is the angle between the polarization axis and the relative momentum $\vec{p}$ of the two nucleons.

A recent experiment [39] performed the first measurement of quasi-elastic electron scattering from a polarized deuterium target in order to measure the tensor analyzing power A_d^T (see also eq. 26):

$$A_d^T = \sqrt{\frac{1}{2}}\frac{\sigma_+(p_m) + \sigma_-(p_m) - 2\sigma_0(p_m)}{\sigma_+(p_m) + \sigma_-(p_m) + \sigma_0(p_m)}$$

$$\stackrel{PWIA}{=} -\frac{2R_0(p)R_2(p) + \sqrt{\frac{1}{2}}R_2^2(p)}{R_0^2(p) + R_2^2(p)}d_{0,0}^2(\theta) \tag{37}$$

with $\sigma_0(\sigma_\pm)$ the cross section measured with $m_z = 0$ ($m_z = \pm1$) and $d_{0,0}^2 = \frac{3}{2}\cos^2\theta - \frac{1}{2}$.

Measurement of A_d^T gives access to R_0, and R_2 and hence to the momentum density distributions of eq. 37. Fig. 18 shows A_d^T as a function of missing momentum for approximately parallel kinematics. In the PWIA, the asymmetry vanishes if there is no D-state component to the ground state wave function. The sizable value of A_d^T indicates that the tensor analyzing power is dominated by the effects of the D-state contribution.

6   A=3 Results

The A=3 ground-state wave functions can be accurately calculated from a realistic NN interaction, including a consistent treatment of mesonic degrees of freedom. Properties of the A=3 systems like the binding energies, charge radii and electromagnetic form factors have been computed with great accuracy by using either numerical solutions of the Faddeev equations, or variational methods. There is also worldwide interest in polarized ^{3}He as an effective neutron target. We review below the experimental evidence on unpolarized electron scattering from the A=3 systems.

6.1   Unpolarized Elastic Scattering

Experimental studies of the electromagnetic structure of ^{3}He and ^{3}H started more than 30 years ago and by the end of the past century all the trinucleon form factors were accurately known up to momentum transfers q^2 of the order of 30 fm^{-2}. An excellent review of the data was compiled by Amroun et al. [40]. They performed a combined analysis of the world data for these nuclei yielding a complete experimental information of the trinucleon electromagnetic form factors, and also providing a separation into the isocalar and isovector form factors.

The experimental results for the magnetic form factors of ^{3}H and ^{3}He from Amroun et al. [40] are shown as the shaded area in Figs. 19 and 20, respectively. These results are compared with several theoretical calculations as shown. The main conclusion is that while the measured ^{3}H magnetic form factor is in excellent agreement with theory over a wide range of momentum transfers, there is a significant discrepancy between the measured and calculated values of the ^{3}He magnetic form factor in the region of the diffraction minimum.

Recently, a new measurement [42] of the ^{3}He magnetic form factor has been performed at Bates that has extended the data up to a momentum transfer of about 42 fm^{-2}. Fig. 21 shows the experimental spectra for all the beam energies used in the experiment. The extracted values for the magnetic form factor are displayed in Fig. 22 in conjunction with the previous world data and theoretical calculations. The very small cross sections, as low as 10^{-40} cm^2/sr, required the use of a high-pressure cryogenic gas target and a detector system with excellent background rejection capability. No existing theoretical calculation satisfactorily accounts for all the available data.

On the other hand, the charge form factors for the A=3 nuclei are in excellent agreement with the experimental data. In Figs. 23 and 24 the calculated ^{3}H and ^{3}He charge form factors are compared with the experimental data.

Clearly seen is the important role of the two-body charge operator contributions above 3 fm^{-1}.

6.2 Short-range correlations

The total integrated strength of the longitudinal response function $R_L(q,\omega)$ measured in inclusive electron scattering (the so-called Coulomb sum rule $S_L(q)$), is defined as

$$S_L(q) = \frac{1}{Z} \int_{\omega_{el}^+}^{\infty} d\omega S_L(q,\omega) \tag{38}$$

$$S_L(q,\omega) = \frac{R_L(q,\omega)}{|G_{E,p}(q,\omega)|^2} \tag{39}$$

where $G_{E,p}$ is the proton electric form factor and ω_{el}^+ is the energy of the recoiling $A_{nucleon}$ system with Z-protons (the lower integration limit excludes the elastic electron-nucleus contribution). It has long been known that $S_L(q)$ is related to the Fourier transform of the proton-proton distribution function,

$$S_L(q) = \frac{1}{Z}\langle 0|\rho_L^\dagger(\mathbf{q})\rho_L(\mathbf{q})|0\rangle - \frac{1}{Z}|\langle 0|\rho_L(\mathbf{q})|0\rangle|^2$$
$$\equiv 1 + \rho_{LL}(q) - Z|F_L(q)|^2 \tag{40}$$

where $|0\rangle$ is the ground state of the nucleus, $\rho_L(\mathbf{q})$ is the nuclear charge operator, $F_L(q)$ is the longitudinal form factor (divided by $G_{E,p}(q,\omega_{el})$) normalized as $F_L(q=0) = 1$, and a longitudinal-longitudinal distribution function has been defined as

$$\rho_{LL}(q) \equiv \frac{1}{Z} \int \frac{d\Omega_q}{4\pi} \langle 0|\rho_L^\dagger(\mathbf{q})\rho_L(\mathbf{q})|0\rangle - 1. \tag{41}$$

If relativistic corrections and two-body contributions to the nuclear charge operator are neglected, then $\rho_L(\mathbf{q})$ (divided by the electric proton form factor) is simply given by

$$\rho_L(\mathbf{q}) \simeq \sum_{i=1,A} e^{i\mathbf{q}\cdot\mathbf{r}_i}(1+\tau_{z,i})/2 \tag{42}$$

and the resulting longitudinal-longitudinal distribution function can be written as

$$\rho_{LL}(q) = \int d\mathbf{r}_1 d\mathbf{r}_2 j_o(q|\mathbf{r}_1 - \mathbf{r}_2|)\rho_{LL}(\mathbf{r}_1, \mathbf{r}_2) \tag{43}$$

with

$$\rho_{LL}(\mathbf{r}_1, \mathbf{r}_2) = \frac{1}{4\pi} \sum_{i \neq j} \langle 0 | \delta(\mathbf{r}_1 - \mathbf{r}_i)\delta(\mathbf{r}_2 - \mathbf{r}_j)(1 + \tau_{z,i})$$
$$\times (1 + \tau_{z,j}) | 0 \rangle \tag{44}$$

$$\int d\mathbf{r}_1 d\mathbf{r}_2 \rho_{LL}(\mathbf{r}_1, \mathbf{r}_2) = Z - 1. \tag{45}$$

Note that, within this approximation and in the large-momentum-transfer limit, the longitudinal cross section is due to the incoherent contributions from the Z protons. In this case, the longitudinal-longitudinal distribution function gives the probability of finding two protons at positions $\mathbf{r}_1$ and $\mathbf{r}_2$, regardless of their spin-projection states. Such a quantity is, therefore, sensitive to the short-range correlations induced by the repulsive core of the NN interaction [43]. The extracted experimental values for ^{3}He[44] and ^{4}He[45,46] are shown in Figs. 25 and 26, respectively. The large error bars on the experimental $\rho_{LL}(q)$ reflect predominantly systematic uncertainties associated with the tail contribution. The agreement between the experimental analyses and the results of calculations in which both one and two-body terms are included in $\rho_L(\mathbf{q})$ is rather good.

6.3 Unpolarized Exclusive Scattering

As an example we present the recent studies of the reaction $^3He(e, e'p)$ carried out by Florizone et al.[41]. Fig. 27 shows the cross section data and extracted momentum distribution for the two-body breakup process compared to various calculations. Fig. 28 shows the experimentally extracted momentum distributions for the continuum, integrated over the missing energy range $E_m = 7 - 20$ MeV. Previous results for this reaction are discussed in Ref.[41].

7 Summary

The purpose of these lectures was to address the physics of the few-body nuclei from the point of view of electron scattering. The formalism for electron scattering was reviewed as well as most of the experimental evidence for the A=2 and A=3 systems. Enormous progress has been achieved for the last 3 decades. Theoretically, these systems can be precisely calculated over a wide range of momentum transfers. Precise experimental data also extends to a wide range of momentum transfers. One of the best references to this rich field is the recent review by Carlson and Schiavilla [16], which we have used extensively in this paper. Ongoing studies in few-nucleon physics are focused on the inclusion of spin degrees of freedom using polarized beams and targets, or related techniques. One of the future goals is to obtain precise information connected with small amplitudes of the ground-state wave function.

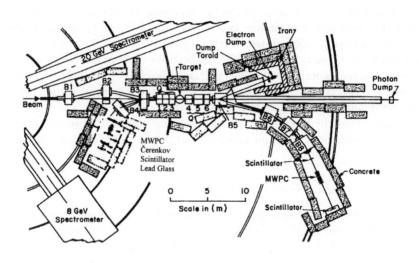

Figure 2: A schematic diagram of the double-arm spectrometer system. The elements B1-B8 are dipole bending magnets and Q1-Q6 are quadrupoles. Also shown are the detector packages for the two spectrometers and the extensive shielding. Reproduced from Bosted *et al.* [6].

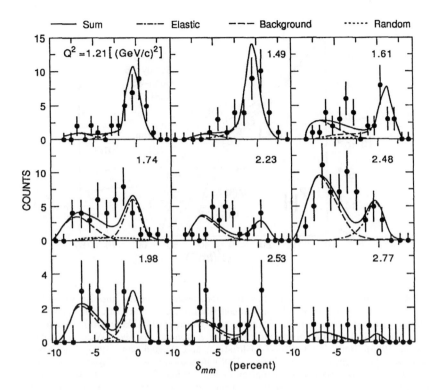

Figure 3: Number of e-d coincidences as a function of missing momentum δ_{mm} at each Q^2 value of this experiment. Also shown are the two-parameter fits using Monte Carlo generated shapes for the elastic peak and background reaction, assumed to be $\gamma d \to d\pi^0$. The dot-dashed curves are the elastic contributions, the dashed curves represent the background reaction, and the dotted curves are from measured random e-d coincidences. The solid curves are the sums. All spectra were taken with a 20-cm long target, except at $Q^2 = 1.20(\text{GeV}/\text{c})^2$ (10-cm target) and $Q^2 = 2.48(\text{GeV}/\text{c})^2$ (40-cm target). Reproduced from Bosted et al. [6]

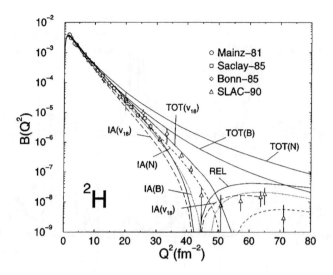

Figure 4: The deuteron $B(Q^2)$ structure function, obtained in the impulse approximation (IA) and with inclusion of two-body current contributions and relativistic corrections (TOT), compared with data from Simon, Schmitt, and Walther(1981; Mainz-81), Cramer *et al.* (1987). Theoretical results corresponding to the Argonned v_{18} (v_{18}; Wiringa, Stoks, and Schiavilla, 1995), Bonn B (B; Plessas, Christion, and Wagenbrunn, 1995), and Nijmegan (N; Plessas, Christion, and Wagenbrunn, 1995) interactions are displayed. Also shown is the relativistically covariant full calculation of Van Orden, Devine, and Gross (1995). The Höhler parameterization is used for the nucleon electromagnetic form factors. Reproduced from Carlson and Schiavilla. [16]

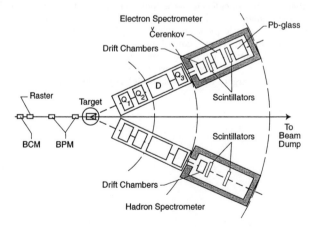

Figure 5: Plan view of the Hall A Facility of JLAB as used in this experiment. Shown are the beam monitoring devices, the cryotarget, the two magnetically identical spectrometers (consisting of quadrupoles Q_1, Q_2, Q_3, and the dipole D), and the detector packages. Reproduced from Alexa *et al.* [7].

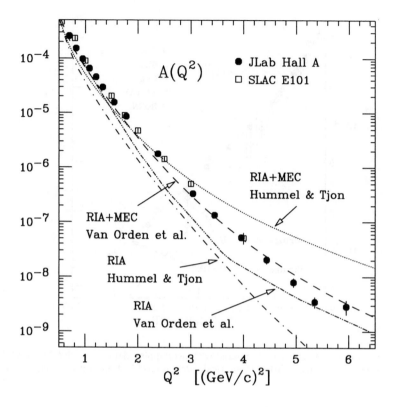

Figure 6: The deuteron elastic structure function $A(Q^2)$ from this experiment compared to RIA [17,18] theoretical calculations. Also shown are previous SLAC [19] data. Reproduced from Alexa *et al.* [7].

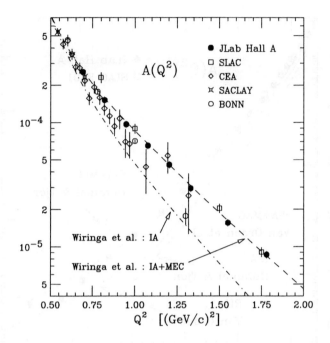

Figure 7: The present $A(Q^2)$ data compared with overlapping data from CEA [20], SLAC [19], Bonn [21], Saclay [22], and IA + MEC theoretical calculations [23]. Reproduced from Alexa *et al.* [7].

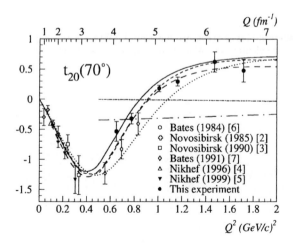

Figure 8: t_{20} at $\theta_e = 70°$ compared to theoretical predictions. Reproduced from Abbott *et al.* [15]

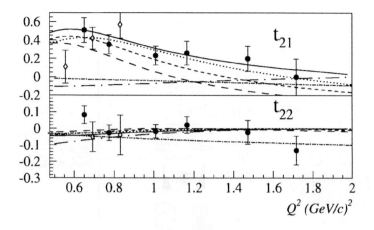

Figure 9: t_{21} and t_{22} at $\theta_e = 70°$ Reproduced from Abbott *et al.* [15]

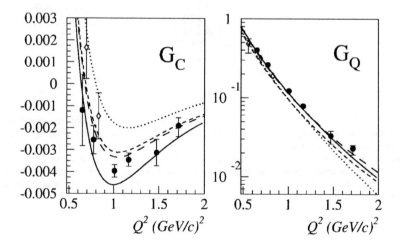

Figure 10: Monopole (G_C) and quadrupole (G_Q) charge form factors of the deuteron. Reproduced from Abbott *et al.* [15]

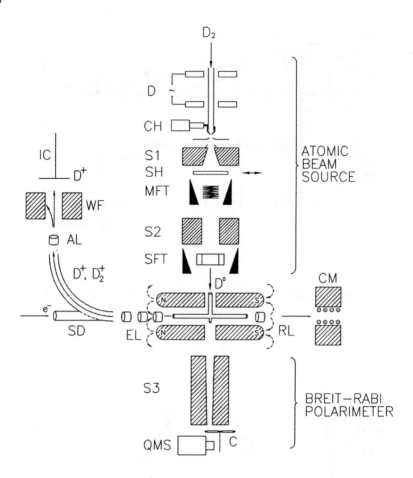

Figure 11: Schematic outline of the atomic beam source, Breit-Rabi polarimeter, internal target, and ion-extraction system. All components, except the target holding field and correction magnets, are inside the vacuum system. D: rf dissociator; CH: cold head; S1, S2, S3: sextupole magnets; MFT, SFT: medium-and strong-field transition units; SH: shutter; C: chopper; QMS: quadrupole mass spectrometer; CM: correction magnet; RL: repeller lens; EL: triplet of ion-extraction lenses; SD: spherical detector; AL: electrostatic lens; WF: Wien filter; IC: ion collector. Reproduced from Ferro-Luzzi *et al.* [12]

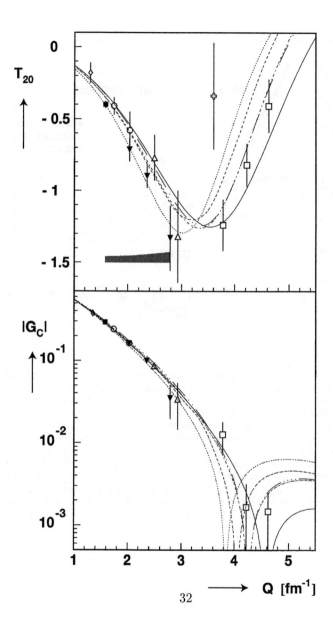

32

Figure 12: Extracted values(solid triangles) of $T_{20}(70°)$ (top) and G_C (bottom) as a function of Q compared to the world data and selected calculations. Data: solid triangles (present experiment), others (see Bouwhuis *et al.* [13]). The shaded area indicates the size of the systematic errors from the present experiment. Reproduced from Bouwhuis *et al.* [13]

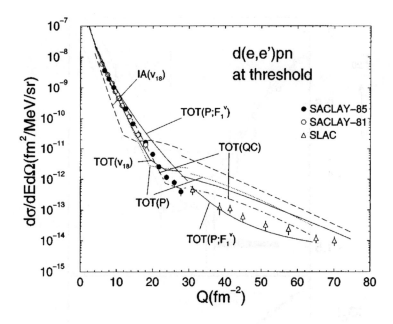

Figure 13: The cross sections for backward electrodisintegration of the deuteron near threshold, obtained in the impulse approximation (IA) and with inclusion of two-body current contributions and relativistic corrections (TOT), compared with data from SLAC-65 [24], Saclay-81 [25], Saclay-85 [26], and SLAC-90 [27]. Theoretical results corresponding to the Argonne [28] v18 (v_{18}), Paris [29] (P) and r-space version C of the Bonn [29] (QC) interactions are displayed. The dipole parameterization (including the Galster factor for $G_{E,n}$) is used for the nucleon electromagnetic form factors. In particular, the Sachs form factor G_E^V (Q^2) is used in the isovector model-independent two-body current operators. For the Paris interaction, the results obtained by using the Dirac form factor F_1^V (Q^2) in these two-body currents are also shown [curve labeled TOT(P;F_1^V)]. Data and theory have been averaged between 0 and 3 MeV for $Q^2 < 30$ fm^{-2} and between 0 and 10 MeV for $Q^2 > 30$ fm^{-2} pn relative energy, corresponding to the break in the curves. Reproduced from Carlson and Schiavilla.[16]

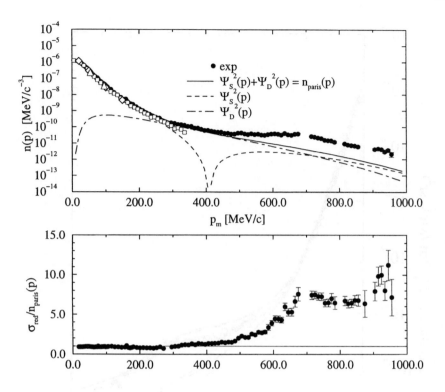

Figure 14: Top: The reduced cross sections compared to the momentum distribution calculated with the Paris potential. The open squares are data from Bernheim et al. [30], the open diamonds from Ducret et al. [32] and the open triangles are from Jordan et al. [31]. The dashed line shows the contribution of the S-state and the dash-dotted line shows the one from the D-state. Bottom: The ratio between the reduced cross sections and the Paris momentum distribution. Reproduced from Blomqvist et al. [34].

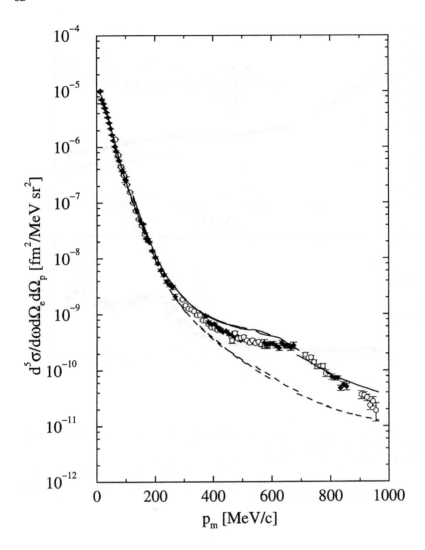

Figure 15: Comparison of the measured D(e,e'p)n cross section to the calculation by Arenhövel with (solid curve) and without (dashed curve) MEC and IC. Reproduced from Blomqvist et al. [34].

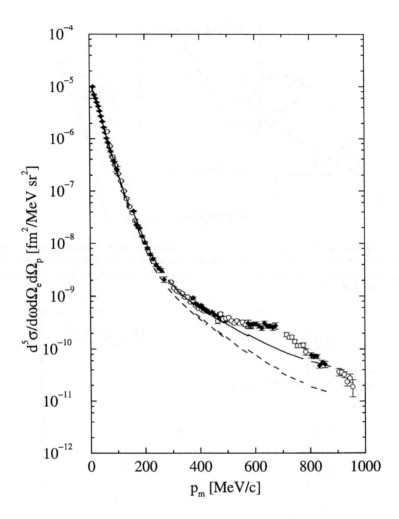

Figure 16: Comparison of the measured D(e,e'p)n cross section to the calculation with Schiavilla's code with (solid curve) and without (dashed curve) MEC and IC. Reproduced from Blomqvist et al. [34].

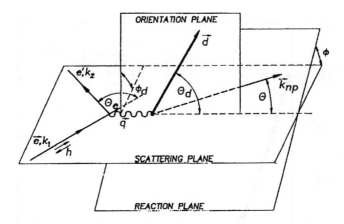

Figure 17: Geometry of exclusive electron-deuteron scattering with polarized electrons and oriented deuteron target. Relative n-p momentum is denoted by $\vec{k}_{np}$ characterized by angles θ and ϕ and deuteron orientation axis by $\vec{d}$ characterized by angles θ_d and ϕ_d. Reproduced from Arenhövel et al.[35]

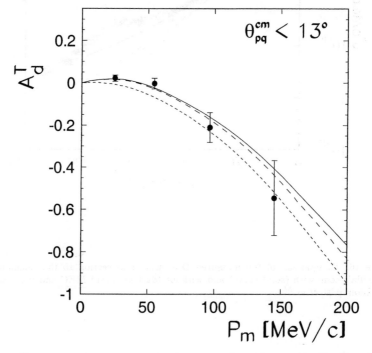

Figure 18: A_d^T as a function of p_m for parallel kinematics (i.e. $\theta_{pq}^{cm} < 13°$). The short-dashed curve represents the result for PWBA, in the long-dashed curve also FSI-effects are included, and the solid curve represents the full calculation. Reproduced from Zhou et al.[39]

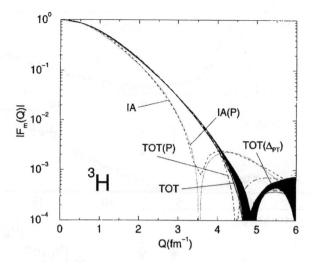

Figure 19: The magnetic form factors of ^{3}H, obtained in the impulse approximation (IA) and with inclusion of two-body current contributions and D admixtures in the bound-state wave function (TOT), compared with data (shaded area) from Amroun et al. (1994). Theoretical results correspond to the Argonne v18 two-nucleon and Urbana IX three-nucleon (Schiavilla and Viviani, 1996) and Paris two-nucleon (P; Strueve et al., 1987) interactions. They use, respectively, correlated hyperspherical harmonics and Faddeev wave functions and employ the dipole parameterization (including the Galster factor for $G_{E,n}$) for the nucleon electromagnetic form factors. Note that the Sachs form factor $G_E^V(Q^2)$ is used in the isovector model-independent two-body current operators for the Argonne-based calculations, while the Dirac form factor $F_1^V(Q^2)$ is used in the Paris-based calculations. Also shown are the Argonne results [curve labeled TOT(Δ_{PT})] obtained by including the two-body currents associated with intermediate excitation of a single D isobar in perturbation theory. Reproduced from Carlson and Schiavilla. [16]

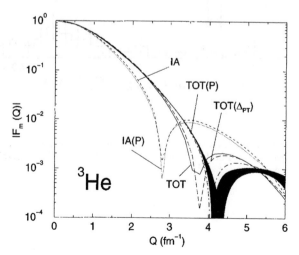

Figure 20: Same as in Fig. 19, but for ^{3}He. Reproduced from Carlson and Schiavilla. [16]

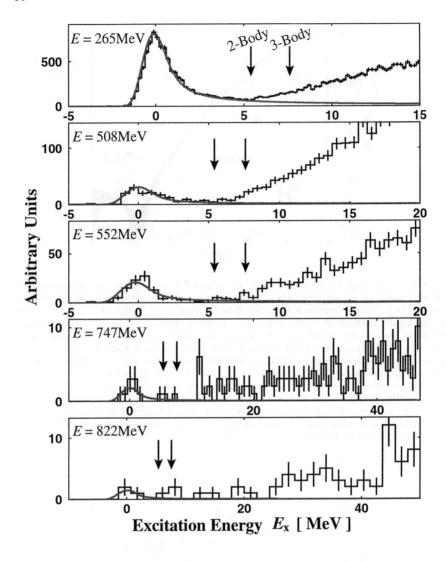

Figure 21: Background-corrected spectra measured for ^{3}He, showing the clear separation of the elastic peaks from the two-body (5.4 MeV) and three-body (7.6 MeV) breakup thresholds. Reproduced from Nakagawa *et al.* [42]

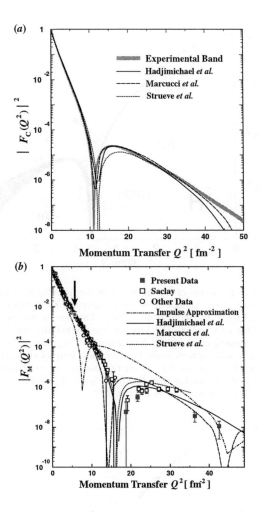

Figure 22: (a) Calculated form factors for elastic charge scattering from ^{3}He and comparison to sum-of-Gaussian fits to experimental values. (b) Elastic magnetic form factor of ^{3}He. The arrow indicates the present measurement at $Q^2 = 5.75 fm^{-2}$. Reproduced from Nakagawa *et al.* [42]

38

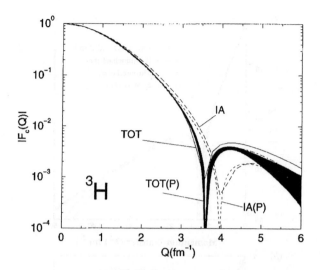

Figure 23: The charge form factors of ^{3}H, obtained in the impulse approximation (IA) and with inclusion of two-body charge contributions and relativistic corrections (TOT), compared with data (shaded area) from Amround et al. (1994). Theoretical results correspond to the Argonne v18 two-nucleon and Urbana IX three-nucleon (Schiavilla and Viviani, 1996) and Paris two-nucleon (P; Strueve et al., 1987) interactions. They use, respectively, correlated hyperspherical harmonics and Faddeev wave functions and employ the dipole parameterization (including the Galster factor for $G_{E,n}$) for the nucleon electromagnetic form factors. Note that the Paris-based calculation also includes Δ-isobar admixtures in the ^{3}H wave function. Reproduced from Carlson and Schiavilla. [16]

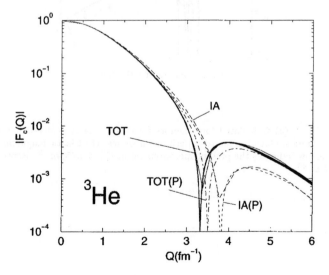

Figure 24: Same as in Fig. 23, but for ^{3}He. Reproduced from Carlson and Schiavilla. [16].

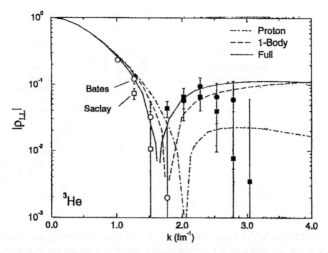

Figure 25: Experimental and theoretical longitudinal-longitudinal distribution functions in ^{3}He. Circles (squares) denote Bates (Saclay) data; solid symbols denote negative values. The curves labeled proton, 1-body, and full show theoretical results obtained from the Faddeev wave functions by including in ρ_L the proton, one-body, and one- plus two-body contributions, respectively. Reproduced from R. Schiavilla, R.B. Wiringa, and J. Carlson [43]

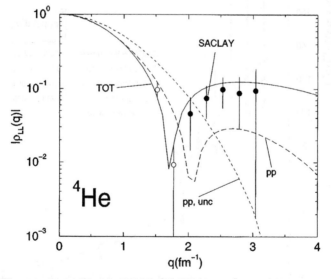

Figure 26: The experimental longitudinal-longitudinal distribution function (LLDF) of ^{4}He, obtained from the measured charge form factor and tail-corrected Coulomb sum-rule data (Saclay), compared with theory (Schiavilla, Wiringa, and Carlson, 1993). The curve labeled pp only takes into account the proton contributions to the nuclear charge operator, while that labeled TOT also includes contributions from two-body charge operators and relativistic corrections. Also shown is the LLDF for uncorrelated protons (curve labeled pp,unc). Note that the empty and filled circles denote positive and negative values for the experimental LLDF. Reproduced from Carlson and Schiavilla. [16]

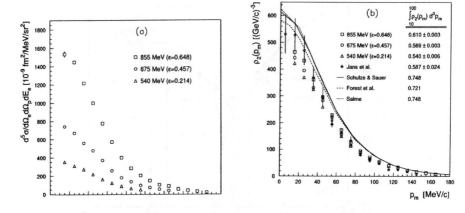

Figure 27: (a) ^{3}He(e,e'p)^{2}H cross sections as a function of missing momentum. (b) extracted momentum distribution for the two-body breakup process are compared to various calculations. All uncertainties are statistical only. Reproduced from Florizone et al. [41]

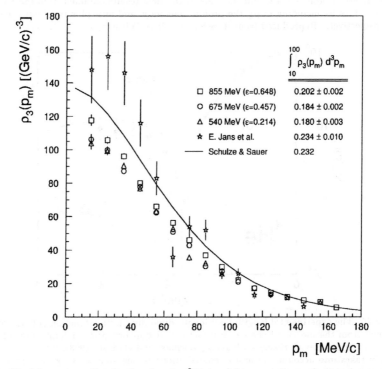

Figure 28: Momentum distribution for the $^3He(e, e'p)n, p$ continuum integrated over the range $E_m = 7 - 20$ MeV. All uncertainties are statistical only. Reproduced from Florizone et al. [41]

8 References

1. Electron Scattering by J. D. Walecka, Lectures Given at Argonne National Laboratory, January 1984, ANL-83-50.
2. T. deForest and J. D. Walecka, Advances in Physics, **15**, 1, 1966.
3. T. W. Donnelly in Modern Topics in Electron Scattering, Editors B. Frois and I. Sick, World Scientific (1991), and references therein.
4. R. Machleidt, Advances in Nuclear Physics, **19**, 189 (1989).
5. J. L. Forest *et al.*, Phys. Rev. C **54**, 646 (1996).
6. P.E. Bosted *et al.*, Phys. Rev. C. **42**, 38 (1990)
7. L.C. Alexa *et al.*, Phys. Rev. Lett. **82**, 1374 (1999)
8. R.G. Arnold *et al.*, Phys. Rev. Lett. **58**, 1723 (1987);
9. D. Abbott *et al.*, Phys. Rev. Lett. **82**, 1379 (1999)
10. M. E. Schulze *et al.*, Phys. Rev. Lett. **52**, 597 (1984)
11. M. Garçon and J.W. Van Orden, preprint. arXiv:nucl-th/0102049(2001)
12. M. Ferro-Luzzi *et al.*, Phys. Rev. Lett. **77**, 2630 (1996)
13. M. Bouwhuis *et al.*, Phys. Rev. Lett. **82**, 3755 (1999)
14. K. Gustafsson, Ph. D. thesis, University of Maryland, 2000.
    Hypertext link: *http://hallcweb.jlab.org/publications/theses*
15. D. Abbott *et al.*, Phys. Rev. Lett. **84**, 5053 (2000)
16. J. Carlson and R. Schiavilla, Rev. Mod. Phys. **70**, 743 (1998)
17. J. W. Van Orden, N. Devine, and F. Gross, Phys. Rev. Lett. **75**, 4369 (1995) and references therein.
18. E. Hummel and J. A. Tjon, Phys. Rev. Lett. **63**, 1788 (1989); Phys. Rev. C. **42**, 423 (1990)
19. R.G. Arnold *et al.*, Phys. Rev. Lett. **35**, 776 (1975)
20. J. E. Elias *et al.*, Phys. Rev. **177**, 2075 (1969)
21. R. Cramer *et al.*, Z. Phys. C **29**, 513 (1985)
22. S. Platchkov *et al.*, Nucl. Phys. **A510**, 740 (1990)
23. R. B, Wiringa *et al.*, Phys. Rev. C. **51**, 38 (1995); R. Schiavilla and D. O. Riska, Phys. Rev. C. **43**, 38 (1991)
24. A. E. Cox *et al.*, Nucl. Phys. **74**, 497 (1965)
25. M. Bernheim *et al.*, Phys. Rev. Lett. **46**, 402 (1981)
26. S. Auffret *et al.*, Phys. Rev. Lett. **55**, 1362 (1985)
27. R.G. Arnold *et al.*, Phys. Rev. C **42**, 1 (1990)
28. R. Schiavilla, unpublished (1996)
29. W. Leidemann *et al.*, Phys. Rev. C **42**, 826 (1990)
30. M. Bernheim *et al.*, Nucl. Phys. A **365**, 349 (1981)
31. D. Jordan *et al.*, Phys. Rev. Lett. **76**, 1579 (1996)
32. J.E. Ducret *et al.*, Phys. Rev. C **49**, 1783 (1994)

33. T. de Forest Jr. Nucl. Phys. A **392**, 392 (1983)
34. K. I. Blomqvist *et al.*, Phys. Lett. B **424**, 33 (1998)
35. H. Arenhövel *et al.*, Phys. Rev. C **46**, 455 (1992)
36. S. Dolfini *et al.*, Phys. Rev. C **51**, 3479 (1995)
37. Z.-L. Zhou *et al.*, preprint. arXiv:nucl-ex/0105006 (2001)
38. J. L. Forest *et al.*, Phys. Rev. C **54**, 646 (1996)
39. Z.-L. Zhou *et al.*, Phys. Rev. Lett **82**, 687 (1999)
40. A. Amroun *et al.*, Nucl Phys. A **579**, 579 (1997)
41. R. E. Florizone *et al.*, Phys. Rev. Lett. **83**, 2308 (1999)
42. I. Nakagawa *et al.*, Phys. Rev. Lett. **86**, 5446 (2001).
43. R. Schiavilla, R.B. Wiringa, and J. Carlson, Phys. Rev. Lett. **70**, 3856 (1993).
44. D. H. Beck, Phys. Rev. Lett. **64**, 268 (1990).
45. A. Zghiche et al., Nucl. Phys. A **572**, 513 (1993);A **584**, 757(1993)(E).
46. K.F. von Reden et al., Phys. Rev. C **41**, 1084 (1990).

STRANGENESS IN NUCLEAR PHYSICS

G.B. FRANKLIN

Carnegie Mellon University,
Department of Physics,
Pittsburgh, PA, USA
E-mail: gbfranklin@cmu.edu

The study of nuclear systems containing one or more strange-quarks provides a means of expanding our understanding of the baryon-baryon interaction. The connection between the baryon-baryon force and the hypernuclear structure is discussed along with calculational techniques used to identify hypernuclear states. A brief discussion of the S=-2 systems, including the H-Dibaryon and $\Lambda\Lambda$ hypernuclei is presented.

1. Introduction

1.1. *Strange Sector Physics*

Strange sector physics involves the use of the s-quark to delineate the role of quark flavor in hadronic matter. The whitepaper, "HADRON PHYSICS in the 21st CENTURY"[1] attempts to elucidate the fundamental issues in non-perturbative QCD. Three of the issues listed are closely tied to strange sector physics:

- What is the connection between QCD and the strong force as manifested in nature?
- What is the mechanism of confinement?
- What is the role of strangeness in hadronic matter?

While perturbative QCD is successful at describing the strong-force between quarks interacting at high relative momentum, the universe is dominated by the low-momentum interactions, e.g. the force binding the nucleons within a nucleus and the quarks within a nucleon. Since the strong force was "invented" to explain the binding within a nucleus, it is somewhat ironic that the QCD Lagrangian cannot be used to directly solve this problem due to the non-perturbative nature of QCD at low energies. Instead, we rely on models to make the connection between QCD and the strong force as it manifests itself in the interaction between two baryons.

A major component of strange-sector physics involves the study of the structure and decay mode of *hypernuclei*, nuclear systems with one or more hyperons. In most cases hypernuclear structure data are combined with nuclear physics techniques developed for conventional nuclei to extract information pertaining to the baryon-baryon interaction. Since hyperon-nucleon low energy scattering data are quite sparse, studies of the binding and level-structures of hypernuclei have provided much of what we know about the ΛN, ΣN, ΞN, and $\Lambda\Lambda$ interactions. These systems provide novel testing grounds to test and expand our models of nuclear physics

Strangeness sector physics also includes the study of light-quark hadrons. Many theoretical models of hadrons assume that $SU(3)_{flavor}$ symmetry is at least approximately conserved since the s-quarks, although heavier than the u- and $d-$ quarks, are much lighter than the hadron masses. This approach seeks a unified understanding of hadrons constructed from u, d and s valence quarks. (The physics of hadrons containing heavy quarks allow different approaches.) Many of the models designed to explain the existing baryon and meson mass spectra also predict the existence of exotic particles such has the H-Dibaryon and the pentaquark. Thus strange-sector physics provides a means of testing our understanding of the mechanism of quark confinement (item 2 above).

Finally, we note that strangeness plays a role in hadronic matter. We now believe that "neutron" stars contain a significant fraction of strange hadrons. The determination of the neutron-star equation of state requires an understanding of the hyperon-baryon potentials at high densities; this is an example of knowledge of hypernuclear physics feeds into other sectors. It is also possible that some neutron stars have a central core of strange-quark matter. The production of strangeness is considered a possible signature for the formation of quark-gluon plasma in relativistic heavy-ion collisions.

The following discussions will concentrate on the use of hypernuclear physics as a tool for exploring the baryon-baryon force. Section II will review the nuclear physics concepts needed to connect measurements of hypernuclear cross sections and energy levels to models of the baryon-baryon force. Section III will discuss experiments which have determined the Λ-Nucleus potential and Section IV will discuss $S = -2$ systems and Section V will discuss studies of the weak ΛN interaction.

1.2. *Extending the NN interaction to the strangeness sector*

The decades of NN data can be understood using a potential description. At a more fundamental level, this potential is, in turn, understood in terms of one-boson-exchange (OBE) at long distances and quark-based models at short distances. One can develop pictures which are self-consistent and constrained by the data at this level.

To connect to QCD, we need to improve our understanding of the short-range regime and be able to calculate the coupling constants used in the OBE calculations. A major barrier is the fact that, despite decades of work in the s=0 sector, there is no unique prescription at this level of understanding; competing models and parameter sets can be used to describe the wealth of data equally well.

When we move to baryons with one and two s-quarks, the object is not to start all over again but to use the studies of ΛN and ΞN interactions to refine our understanding of ordinary matter. If our models of the NN interaction are accurate, they should be able to describe the generalized baryon-baryon interaction.

For example, the OBE diagrams used in medium and long-range portion of NN interaction calculations can be transformed using $SU(3)_{flavor}$ considerations. In some approaches, the short range description introduces an opportunity to "do some fiddling", but this generally has only small effects. In most approaches, there is little wiggle room once the NN-sector is fixed.

The strangeness sector data, however, are limited. Most of what we know is in the $S = -1$ sector and comes from hypernuclear excitation spectra. However, even limited information from systems with one or two strange quarks can make a significant contribution. For example, the well known Nijmegan D and F potentials (used in many nuclear models) do equally well in describing NN data, but don't even agree on the sign of the residual binding when transformed to a $\Lambda\Lambda$ system. So rudimentary information from such a system can make a significant contribution to our knowledge of the generalized baryon-baryon interaction.

1.3. *NN Interaction Calculations*

1.3.1. *Scattering cross sections using potential theory*

A description of the NN interaction based on a potential can be used to calculate scattering cross sections simply by solving Schroedinger's Equation with the appropriate boundary conditions. Let H_0 be the free Hamiltonian (describing two non-interacting nucleons) and V be the potential that describes their interaction. Schroedinger's equation can then be written:

$$(H_0 + V)|\Psi_{\vec{k}}\rangle = E_k|\Psi_{\vec{k}}\rangle. \tag{1}$$

where $|\Psi_{\vec{k}}\rangle$ is the wavefunction that describes the two-body system.

Since $|\Psi_k\rangle$ represents an unbound pair of nucleons, we wish to find a solution to Eq. 1 with appropriate bounding conditions. In the case of scattering, we want an equation that corresponds to an incident plane wave of relative momentum k and an outgoing scattering wave that may have some angular dependence $f(\theta)$ so that at large r (outside the range of the NN potential) we have:

$$\text{as } r \to \infty, \quad |\Psi_{\vec{k}}\rangle \to e^{i\vec{k}\cdot\vec{r}} + \frac{e^{ikr}}{r}f(\theta). \tag{2}$$

The scattering amplitude $f(\theta)$ has units of one over distance and gives the amplitude of the outgoing particles at angle θ. The differential cross section is then given by $d\sigma/d\Omega = |f(\theta)|^2$.

We want to put Eq. 1 into a form that can be solved using an iterative solution. As a first attempt, we start by rearranging the terms and then dividing by $(E_k - H_0)$.

$$(E_k - H_0)|\Psi_{\vec{k}}\rangle = V|\Psi_{\vec{k}}\rangle \tag{3}$$

$$|\Psi_{\vec{k}}\rangle = \frac{1}{(E_k - H_0)}V|\Psi_{\vec{k}}\rangle \tag{4}$$

We can understand the meaning of having the operator H_0 in the denominator of Eq. 4 by writing the state $|\Psi_{\vec{k}}\rangle$ in terms of the set of eigenstates of the free 2-nucleon system, $|\Phi_{\vec{k'}}\rangle$, where $H_0|\Phi_{\vec{k'}}\rangle = E'_k|\Phi_{\vec{k'}}\rangle$.

$$|\Psi_{\vec{k}}\rangle = \sum_{\vec{k'}} \frac{1}{(E_k - H_0)}|\Phi_{\vec{k'}}\rangle\langle\Phi_{\vec{k'}}|V|\Psi_{\vec{k}}\rangle \tag{5}$$

$$|\Psi_{\vec{k}}\rangle = \sum_{\vec{k'}} |\Phi_{\vec{k'}}\rangle\frac{1}{(E_k - E_{k'})}\langle\Phi_{\vec{k'}}|V|\Psi_{\vec{k}}\rangle \tag{6}$$

Note that the sum over states is really an integral over the unbound, plane-wave states that describe two non-interacting nucleons with relative momentum k'. It is now apparent that there is a problem with Eq. 4 since the denominator goes to zero for values of k' that give $E_k = E_{k'}$. To understand the proper form of Eq. 4, we temporarily consider time-dependent solutions to Schroedinger's Equation built from the time-independent solutions $|\Psi_{\vec{k}}\rangle$ and form a wave-packet describing particles with a range of momentum, k, and angles, $d\Omega$.

$$|\Psi_k(t)\rangle = \frac{1}{(2\pi)^3} \int k^2 dk d\Omega A(\vec{k})|\Psi_{\vec{k}}\rangle e^{-iE_k t/\hbar} \tag{7}$$

We want to find the stationary state solutions for $|\Psi_{\vec{k}}\rangle$ that form a wavepacket with appropriate boundary conditions when they are used in Eq. 7. For an interaction around $t = 0$, we want a wavepacket that describes incident particles with momentum $\sim \vec{k}$ at negative times and contains a term corresponding to a scattered wave at positive t. Consider the following modified form of Eq. 4.

$$|\Psi_{\vec{k}}^{+}\rangle = |\Phi_{\vec{k}}\rangle + \frac{1}{(E_k - H_0 + i\epsilon)}V|\Psi_{\vec{k}}^{+}\rangle \tag{8}$$

which has the form

$$|\Psi_{\vec{k}}^{+}\rangle = |\Phi_{\vec{k}}\rangle + G^{+}V|\Psi_{\vec{k}}^{+}\rangle. \tag{9}$$

The wavepacket propagating in time is then given by plugging Eq. 9 into Eq. 7.

$$|\Psi(t)\rangle = \frac{1}{(2\pi)^2} \int k^2 dk d\Omega A(\vec{k}) \left(|\Phi_{\vec{k}}\rangle + G(E_k)^{+}V|\Psi_{\vec{k}}^{+}\rangle\right) e^{-iE_k t/\hbar} \tag{10}$$

As $t \to -\infty$, the integral over E_k can be closed above the real axis and in this limit $|\Psi_{\vec{k}}^{+}\rangle \to |\Phi_{\vec{k}}\rangle$. After the interaction occurs (corresponding to $t \to +\infty$) the integral must be closed below the real axis and the pole at $E_k = H_0 - i\epsilon$ contributes a delta function, $-2\pi i \delta(E_k - H_0)$, which enforces energy conservation. Thus solving Eq. 8 for $|\Psi_{\vec{k}}^{+}\rangle$ gives us stationary states that correspond to a particle with initial momentum $\vec{k}$. At later times, the amplitude for finding the particle with momentum $\vec{k'}$ is then

$$\langle\Phi_{\vec{k'}}|\Psi_{\vec{k}}^{+}\rangle \to \langle\Phi_{\vec{k'}}|\Phi_{\vec{k}}\rangle + \frac{-2\pi i}{(2\pi)^3} \int d^3 k A(\vec{k})\delta(E_k - E_{k'})\langle\Phi_{\vec{k'}}|V|\Psi_{\vec{k}}^{+}\rangle. \tag{11}$$

If the wavepacket amplitude $A(\vec{k})$ is non-zero only over a small region around $\vec{k_0}$, the quantity containing the state-functions can be factored out of the integral. Calculating the probability that a detector with finite momentum acceptance detects the particle involves squaring the amplitude and integrating over the acceptance. The cross section is obtained after dividing by the incident flux. (See Rodberg[2] Chpt.7 for details.) The result is

$$\frac{d\sigma}{d\Omega} = \frac{2\pi}{\hbar v_i}\left| < \Phi_{\vec{k'}}|V|\Psi_{\vec{k}}^{+}\rangle > \right|^2 \rho_f(E) \tag{12}$$

1.3.2. *T-matrix formalism using potentials and meson exchange models*

The connection between a potential-model description of the baryon-baryon interaction and a meson-exchange model can be obtained by equating the *T-matrices* that are generated from the two approaches. We begin by continuing the previous section's discussion of potential scattering to introduce the concept of the *transition matrix* or *T-matrix*.

To calculate the scattering amplitude $\langle \Phi_{\vec{k}'} | V | \Psi_{\vec{k}}^+ \rangle$ we need to determine $V | \Psi_{\vec{k}'} \rangle$. This gives the description of the scattered state. As an alternative, we can express the scattering completely in terms of an operator, T that transforms the initial plane-wave state into the scattered state.

$$T|\Phi_{\vec{k}}\rangle \equiv V|\Psi_{\vec{k}}^+\rangle \tag{13}$$

This operator, the T-matrix, can be found by multiplying the Lippmann-Schwinger Equation (Eq. 8) by V and using Eq. 13 to replace $V|\Psi_{\vec{k}}\rangle$.

$$V|\Psi_{\vec{k}}^+\rangle = V|\Phi_{\vec{k}}\rangle + VG^+V|\Psi_{\vec{k}}^+\rangle \tag{14}$$

$$T|\Phi_{\vec{k}}\rangle = V|\Phi_{\vec{k}}\rangle + VG^+T|\Phi_{\vec{k}}\rangle \tag{15}$$

This shows that the operator T can be derived by iterating the potential V and the Green's Function G.

$$T = V + VG^+T \tag{16}$$

$$T|\Phi_{\vec{k}}\rangle = V|\Phi_{\vec{k}}\rangle + VG^+V|\Phi_{\vec{k}}\rangle + VG^+VG^+V|\Phi_{\vec{k}}\rangle + \ldots \tag{17}$$

and the cross section can now be written as

$$\frac{d\sigma}{d\Omega} = \frac{2\pi}{\hbar v_i} \left| \langle \Phi_{\vec{k}'} | T | \Phi_{\vec{k}} \rangle \right|^2 \rho_f(E). \tag{18}$$

In this form, the physics of the interaction is contained in the operator T or, equivalently, the T-matrix $\langle \Phi_{\vec{k}'} | T | \Phi_{\vec{k}} \rangle$.

Although a potential-model description of the NN interaction is convenient for non-relativistic calculations, where does it come from? As mentioned earlier, it is not possible to derive it directly from QCD. Instead, we use a *hadrodynamic* model based on meson exchanges such as the diagrams shown in Fig. 1a. The connection between a meson-based effective field theory and a potential model can be made by equating the T-matrix. For example, we will first consider a model based on one pion exchange. Since the pion is a pseudo scaler, the vertex shown in Fig. 1b has the form

$$g\bar{\Psi}_{k'}\gamma_5\Psi_k = g\left(\chi^T, \frac{\vec{\sigma}\cdot\vec{k}'}{E+M}\chi^T\right)\begin{bmatrix} 0 & -1 \\ -1 & 0 \end{bmatrix}\begin{pmatrix} \chi \\ \frac{\vec{\sigma}\cdot\vec{k}}{E+M}\chi \end{pmatrix}. \tag{19}$$

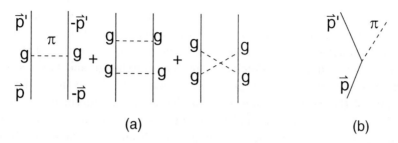

Figure 1. a) Meson-exchange model for calculation of T-matrix. b) π-nucleon vertex.

We are interested in applications in which the nucleons are non-relativistic, $E \approx M$ so that

$$g\bar{\Psi}(k')\gamma_5\Psi(k) \approx g\frac{\vec{\sigma}\cdot(\vec{k}'-\vec{k})}{2M} = g\frac{\vec{\sigma}\cdot\vec{q}}{2M}. \tag{20}$$

This can be used to calculate the T-matrix using Fig. 1a and from Eq. 15 the T-matrix gives the potential, expressed in momentum space, to first order. Thus One Pion Exchange Potential (OPEP) has the form

$$V_{OPEP}(q) = \langle\Phi_{\vec{k}'}|T|\Phi_{\vec{k}}\rangle = \frac{-g^2}{m_\pi^2+q^2}(\vec{\sigma}_1\cdot\vec{q})(\vec{\sigma}_2\cdot\vec{q}). \tag{21}$$

This configuration space potential can be shown to be[3]

$$V_{OPEP}(r) = \frac{1}{(2\pi)^2}\int e^{i\vec{q}\cdot\vec{r}}V_{OPEP}(q)d^3q \tag{22}$$

$$= \frac{g^2}{4M^2}m_\pi^3\left[\vec{\sigma}_1\cdot\vec{\sigma}_2 + S_{12}\left(1 + \frac{3}{m_\pi r}\frac{3}{(m_\pi r)^2}\right)\right]\frac{e^{-m_\pi r}}{m_\pi r} \tag{23}$$

The tensor operator is given by $S_{12} = 3(\vec{\sigma}_1\cdot\hat{r})(\vec{\sigma}_2\cdot\hat{r}) - \vec{\sigma}_1\cdot\vec{\sigma}_2$.

1.3.3. General form of the NN potential

The proceeding section shows the form of the NN-potential derived from a one-pion exchange model. Inclusion of other mesons will add additional terms. It is useful to consider the general form for the interaction between two nucleons by considering combinations of their spin, relative momentum, and relative position that give scalar terms. This gives the form of the general NN potential

$$V = V_C + V_{LS}\vec{L}\cdot(\vec{\sigma}_1+\vec{\sigma}_2) + V_\sigma\vec{\sigma}_1\cdot\vec{\sigma}_2 + V_T S_{12} + V_{\sigma p}(\vec{\sigma}_1\cdot\vec{p})(\vec{\sigma}_2\cdot\vec{p}) \tag{24}$$

Each of the coefficients V_j is in general a function of r, p, L, and *isospin channel*. In principle, they can be determined within a model that includes the one-pion-exchange but the additional diagrams introduce coupling constants and form-factors which need to be determined. There are various models available. In general, the long range ($r > 3fm$) portion of the NN potential is found to be dominated by one π-exchange and has a strong tensor-force component. This force is responsible, for example, for the mixing of the S- and D- states in the deuteron. At intermediate ranges ($0.7fm < r < 2fm$) the models employ a scalar meson or 2π-exchange which gives an attractive potential with a significant $L \cdot S$ (spin-orbit) term. At short range, the potential is determined by vector mesons or a quark-based model. The potential has a strong repulsive core with significant $L \cdot S$ dependence.

1.4. *Connecting to nuclear structure using the shell model*

1.4.1. *Single Particle Model*

We now consider a system of A interacting nucleons. Schroedinger's Equation has the form

$$H\Psi(r_1, r_2, ...r_A) = E\Psi(r_1, r_2, ...r_A). \tag{25}$$

In the extreme single particle model, we make an initial approximation for the multi-particle wavefunction by writing $\Psi(r_1, r_2, ...r_A) = \Phi_1(r_1)\Phi_2(r_2)...\Phi_A(r_A)$ and find the single particle wavefunction $\Phi_j(r)$ by assuming the nucleons see an averaged central potential $V_0(r)$. The solution is then separable and each single-particle wavefunction satisfies

$$-\frac{\hbar^2}{2m}\nabla^2\Phi(r, \theta, \phi) + V_0(r)\Phi(r, \theta, \phi) = \varepsilon\Phi(r, \theta, \phi) \tag{26}$$

If we assume the averaged central potential is spherically symmetric, the single particle solutions have the form $\Phi(r, \theta, \phi) = \phi(r)Y_{lm}(\theta, \phi)$ and separation of variables gives

$$\frac{\hbar^2}{2m}\frac{d^2}{dr^2}[r\phi(r)] + \left[V_0(r) + \frac{\hbar^2}{2m}\frac{l(l+1)}{r^2}(r\phi(r))\right] = \varepsilon r\phi(r) \tag{27}$$

From this we can see the the single particle solutions carry the quantum numbers n, l, and m_l for the spacial solution. We will use m_s for the spin projection. Since the potential is assumed to be spherically symmetric and so far we've left out any spin dependence, the single particle energies, ε_{nl}, depend only on n and l. In labeling these solutions, we usually use spectroscopic notation, $l = 0, 1, 2, 3, 4... \rightarrow s, p, d, f, g...$ The actual energy

Table 1.   Test Ordering of single particle states for a Woods Saxon Potential.

level	1	2	3	4	5	6	7
n	1	1	1	2	1	2	1
l	0	1	1	0	3	2	4
label	1s	1p	1d	2s	1f	2p	1g
# states in shell	2	6	10	2	14	6	18
total # states	2	8	18	20	34	40	58

levels depend on the details of $V(r)$. Table 1 shows the level ordering for a typical Woods-Saxon central potential.

We would expect shell structure (discontinuities in the binding potential, for example) at closed shells. Table 1, implies that this should occur when the number of neutrons or protons equals 2, 8, 18.... Indeed, ^4He and ^{16}O have double closed shells and exhibit this behavior. However, nuclei show that the *magic numbers* which indicate closed shells are at 2, 8, 28, 50, 82...; only the first two magic numbers are in agreement with Table 1. The conclusion is that the spin-independent potential does not adequately describe even the rough systematics of nuclear binding. The spin-dependent interactions of the NN force results in an effective single-particle spin-orbit term which must be added to the spin-independent potential.

$$V = V_C(r) + V_{LS}(r)\vec{L} \cdot \vec{S} \tag{28}$$

The new term mixes states of different m_l and m_s, the quantum numbers that describe the eigenstates become n, l, j, m_j, and the single particle energies now depend on n, l and j. The quantum number j of course refers to the eigenstates of the total angular momentum operator $\vec{J} = \vec{L} + \vec{S}$. The eigenvalues of the $\vec{L} \cdot \vec{S}$ operator are found by the usual technique of squaring $\vec{J}$ to find

$$\vec{L} \cdot \vec{S} = \frac{1}{2}(J^2 - L^2 - S^2)|\Phi_{njlm_j}\rangle \tag{29}$$

$$= \frac{1}{2}\hbar^2[j(j+1) - l(l+1) - s(s+1)|\Phi_{njlm_j}\rangle \tag{30}$$

For example, the six $1p$ states now split into two subshells. There are two states with $j = 1/2$ (denoted $1p_{1/2}$) and four states with $j = 3/2$ (denoted $1p_{3/2}$). Figure 2 (taken from Tipler and Llewellyn[4] shows the level ordering when the spin-orbit term is included. Note that the third shell includes both the 1d and 2s states so that the magic number 18 disappears and the $1f_{7/2}$ forms its own shell, giving a magic number at $n = 28$.

Inclusion of the spin-orbit term brings the shell ordering into qualitative agreement with experiment. The connection between the nuclear potential

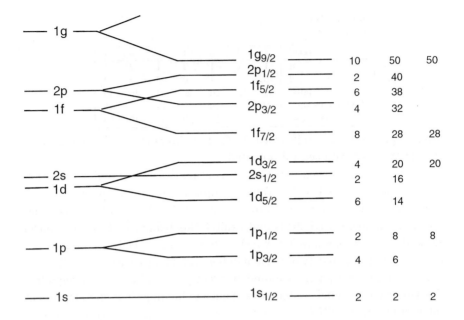

Figure 2. Shift in Woods-Saxon potential single-particle state energy levels when spin-orbit interaction is included.

and the NN 2-body interaction potentials can be made by demanding self-consistency. That is, an initial effective nuclear potential can be used to generate the single-particle states, $\Phi_i(r)$. A new mean nuclear potential seen by a single nucleon can be calculated using the two-body potential and integrating over the other single-particle states. The process can be iterated to convergence.

1.4.2. The G-Matrix Effective Interaction

To make an accurate connection between the the two-body NN interaction and experiment using the nuclear shell model, we need to make corrections for the two-body correlations. With single particle wavefunctions, the amplitude for finding one nucleon at position $\vec{r}_1$ and another at position $\vec{r}_2$ has the form $\Phi_i(\vec{r}_1)\Phi_j(\vec{r}_2)$; the probability of finding a nucleon at $\vec{r}_1$ appears to be independent of the position of the other nucleons. In reality, this cannot be true! (Consider Pauli Blocking and the effects of the short range repulsive core of the NN potential.) If we use terms of the form $V(r_1, r_2)\Phi_i(r_1)\Phi_j(r_2)$ in our Hamiltonian, we would overestimate the effects of the hardcore repulsion in computing the average potential since the correct treatment involves a correlated wavefunction $V(r_1, r_2)\Psi(r_1, r_2)$ and $\Psi(r_1, r_2) \to 0$ as $r_1 \to r_2$ for a strongly repulsive short range force. The solution is to replace the actual NN potential, $V(r_1, r_2)$ with an effective potential using *G-matrix* theory so we can replace $V(r_1, r_2)\Psi(r_1, r_2)$ with $G(r_1, r_2)\Phi_i(r_1)\Phi_j(r_2)$. The details won't be covered here, but it is interesting to note in the derivation we can write the correlated two-particle wavefunction in terms of our single-particle states

$$\Psi(r_1, r_2) = \Phi_1(r_1)\Phi_2(r_2) + \sum_{\substack{i,j \\ \text{not filled}}} \alpha_{ij}\Phi_i(r_1)\Phi_j(r_2) \qquad (31)$$

The fact that some single particle states are filled plays a role in determining the G-matrix because the summation can be shown to be only over the unfilled states. This can be understood in a semi-classical model by considering two nucleons described by plane-wave states. In free space, their scattering amplitude is determined by the NN potential alone. However in a nucleus they must scatter to unoccupied states and so the interaction is effected by the other nucleons which fill the states below the fermi level. The correlated 2-body residual interaction must proceed through intermediate two-particle two-hole states as illustrated in Fig. 3. The corrections to the uncorrelated product wavefunctions are suppressed by the energy denominator. This is particularly true for closed shells.

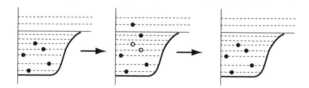

Figure 3. Representation of interaction of two nucleons scattering within the nucleus to an intermediate two-particle two-hole state.

2. S=-1 Hypernuclei

2.1. *Production Considerations*

Most experiments involving the production of nuclei containing one strange quark are done with kaon, pion, or electron beams although other beams such as heavy ions and antiprotons are sometimes utilized. Figure 4 shows that K^- mesons can be used to create Λs (or Σs) using a *strangeness exchange* reaction. Pions and electrons are used to create $s\bar{s}$ quark pairs; this is called *associated strangeness production*.

Figure 5 shows the momentum transfers involved for the (K, π), (π, K), and (γ, K) reactions. Note that the (K, π) reaction has a *magic momentum* for Λ production; the Λ is produced at rest when a 530 MeV/c momentum kaon is used. This would imply a high probability for the Λ to stick in the nucleus and form a hypernucleus. In contrast, the (γ, p) and (π, n) reactions involve momentum transfers above 400 Mev/c so we would expect smaller production cross sections.

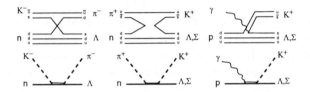

Figure 4. Production diagrams for elementary production of Λ and Σ hyperons.

Most of the early hypernuclear spectroscopy was done using a kaon beam to produce Λ–hypernuclei. A typical production reaction can be written as

$$K^- +{}^{12}C \rightarrow \pi^- +{}_{\Lambda}^{12}C. \tag{32}$$

It is worth noting that the *12* in ${}_{\Lambda}^{12}C$ denotes the total baryon number (including the Λ) and the C denotes the charge Z=6 so it is really a Λ bound to a ${}^{11}C$ core. The symbol ${}_{\Sigma}^{12}C$ refers to a system with the same quantum numbers ($S = -1$,$B = 12$, and $Z = 6$), but a mass that indicates that it is 11-nucleon system bound to a Σ. It could be an admixture of $\Sigma^0 - {}^{11}C$ and $\Sigma^- - {}^{11}N$ states.

Some of the early hypernuclear work done at BNL is shown in Fig. 6.[5,6] These experiments were performed using the 28 GeV/c proton beam of the AGS accelerator. The kaons were produced when the proton beam struck a tungsten target. (Any target that won't melt will do.) Secondary particles produced from this target pass through a magnetic spectrometer which utilizes quadrupole fields to contain and focus the secondary beam and a dipole to select the desired momentum region, in this case around 700 MeV/c. Unfortunately, the kaon production rate is very small compared to the pion production rate. To alleviate this problem, *separated*

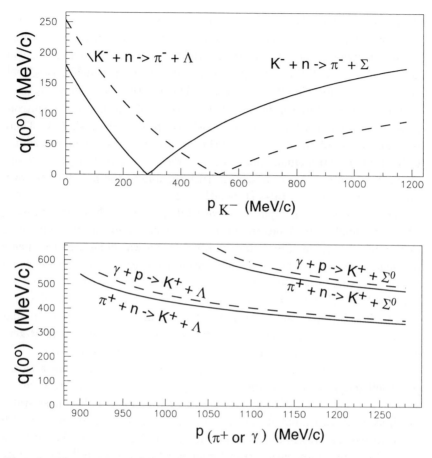

Figure 5. Momentum transfer as a function of incident beam momentum for the (K, π), (π, K), and (γ, K) reactions.

beams are used for these experiments. The secondary beam passes through an *electrostatic separator*, a velocity selector with crossed E and B fields, The fields are tuned so that the 700 MeV/c kaons pass nearly undisturbed through this region due to the cancellation of the $e\vec{E}$ and $e\vec{v} \times \vec{B}$ forces and pass through a narrow slit in a tungsten collimator. The pions are deflected a few millimeters and most are ranged out in the collimator. Although the beam that reaches the experiment's target (in this case ^{12}C) is kaon enriched by the separator, the ratio of kaons to pions is at best 1 to 10 in this momentum region and in most cases significantly worse.

The excitation spectra shown in Fig. 6 can be easily interpreted using the shell model. Fig. 6a shows the excitation spectrum for $^{12}_{\Lambda}C$. Since the experiment was done using the (K^-, π^-) reaction, the hypernucleus was formed when a neutron in the ^{12}C was transformed into a Λ. The dominant peak near 10 MeV excitation energy corresponds events in which a $p_{3/2}$ shell neutron is transformed into a $p-$shell Λ. The notation $(p_{n3/2}^{-1}p_{\Lambda 3/2,1/2})_{0+2+}$ signifies that the state can be modeled as a core ^{11}C, in this case a neutron-hole in the originally filled ^{12}C $p_{3/2}$ shell, coupled to the p-shell Λ. There is a multiplicity of possible states of this nature since the Λ spin can couple to its L=1 orbital angular momentum to form a $j = 3/2$ or $j = 1/2$ single particle state. This state can, in turn, couple to the $p_{3/2}$ hole to form positive parity nuclear states states 0^+ or 2^+.

This state is an example of a *substitutional state*; these are states in which the final state Λ lies in the same sub-shell and has the same spin as the initial nucleon. Since this (π, K) reaction was done near the magic momentum shown in Fig. 5, the momentum transfer is small and we would expect this substitutional state to dominate. On the other hand, since the Λ is not Pauli-blocked, the ground state of $^{12}_{\Lambda}C$ is reached by transforming a $p-$shell neutron to an $s-$shell Λ. This requires a transfer of angular momentum and so the cross section for forming the ground state at zero excitation energy is small when the momentum transfer is near zero.

Figure 6b and 6c show the results for a ^{13}C target (which has a closed $p_{3/2}$ shell and a valence neutron in the $p_{1/2}$-shell). Near zero degrees we can see the $p_{n1/2}^{-1}p_{\Lambda 1/2}$ substitutional state as well as other higher mass states (such as s-substitutional states, for example.) At 15 degrees, we can still see a trace of a state around 10 MeV but calculations indicate that at these angles the larger momentum transfer should enhance the $p_{n1/2}^{-1}p_{\Lambda 3/2}$ state and reduce the $p_{n1/2}^{-1}p_{\Lambda 1/2}$ substitutional state. An attempt to measure the shift of the peak position as a function of angle placed an upper limit on the spin-orbit splitting for Λ hyperons of 0.4 MeV, significantly less than the spin-orbit splitting required to understand conventional nuclei. The Λ potential depths extracted from this data is around 28 MeV, about half the nuclear well depth of the nucleon.

2.2. *Calculation of Cross Sections and Identification of States*

The qualitative approach to identifying the observed states discussed in the preceding section can be made more quantitative by comparing the angular dependence with theoretical predictions. These calculations are

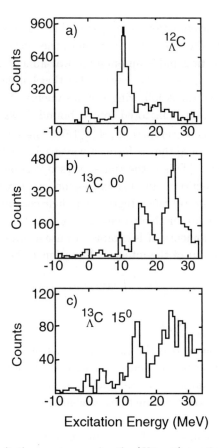

Figure 6.   Excitation spectrum using the (K^-, π^-) reaction at the AGS.

usually done using a Distorted Wave Impulse Approximations (DWIA). The differential cross section for a given final state can be written using the T-matrix formalism given in Eq. 18. The calculation of the T-matrix can be done in terms of an operator that represents the elementary production mechanism, for example $t_{KN \to \pi \Lambda}$

$$T(k_i, k_f) = \int \Psi_\Lambda^*(\vec{r}) \Psi_{k_f \pi}^{(-)}(\vec{r}) t_{KN \to \pi \Lambda} \Psi_n(\vec{r}) \Psi_{k_i K}^{(+)}(\vec{r}) d^3 r \qquad (33)$$

In this form, we have replaced the plane waves $|\Phi_{\vec{k}}\rangle$ and $|\Phi_{\vec{k'}}\rangle$, which represent the initial and final state solutions to the free Hamiltonian for structureless particles, with the state functions representing the meson-nucleon system before and after the 2-body transition occurs. In configuration space, the initial state nucleon wavefunction is given by $\Psi_n(\vec{r})$ and the initial state kaon wavefunction is $\Psi^{(+)}_{k_i K}(\vec{r})$. The kaon wavefunction is, in general, *distorted* in the sense that it includes the effect of the initial-state kaon-nucleon interaction and is an outgoing wavefunction in the sense that it can viewed as a component of a wave packet that has momentum centered around k at negative times following the arguments of section 1.3.1. The final-state wavefunction is factored into a bound Λ wavefunction, $\Psi^*_\Lambda(\vec{r})$ and a final-state pion wavefunction., $\Psi^{(-)}_{k_f \pi}(\vec{r})$. The latter is an *incoming* wavefunction in the sense that is forms a wavepacket with central momentum k_f at late times. (This is the time-reversed version of the solution to an outgoing wavefunction.) In this expression, we have factored out the spectator nucleons and set the integral over their coordinates to one.

If the elementary production t-matrix, $t_{KN \to \pi \Lambda}$, has only a slowly varying dependence on momenta, it is common to use a suitably averaged value and factor the term out of the integral. Equation 18 becomes

$$\frac{d\sigma}{d\Omega} = \frac{2\pi}{\hbar v_i} |t_{KN \to \pi \Lambda}|^2 \rho_f(E) \left| \int \Psi^*_\Lambda(\vec{r}) \Psi^{(-)}_{k_f \pi}(\vec{r}) \Psi_n(\vec{r}) \Psi^{(+)}_{k_i K}(\vec{r}) d^3 r \right|^2 \quad (34)$$

$$= \alpha \frac{d\sigma}{d\Omega}_{KN \to \pi \Lambda} N_{eff}. \quad (35)$$

The kinematic factor α is the ratio of the density of final states for hypernuclear production to the density of final states for $KN \to \pi \Lambda$. The overlap between the initial and final nuclear states, weighted by the incoming and outgoing meson wavefunctions, is contained in the factor N_{eff}, the *number of effective nucleons*.

The calculation can be further simplified using a Plane Wave Impulse Approximation (PWIA). In this case we replace the distorted incident and final-state meson wavefunctions with plane waves,

$$\Psi^{(-)}_{k_f \pi}(\vec{r}) \Psi^{(+)}_{k_i K}(\vec{r}) \approx e^{-i\vec{k}_f} e^{i\vec{k}_i} = e^{i\vec{q} \cdot \vec{r}} \quad (36)$$

where $\vec{q} = \vec{k}_K - \vec{k}_\pi$ is the momentum transfer to the nuclear system.

The expression for the effective number of nucleons then becomes

$$N_{eff} = \left| \int \Psi^*_\Lambda(\vec{r}) \Psi_n(\vec{r}) e^{i\vec{q} \cdot \vec{r}} d^3 r \right|^2. \quad (37)$$

As an aside, it is interesting to transform the *form factor* given by Eq 37 into momentum space by writing the configuration space wavefunctions for the Λ and the nucleon in terms of their momentum space descriptions:

$$\Psi_\Lambda(\vec{r}) = \frac{1}{(2\pi)^3} \int \tilde{\Psi}_\Lambda(\vec{k}_\Lambda) e^{i\vec{k}_\Lambda \cdot \vec{r}} d^3 k_\Lambda \quad (38)$$

$$\Psi_n(\vec{r}) = \frac{1}{(2\pi)^3} \int \tilde{\Psi}_n(\vec{k}_n) e^{i\vec{k}_n \cdot \vec{r}} d^3 k_n. \tag{39}$$

The integral over d^3q turns the exponential into a delta function which forces momentum conservation, $\vec{k}_\Lambda = \vec{k}_n + \vec{q}$. Equation 37 then becomes

$$N_{eff} = \frac{1}{(2\pi)^6} \left| \int \tilde{\Psi}_\Lambda(\vec{k}_n + \vec{q}) \tilde{\Psi}_n(\vec{k}_n) d^3 k_n \right|^2. \tag{40}$$

Thus we see the form factor is just the amplitude for the initial-state nucleon to have momentum $\vec{k}_n$ times the amplitude for the final-state wavefunction to have momentum $\vec{k}_n + \vec{q}$, integrated over all possible initial state momenta.

Going back to the configuration-space PWIA description of Eq. 37, it can now be seen why the substitutional states are dominant at low momentum transfer. Since, for substitutional states, we would expect the wavefunction for the final-state Λ, $\Psi_\Lambda^*(\vec{r})$ to be similar to the initial state nucleon and N_{eff} could be near one as $q \to 0$.

To calculate the cross sections to specific final states for closed shell nuclei, we model the states using the *particle-hole* notation. That is, the hypernuclear state is modeled as a single-particle neutron-hole labeled (j_n, l_n) corresponding to the total and orbital angular momenta of the initial-state neutron; the lambda is in a (j_Λ, l_Λ) state. For a given state, the neutron-hole couples to the Λ-particle to form a hypernuclear state with total angular momentum J. Assuming a spherically symmetric potential core, the neutron-hole wavefunction can be written in terms of a radial wavefunction, Clebsh-Gordan coefficients, spherical harmonics, and a spinor:

$$\Psi_{n(j_n,l_n,m_{jn})}(\vec{r}) = R_{n(j_n,l_n)}(r) \sum_{m_l,m_s} (j_n m_{jn} | l m_l \tfrac{1}{2} m_s) Y_{lm_l}(\theta_r, \phi_r) \chi_{m_s}. \tag{41}$$

The Λ wavefunction has a similar form. By expanding $e^{i\vec{q}\cdot\vec{r}}$ in terms of spherical Bessel functions, $j_l(kr)$, and spherical harmonics, the radial integrals can be done. Dover[7] applied the methodology used by Hufner *et al*[8] to produce an expression for hypernuclear production at zero degrees from 0^+ initial-state nuclei in terms of a Wigner 6-J coefficient and a form factor that depends only on the momentum transfer

$$N_{eff}(\theta = 0°; 0^\pi \to J^+) = (2J+1)(2j_\Lambda+1)(2j_n+1) \begin{pmatrix} j_\Lambda & j_n & J \\ 1/2 & -1/2 & 0 \end{pmatrix}^2 F_J(q) \tag{42}$$

where the form factor is given by

$$F_J(q) = \left[\int_0^\infty r^2 dr R_\Lambda(r) R_n(r) j_J(qr) \right]^2. \tag{43}$$

This form shows explicitly that a non-zero momentum transfer is needed to reach states with $J \neq 0$ and that an experimental determination of a cross sections' dependence on momentum transfer can be used to determine the J of a specific final state. We would expect to see a maximum in the cross section when the peak in the spherical Bessel function, $j_J(qr)$, is matched to the nuclear size.

To take the distortions of the incident and outgoing meson into account, the factor $j_J(qr)$ can be replaced with distorted waves. This can be determined using nuclear optical potentials and solving for the meson wavefunctions. The primary effect is a reduction in the cross sections due to the absorption of the mesons, particularly in large A nuclei. Since some production cross sections benefit from higher momentum transfers and the absorption of the mesons is an issue, the $(e, e'K^+)$ and (K^-, π) production mechanisms can have advantages over (K^-, π). Table 2 shows the we would expect (π, K) to have a smaller attenuation factor due to the longer mean-free-path of the K^+ as compared to the K^-; the $(e, e'K^+)$ reaction is better yet.

Table 2. Momenta and nuclear mean free paths of mesons used in hypernuclear production

Meson	P (MeV/c)	Mean free path (fm)
K^-	700	2.4
π^-	750	2.4
π^+	1100	1.6
K^+	730	4.5

Figure 7a shows the hypernuclear mass spectrum for $^{12}_{\Lambda}C$ obtained at KEK using the (π^+, K^+) reaction, The collaboration used a superconducting kaon spectrometer (SKS) for this experiment. The reduction in cross section due to the large momentum transfer (as compared to (K, π)) is compensated by the higher fluxes available for pion beams. This, combined with the improved resolution of the SKS spectrometer, yields the clear structure seen in the figure. This technique has allowed access to states in heavier nuclei, as seen in Fig. 7b. This data have allowed systematic studies of the energy levels as a function of nuclear size and the results indicate the model of a hypernucleus as a Λ bound in a Wood-Saxon type potential appear to reproduce the data. The general behavior of the central potential seen by the Λ appears to be understood.

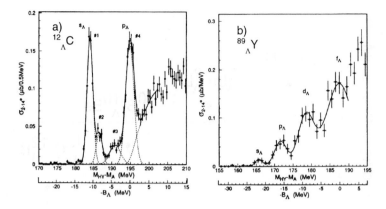

Figure 7. Hypernuclear mass spectra of a) $^{12}_{\Lambda}C$, and b) $^{89}_{\Lambda}Y$ obtained at KEK using the SKS spectrometer and the (π^+, K^+) reaction.

2.3. Details of the Λ Potential

2.3.1. Charge Symmetry Breaking

Calculations have shown that models which use a single-particle Λ coupled to a central potential to represent interactions with the core nucleus do a good job of reproducing the hypernuclear states obtained in the recent (π, K) data on medium and heavy hypernuclei. The gross features of Λ hypernuclei can be understood using formalism developed for nuclear physics in the $S = 0$ sector. Close inspection of the results, however, show interesting contrasts between the NN and ΛN interactions.

As a first example, consider the large charge symmetry breaking that is evident in the binding energies of mirror hypernuclei as shown in Fig. 8. These data are calculated from the published binding energies of light hypernuclei.[9] It can be seen that the breaking of isospin symmetry can be as large as 0.5 MeV in these light hypernuclear systems. This is significantly larger than the isospin breaking observed for conventional nuclei. In addition, the sign of the violation actually changes sign as the atomic number goes from 4 to 12.

It is believed that the isospin violation can be understood by first considering another difference between the NN and ΛN interaction. The exchange of the π-meson plays a significant role in the NN interaction. In contrast, since the Λ is isospin zero so the π cannot couple to the Λ without changing its isospin.. We expect a significant portion of the interaction to be due to 2-pion exchange with a Σ as an intermediate state. The mass

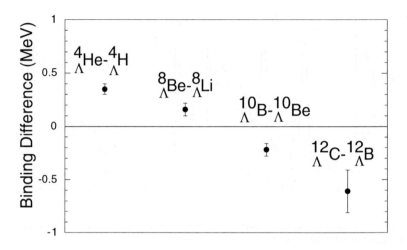

Figure 8.   Binding asymmetries of light hypernuclear mirror nuclei.

differences of the Σ hyperons may account for much of the isospin breaking since there is a 7 MeV difference between the Σ^+ and Σ^- masses.
Fig. 9 shows that Λp and Λn 2-pion exchange interactions involve different Σ charge states. The observed dependence on the size of the nucleus is
thought to be due to the shrinkage of the core-nucleus when the Λ is added.
This shrinkage causes a shift in the Coulomb energy which will be different
for the two different charge states of the isospin pairs plotted in Fig. 8

Since 2-pion exchange can be expected to play a more significant role in
the ΛN interaction than in the NN interaction, we might expect interactions of the type shown in Fig. 9c to be significant. This diagram illustrates
an effective three-baryon force. Indeed, this type of three-baryon interaction is needed to reproduce the correct binding of the hypertriton, $^3_\Lambda H$.

2.3.2. Spin Dependence

Detailed knowledge of the baryon-baryon interaction requires extraction of
the spin-dependence of the ΛN potential. The spin-orbit term in the effective $\Lambda - Nucleus$ interaction was needed to understand nuclear binding-
energy systematics in the $S = 0$ sector. However, early hypernuclear experiments were able to place only upper limits on the spin-orbit terms for Λ
hypernuclei. Models of the ΛN interaction generally "predict" small spin
dependence, but the strength of the terms depend on the model.

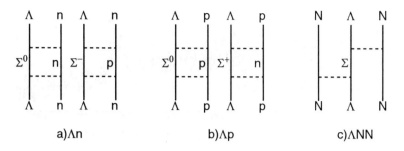

Figure 9. Two pion exchange diagrams for ΛN interactions.

The ΛN potential has the general form:

$$V_{\Lambda N} = V_0(r) + V_\sigma \mathbf{s}_N \cdot \mathbf{s}_\Lambda + V_\Lambda \mathbf{l}_{N\Lambda} \cdot \mathbf{s}_\Lambda + V_N \mathbf{l}_{N\Lambda} \cdot \mathbf{s}_N + $$
$$V_T[\, 3(\boldsymbol{\sigma}_N \cdot \mathbf{r})(\boldsymbol{\sigma}_\Lambda \cdot \mathbf{r}) - \boldsymbol{\sigma}_N \cdot \boldsymbol{\sigma}_\Lambda]$$

Note that, unlike the NN interaction, there are two $L \cdot S$ terms and the ΛN interaction is often written using a symmetric and antisymmetric linear combination of these two terms. The strength of these spin-dependent terms can be determined by measuring the level splittings in carefully selected hypernuclei. However, the first attempts to measure level splitting dominated by spin-orbit contributions produced only upper limits due to insufficient energy resolution and insufficient acceptance. These barriers have been overcome with the commissioning of the *Hyperball*. The Hyperball is a 14-element Ge detector array which was optimized to detect hypernuclear γ-rays. This device required 4 years of R&D, design, and construction. The hyperball was first run at KEK and then moved to the AGS to take advantage of the enormous increase in kaon flux and was used at the AGS during the last running period before the turn-on of RHIC. The U.S. maintained kaon-line, spectrometers, and software used for the tagging of hypernuclei were combined with the Japanese built Hyperball; the results of the first run is shown in Figure 10. The experiment, which can only be performed using a hadron beam, successfully measured the energy splitting of an excited state of $^9_\Lambda Be$ due to the spin-orbit interaction, 31 ± 2 keV, with an unprecedented accuracy.[10]

The JLAB hypernuclear program uses the $(e, e'K^+)$ reaction to produce and tag hypernuclei. This approach involves momentum transfers that are similar to (π^+, K^+) but turn a proton into a Λ instead of a neutron. In addi-

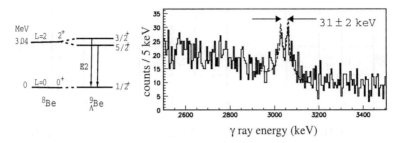

Figure 10. AGS E930 Hyperball results. A kaon beam was used to create $^{9}_{\Lambda}$Be. Hypernuclei produced with the ^{8}Be core in the 2^{+} state (energy diagram on left) were tagged using magnetic spectrometers and the energy spectrum of the coincident γ-rays detected in the Hyperball, shown on right, represents the first direct measurement of the energy splitting due to the spin-orbit term of the ΛN potential.

tion, unlike (π, K), this electroproduction reaction tends to flip the baryon spin. Early results indicate that JLAB experiments will attain better resolution and statistics than the meson-beam based production experiments. Experiment E89-009 has produced an excitation for $^{12}_{\Lambda}B$ with a reasonable background level and excellent resolution. One of the experimental difficulties in this first experiment arises from the limitations of the existing pair of spectrometers used to detect the outgoing electron and the kaon; the two spectrometers are not designed for extreme forward angle coincidence measurements. Since the production cross section for producing hypernuclei is maximum at zero degrees and drops rapidly with laboratory angle, the experiment could not be performed in the optimal kinematic region. A new experiment at JLAB, E01-011 promised more than an order of magnitude increase in production rates when a new spectrometer, specifically designed for this experiment, will be used at JLAB.

3. Strangeness -2 Systems

In the strangeness -2 sector, the possibility of a bound (stable against strong decay) object with baryon number 2 and $S = -2$ has motivated considerable theoretical and experimental interest. This *dibaryon*, called the H, is a six-quark (uuddss) object first shown to be bound with respect to twice the lambda mass in the MIT bag model by Jaffe.[11] Since the original prediction, dozens of calculations of the H-Dibaryon mass have been performed using a variety of confinement models. Most models have predicted a bound H-Dibaryon, although recent work on the lattice do not show binding.

Many of the experimental searches for this predicted particle use the

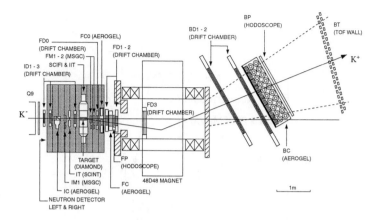

Figure 11. Experimental setup used by AGS experiment E885 to search for the H and $\Lambda\Lambda$ hypernuclei using the (K^+, K^-) reaction.

(K^-, K^+) reaction to create a system with two strange quarks. For example, AGS experiment E885 searched for a signature of H-Dibaryon production using this reaction on a ^{12}C target.[12] This experiment was performed using a 1.8 GeV/c separated kaon beam for the incident kaons. The momenta and trajectories of the outgoing K^+ mesons were measured using spectrometer designed around a "48D48" dipole magnet. The setup is shown in Fig. 11. Kaons were distinguished from pions using time-of-flight measurements and threshold aerogel Cerenkov detectors.

The reconstructed outgoing K^+ momentum was combined with the measured incident K^- momentum to calculate the missing mass for each event. The resulting excitation energy spectrum (which is a missing mass spectrum with an energy offset) is shown in Fig. 12. The zero of this excitation spectrum is defined to be the kinematic onset of the production of Ξ hyperons. The region just below zero corresponds to possible $S = -2$ hypernuclear production and $\Lambda\Lambda$ production. This experiment was particularly sensitive to H production in the region below E=−30 MeV, well below the kinematically allowed Ξ production region. Although there are some background counts, there is no evidence of structure and the rates are well below the H-Dibaryon production rate estimates shown by the dash-dotted lines.

The E885 results are shown in Fig. 13 as an upper limit for H-dibaryon production along with the results from KEK-E224[13] (which also studied the

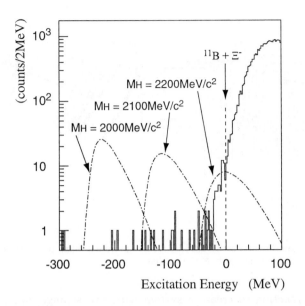

Figure 12. Excitation energy spectrum from E885. The dash-dotted lines show the calculated peaks for h production which correspond to an assumed H mass of 2000 MeV/c^2, 2100 MeV/c^2, and 2200 MeV/c^2.

$^{12}C(K^-, K^+)$ reaction and an earlier AGS experiment, E836[14], which used a ^3He target. It can be seen that these experiments have produced upper limits which are a factor of 50 or more below predicted production rates over a large mass region although there is large theoretical uncertainties in the rate calculations. There is little experimental data in the mass region very close to $2m_\Lambda$ and in the unbound region due to the background of quasi=free Ξ production.

Measurements of the ground states of $\Lambda\Lambda$ hypernuclei allows investigation of the $\Lambda\Lambda$ interaction. In addition, they have implications on the possible existence of the H-Dibaryon. Over the past 40 years, there have been four publications which have reported binding energies of nuclei with two bound Λs extracted from emulsion events which are interpreted as the formation and decay of double-Λ hypernuclei.

The first $\Lambda\Lambda$ hypernuclear event was reported by Danysz[15,16] in 1963. The event provided a measurement of the residual binding $\Delta B_{\Lambda\Lambda}$ by subtracting the the measured total binding of the 2 Λs from twice the binding of a Λ in the corresponding $S = -1$ hypernucleus. The event indicated a net attractive interaction between the Λs.

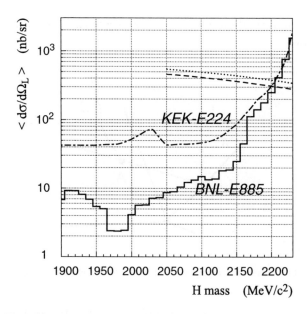

Figure 13. The 90% C.L. upper limits on the direct H production for BNL-E885 (^{12}C target), KEK-E224 (^{12}C target), and BNL-E836 (^3He target). The dashed line shows a theoretical calculation for production off of ^{12}C based on the model of Aerts and Dover and the dotted line represents Aerts and Dover's original calculation for a ^3He target.

In the following years, considerable effort was put towards finding additional $\Lambda\Lambda$ hypernuclear events. The techniques used for finding events in emulsion stacks limited the usable kaon luminosity and analysis shows the original event was very fortuitous when the kaon flux, stopping rate, and formation rate is considered. An event reported in 1966 described a $^6_{\Lambda\Lambda}He$ hypernucleus but the limited information contained in the report prevented verification.[17]

Experimenters at KEK developed a hybrid emulsion/tracking experiment that found the $\Lambda\Lambda$ event illustrated in Fig 14.[18] The event can be interpreted as a Ξ hyperon stopping in the emulsion, forming a $^{10}_{\Lambda\Lambda}Be$ hyperfragment which recoils and undergoes a mesonic decay at point B. This results in a recoiling $^{10}_{\Lambda}B$ hypernucleus which undergoes a multi-body nonmesonic decay at point C. This interpretation leads to a residual $\Lambda\Lambda$ binding of -4.9 MeV (repulsive). Unfortunately, the emulsion event is also consistent with a second interpretation in which an excited state of $^{14}_{\Lambda\Lambda}C$ is initially formed. This interpretation is consistent with a residual $\Lambda\Lambda$ binding of 4.8 MeV (attractive). There are experimental reasons to favor the former interpretation and theoretical arguments that favor the latter interpretation.

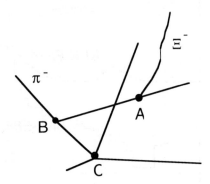

Figure 14. Schematic representation of the formation and decay of $^{10}_{\Lambda\Lambda}Be$

A measurement of the ground state mass of a double-Λ hypernucleus via its weak decy can be used to deduce a lower limit on the possible mass of the H-Dibaryon. If one assumes that the $S = -2$ hypernucleus reaches its ground state through strong and electromagnetic transitions before undergoing the weak mesonic decay, the range of the π^- meson and recoiling hyperfragment serve as a direct measure of the ground state of an $S = -2$ system. For example, if the H-Dibaryon exists and is bound by more than 30 MeV, then the transition $\Lambda\Lambda \to H$ would be kinematically allowed within the $^{10}_{\Lambda\Lambda}Be$ system and result in a lighter final-state mass.

Last year, a second generation hybrid-emulsion experiment published an event identified as the formation and decay of $^6_{\Lambda\Lambda}He$. This important event indicates that the residual binding between the two Λs is $\Delta B_{\Lambda\Lambda} = 1.01 \pm 0.30$ MeV. Assuming the arguments in the preceeding paragraph, the lower limit on the H-Dibaryon mass is 2223.7 MeV/c^2 at the 90% confidence level. It now appears that if the H-Dibaryon does exist, it is unbound or has a binding energy of less than 6 MeV.

References

1. B. Nefkens, editor, "Hadron Physics in the 21st Century," (unpublished), (1998).
2. L.S. Rodberg and R.M. Thaler, "The Quantum Theory of Scattering, Academic Press(1967).

3. M.A. Preston and R.K. Bhaduri, "Structure of the Nucleus", Addison-Wesley (1975).
4. P.A. Tipler and R.A. Llewellyn, "Modern Physics", Freeman and Company (2000).
5. R.E. Chrien, *et al.* Phys. Lett. B **89B**, 31 (1979) .
6. M. May, *et al.* Phys. Rev. Lett. **47**, 1106 (1981).
7. C. Dover, BNL-41894 (1988).
8. J. Hufner *et al.*, Nucl Phys. A234 (1974).
9. J. Tulie, Nuclear Wallet Cards, National Nuclear Data Center (2000).
10. H. Akikawa *et al*, Phys. Rev. Lett. **88**, 082501 (2002).
11. R.L. Jaffe, Phys. Rev. Lett. **38**, 195 (1977).
12. K. Yamamoto *et al.*, Phys. Lett. B*478*, 401 (2000).
13. J.K. Ahn, *et al.*, Phys. Lett. B **378** 53 (1996).
14. R.W. Stotzer *et al.*, Phys. Rev. Lett. **78**, 3646 (1997).
15. M. Danysz *et al.*, Nuc. Phys. **49** 121 (1963).
16. R.H. Dalitz *et al.*, Proc. R. Soc. London A**426** 1 (1989).
17. D.J. Prowse, Phys. Rev. Lett. **17**, 782 (1966).
18. S. Aoki et al., Prog. Th. Phys. **85**, 1287 (1991).
19. H. Takahashi *et al.*, Phys. Rev. Lett. **87**, 212502 (2001).

PION ELECTROPRODUCTION AND THE SEARCH FOR NUCLEAR PIONS

D. GASKELL

Nuclear Physics Laboratory, University of Colorado at Boulder,
Campus Box 390,
Boulder CO 80309, USA
E-mail: gaskell@auric.colorado.edu

Pion exchange plays a significant role in conventional models of the nuclear force. In these models, the pion contributes significantly at long and intermediate distances – this gives rise to an enhancement of the virtual pion cloud of nuclei relative to that of the free nucleon. However, several experiments that should be sensitive to nuclear pion currents, including deep inelastic scattering, $(\vec{p}, \vec{p}')$ and $(\vec{p}, \vec{n})$ scattering, and Drell–Yan experiments, yield inconclusive results. In this talk, I will discuss pion–electroproduction as a probe of the nuclear pion field and show recent results from both Jefferson Lab and Mainz. Finally, I will discuss potential future measurements that may shed more light on this issue.

1   The Elusive Pion Excess

Since Yukawa's suggestion [1] that there exists an exchange particle that mediates the nucleon–nucleon force, analogous to the way in which the photon mediates the electromagnetic force, the pion has played a key role in descriptions of the nucleon–nucleon interaction. Although modern potentials control the pion contribution at short distances via form–factors and include contributions from other mesons, pion exchange is still the dominant contribution to the long–range nuclear force.

Since the nucleon–nucleon interaction gives rise to tightly bound nuclear systems, it seems quite intuitive that if pion exchange plays a significant role in that interaction, then one might expect the "number" of virtual pions in nuclei to be greater than that in the free nucleon. Indeed, this can be quantified using standard nuclear models (i.e., either a mean field theory, or a theory that relies on complex nucleon–nucleon potentials) and a pion number operator. The result of such a calculation is shown in Fig. 1 where the pion excess density (i.e., the number of pions in nuclei after subtracting away the intrinsic pion field of each nucleon) for various nuclei is plotted. One can see that there is in fact a negative pion excess at low virtual pion momentum, k, due to Pauli blocking of the nucleons in the final state, but when integrated over all k, there is a net positive excess. For example, the integrated pion excess (per nucleon) is predicted to be about 0.14 for Fe. The question that

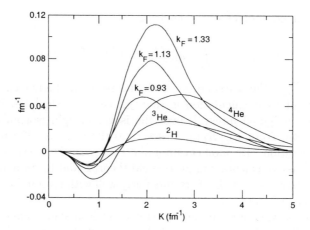

Figure 1. Pion excess distribution as calculated by Friman *et al.* [2] The excess is calculated as a function of the virtual pion momentum, k.

follows then, is in what way can we gain access experimentally to signatures of this pion excess?

1.1 The EMC Effect

To first order, deep inelastic scattering (DIS) is viewed as the exchange of a virtual photon from a high energy electron or muon with a target nucleon or nucleus. At large energy (ν) and four–momentum ($-Q^2$) transfers, DIS is interpreted as scattering from the quarks in the target. The cross section, then, can be written in terms of structure functions, F_1 and F_2, which basically describe the charge-weighted quark distributions. Since these high energy probes can also scatter from the quarks in pions, one might expect to see some effect from the pion excess in the DIS cross section for nuclear targets as compared to the free nucleon.

The European Muon Collaboration reported just such an effect when they measured the F_2 structure function in iron and compared it to the same for deuterium [3]. In particular, an enhancement in the F_2 structure function was seen for $x < 0.2$ (where $x = Q^2/m_p\nu$, and can be interpreted as the fraction of the nucleon momentum carried by the struck quark). This was quickly interpreted as an effect of the pion excess in nuclei [4]. In a calculation of Ericson and Thomas, the change in F_2 due to the pion excess, δF_2^π, was written terms of a convolution of the pion structure function, F_2^π, with the distribution of extra pions in the nucleus,

$$\delta^\pi F_2^A(x) = \int_x^1 dy[f_\pi^A(y) - f_\pi^N(y)]F_2^\pi(x/y), \qquad (1)$$

where f_π^A (f_π^N) is the distribution of virtual pions in the nucleus (nucleon). A key point of their calculation is that f_π^A can be calculated in terms of (form–factor) weighted integrals over the isovector spin–isospin nuclear response function, $R_L(q,\omega)$. $R_L(q,\omega)$ describes the response of a nucleus to a pion–like excitation with momentum q and energy ω. This response function is relatively easy to calculate in mean–field theories and Ericson and Thomas were able to reproduce the EMC results fairly well with reasonable parameters for their model of $R_L(q,\omega)$.

Unfortunately, later experiments found a much smaller enhancement of F_2 at low x [5,6]. While the calculations of Ericson and Thomas could be adjusted somewhat to better agree with this smaller enhancement, this reduces the pion excess and hence reduces the role of pion exchange in nuclei.

1.2 Polarization Transfer Experiments

Since the predicted enhancement in the F_2 structure function came about from the spin–isospin nuclear response function, $R_L(q,\omega)$, it seems reasonable to try to probe that quantity directly (rather than relying on multidimensional weighted integrals over R_L). This can be accomplished in $(\vec{p},\vec{p}')$ and $(\vec{p},\vec{n})$ experiments in which a polarized proton scatters from a nucleus and the outgoing proton or neutron polarization is measured. The $(\vec{p},\vec{p}')$ reaction measures a linear combination of the isovector and isoscalar response, while the $(\vec{p},\vec{n})$ reaction is purely isovector and hence a cleaner probe, but is more challenging technically.

An early $(\vec{p},\vec{p}')$ experiment [7], and later, a $(\vec{p},\vec{n})$ experiment [8] both failed to see any effect from nuclear pions. However, it should be pointed out that these experiments measured the ratio of longitudinal to transverse responses, R_L/R_T. It was assumed that there should be no nuclear dependence to the transverse response, hence a measurement of the nuclear dependence of R_L/R_T should be sensitive to an A dependence in the longitudinal response. However, further analysis of the $(\vec{p},\vec{n})$ results indicated that the longitudinal response alone was not inconsistent with the effects predicted from the pion excess, and that the lack of effect seen in the R_L/R_T ratio may have been due to an enhancement of the transverse response [9]. Unfortunately, the systematic uncertainties on the individual responses were too large to make a strong statement either way.

1.3 The Drell–Yan Process

Similar to DIS, the Drell–Yan process is also sensitive to quark distributions in nuclei and nucleons. In this case, a quark from a high proton beam combines with an antiquark from the target to form a virtual photon, which then becomes a lepton, anti–lepton pair. Since the valence structure of the pion is quark, anti–quark (while the valence structure of the nucleon involves only quarks), the Drell–Yan process, which probes anti–quarks in the target, should be more sensitive to contributions from nuclear pions.

Experiment E772 at Fermi National Accelerator measured the ratio of Drell–Yan cross sections between various nuclei and deuterium [10], and saw little evidence for any nuclear effects. The ratios were inconsistent with predictions based models including nuclear pion effects.

1.4 Discussion

The results of the experiments discussed above do not give a clear signature of the presence of "extra" pions in nuclei. However, one must keep in mind the limitations of these experiments. First, both the DIS and Drell–Yan experiments do not probe the pion field directly. These experiments probe quark distributions, and while pions in the nucleus should modify these distributions, the relation between quark distributions and the pion excess is not without some ambiguity. Further, "shadowing" (where one sees a suppression of the F_2 distribution in nuclei) at low x complicates the picture even more. Issues with the polarization transfer experiments have already been discussed - namely that there seems to be an unanticipated enhancement of the transverse nuclear response masking signatures of an enhancement of R_L, although the precision of these results is not sufficient to make a definite statement about the presence or lack of nuclear dependence in R_L.

The have also been several theoretical attempts to explain the apparent absence of nuclear pions. One such explanation supposes that the enhancement of $R_L(q, \omega)$ occurs at large energy transfers, ω [11]. Since the polarization transfer experiments carried out thus far did not sample large ω they were not sensitive to this region. Furthermore, the DIS and Drell–Yan experiments cannot probe this region of the response due to kinematic constraints. A second explanation supposes that chiral symmetry is partially restored in nuclei, leading to medium modifications of particle masses and coupling constants. This Brown–Rho rescaling, then, reduces the effects one might expect from the presence of excess pions [12].

Of course the simplest explanation may be that there just are not as many pions in the nucleus as models predict. In particular, as noted in Ref. [2], a large fraction of the pion excess comes from pions coupling to virtual Δ's in the ground state nuclear wave function. It turns out that many nuclear properties are not terribly sensitive to the presence of Δ's and a reduction of their contribution does not have a large impact the overall success of these potentials. If the pion excess is smaller than originally predicted, a more sensitive probe of the nuclear pion filed may be needed.

2 Pion Electroproduction as a Probe of the Nuclear Pion Field

Perhaps the most direct probe of the pion field of the nucleus (and nucleon) is pion electroproduction. The cross section for this process can be written in terms of a virtual photon flux (Γ) and the virtual photon cross section,

$$\frac{d^5\sigma}{d\Omega_e dE_e d\Omega_\pi} = \Gamma \left[\frac{d\sigma_T}{d\Omega_\pi} + \epsilon \frac{d\sigma_L}{d\Omega_\pi} \sqrt{2\epsilon(1+\epsilon)} \frac{d\sigma_{LT}}{d\Omega_\pi} \cos\phi + \epsilon \frac{d\sigma_{TT}}{d\Omega_\pi} \cos 2\phi \right] . \quad (2)$$

The virtual photon cross section above has been broken down into contributions from transverse (T) and longitudinal (L) virtual photons, as well as interference terms (LT, TT). The Born level processes that contribute to pion electroproduction are shown in Fig. 2. What is important to note here is that the first diagram (the pole process), in which a virtual pion is emitted from the nucleon and is knocked on–shell via its interaction with the virtual photon, dominates the longitudinal cross section in forward kinematics (i.e. the pion emitted along the virtual photon direction). If one integrates over ϕ, the interference terms disappear and the longitudinal cross section can be extracted via a Rosenbluth separation in which one measures the cross section at two or more values of ϵ (the virtual photon longitudinal polarization). One then plots the virtual photon cross section as a function of ϵ and fits the linear dependence – the slope giving the longitudinal cross section and the intercept the transverse.

In the limit of pole dominance, a measurement of the longitudinal pion electroproduction cross section in nuclei (compared to the free nucleon) should give some insight into the role of pions in the nucleus.

3 Recent Pion Electroproduction Results

In this section, I will discuss the results of two pion electroproduction experiments and their impact on the search for the effects of nuclear pions. The first, Jefferson Lab experiment E91003, focused on quasifree production of charged pions from ^{1}H, ^{2}H, and ^{3}He. In quasifree production, one views the process as occurring on one of the bound nucleons in the nucleus. The final

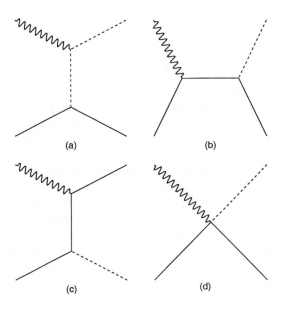

Figure 2. Born diagrams for pion electroproduction. Solid lines denote nucleons, wavy lines denote virtual photons, and dashed lines are pions. In forward kinematics, the pion pole process (a) dominates the longitudinal cross section.

state is then a continuum state with the "struck" nucleon is no longer bound. Alternatively, one can investigate coherent pion electroproduction, in which the nucleus remains intact and the process can, to some extent, be viewed as occurring on the nucleus as a whole. The interesting results of such an experiment carried out at the MAMI accelerator at Mainz then motivated a similar analysis of the E91003 data.

3.1 Quasifree Pion Electroproduction

As mentioned above, in quasifree pion electroproduction the "struck" nucleon is knocked out of the nucleus and the final state is no longer bound. Because of this, the pion electroproduction cross section as written in Eq. 2 has another degree of freedom due to the relative momentum between the recoiling nucleon and the target remnants. Hence, the virtual photon cross section can now be written (ignoring the LT and TT terms),

$$\frac{d\sigma}{d\Omega_\pi dM_x} = \frac{d\sigma_T}{d\Omega_\pi dM_x} + \epsilon \frac{d\sigma_L}{d\Omega_\pi dM_x} , \tag{3}$$

where M_x is the missing mass of the recoiling system ($M_x^2 = (q + P_A - p_\pi)^2$, where q, P_A, and p_π are the momentum four–vectors for the virtual photon, target, and pion respectively), which in the quasifree case is no longer fixed. To compare with the free nucleon cross section, one would ideally integrate the cross section over all M_x.

The primary aim of Jefferson Lab experiment E91003 was to measure the longitudinal cross section for quasifree charged pion (π^+ and π^-) electroproduction from light nuclei (^{2}H and ^{3}He) and compare to hydrogen to look for signatures of the pion excess. The kinematics were chosen so as to correspond to the region where one expects a suppression ($k \approx 1.0$ fm^{-1}) and an enhancement ($k \approx 2.4$ fm^{-1}) of the virtual pion field (see. Fig. 1). For each target, data were taken at two ϵ, and the longitudinal and transverse cross sections extracted.

Unfortunately, once one has the longitudinal cross section for each target, one can not immediately use the raw target ratios to look for signs of the pion excess. Naively, one would assume that in the absence of nuclear pions, the quasifree integrated cross sections would be the same as that for the free nucleon. However, there are several effects that have nothing to do with nuclear pions that could cause the nuclear results to differ from the nucleon result.

First, the experiment was not able to measure the full quasifree M_x distribution at all settings. This is illustrated in Fig. 3 where the low ϵ virtual photon π^+ cross section from deuterium is shown at both values of k sampled in the experiment. While the M_x acceptance was quite good for the low k data, only about 80% of the M_x distribution from deuterium was measured for the high k data. In addition to this incomplete coverage of the quasifree peak, there are effects that simply come from the kinematic threshold for pion electroproduction. The total photon–nucleon center–of–mass energy (commonly denoted W) necessary to produce a charged pion is ≈ 1.08 GeV. At the high k kinematics, the total center–of–mass energy was about 1.15 GeV for the free nucleon; rather close to threshold. A nucleon bound in a nucleus has a corresponding Fermi momentum. This means that in a quasifree picture of pion electroproduction, the total energy in the center of mass of the virtual photon and the bound nucleon is no longer single valued due to the nucleon motion, and in particular, W may dip below pion electroproduction threshold for certain regions of the bound nucleon wave function. Hence, the quasifree ratio

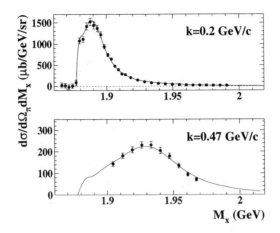

Figure 3. M_x acceptance for ^{2}H$(e, e'\pi+)nn$ data at the low (top) and high (bottom) k kinematics accessed in Jefferson Lab experiment E91003 [13]. The points are data and the curves are quasifree calculations.

may differ from unity due to simple threshold effects. To complicate matters further, the Fermi motion of the nucleon will also affect the fundamental $\gamma^* - N$ cross section for pion electroproduction impacting the comparison with the free nucleon.

In order to disentangle the potential effects of nuclear pions from the complications described above, experimental ratios (R^{exp}) of ^{3}He and ^{2}H to ^{1}H and ^{3}He to ^{2}H were formed. These ratios were then compared to quasifree calculations of the target ratios ($R^{\mathrm{q.f.}}$). These calculations started with some model of the free nucleon cross section, which was then convolved with realistic nucleon distributions for the ^{3}He and ^{2}H targets. Since the calculation itself involves ratios, it is not very sensitive to the details of the input pion electroproduction model. A comparison of R^{exp} to $R^{\mathrm{q.f.}}$ should therefore reveal any nuclear effects *not* considered above, including nuclear pions. The results of this comparison are shown in Fig. 4. Clearly, the results at high k are consistent with no enhancement while, on the other hand, the low k results show some obvious suppression. At this point, it is worth pointing out that nothing has been said about final–state interactions (FSI) between the outgoing pion the recoiling nucleons. One of the motivating reasons for doing the experiment on light nuclei (rather than some heavy nuclei where the effect should be larger) was to suppress complications from large FSI. In

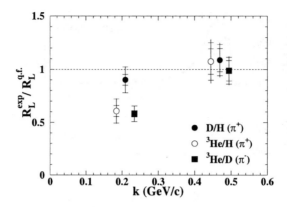

Figure 4. Longitudinal cross section ratios compared to quasifree calculations [13]. Inner error bars are statistical, middle errors are statistics+systematics added in quadrature, and outer errors include the additional uncertainties from the quasifree calculations.

fact, for most of the points in Fig. 4, the FSI effects are estimated to be less than 5%. The one exception is the high k, ^{3}He to H ratio, where the effect may be as large as 33%. Even accounting for this effect leads to a ratio of $R^{\text{exp}}/R^{\text{q.f.}} \approx 1.43 \pm 0.28$, which is about a 1.5σ effect (although this ignores the uncertainty involved is the estimate if the $\pi - N$ FSI).

One may ask then if the above result rules out the presence of excess pions in the nucleus? To answer this question, we must make some connection between the pion excess and the longitudinal ratios shown in Fig. 4. The simplest approximation is to assume that the pion electroproduction cross section is proportional to the number of virtual pions. Using the pion excess results of Friman *et al.* [2] and the formalism of D. Koltun [11], such an estimate yields a suppression of 1% (14%) for the ^{2}H (^{3}He) to ^{1}H ratios at low k and an enhancement of 6% (13%) at high k. Clearly, the results at high k are consistent with these estimates, although the low k results are a bit more suppressed (although this may in part be due to the exclusion of the coherent final state which is implicitly included in the calculation of the pion excess). The results of this experiment, then, seem to lack the neccesary precision relative to the estimated size of the effect to make any definite statements about the pion excess. However, the presence of the Pauli blocking suppression at low k, which should be present whether there are nuclear pions or not, gives one some confidence that the method is fundamentally sound and that one may be able make a more definitive statement about nuclear pions with improved experimental parameters (i.e. kinematics that have better M_x coverage, using a ^{4}He target where the signal may be twice as large),

3.2 Coherent Pion Electroproduction

Since coherent pion electroproduction leaves the nucleus intact, there is an inherent reduction in the pion electroproduction cross section (relative to the free nucleon) due to the overlap of the initial and final state nuclear wave functions. At first glance, this would seem to complicate the experimental extraction of nuclear pion effects. In fact, as will be shown, one can make some rather simple approximations to try and deal with these effects and get a handle on modifications to the pion electroproduction cross section.

A detailed calculation [14] of the ratio of the coherent cross section to that from hydrogen showed that the dominant factor, as mentioned above, is the overlap of the initial (^{3}He) and final (^{3}H) state nuclear wave functions. This overlap is described by the square of the nuclear form factor, $F^2(k)$, where k is the 3–momentum transfer to the nucleus. Including an additional factor, ρ, that comes from the difference in the density of final states (i.e. the pion momentum is different in the coherent process from than in production from the free nucleon), the cross section for coherent production can be approximated,

$$\sigma(^3He) = \rho F^2(k)\sigma(H). \tag{4}$$

Here it is important to note that ρ is typically on the order of unity and that the dominant modification comes from the form factor, which drops off rapidly with k. Of course, more detailed calculations are helpful, but the above already gives one some insight into the size of the effect one might expect in the absence of medium modifications to the electroproduction cross section.

3.3 Mainz Results

The A1 Collaboration using the Mainz Microtron performed a series of measurements of charged pion electroproduction from ^{3}He. The high resolution of their experimental setup allowed them to cleanly measure the coherent ^{3}He$(e, e'\pi^+)^3$H process in which a π^+ is produced and the helium–3 nucleus (two protons and a neutron) becomes a mirror triton nucleus (one proton and two neutrons). Laboratory cross sections were measured for three values of ϵ at each of two Q^2 values and longitudinal and transverse cross sections were extracted via Rosenbluth separations [15,16].

These separated cross sections for the ^{3}He$(e, e'\pi^+)^3$H process were compared to distorted wave impulse approximation (DWIA) calculations that

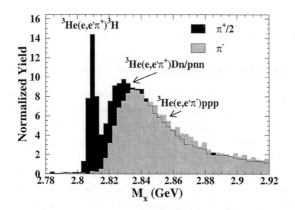

Figure 5. Missing mass distributions for π^+ and π^- production from ^{3}He the E91003 low k kinematic setting [18]. The π^+ yield has been divided by two for comparison with the π^- yield. The ^{3}H final state is clearly visible in the π^+ data.

started with a unitary isobar model of pion electroproduction from the proton [17] and included final state interactions between the outgoing pion and recoiling nucleus. The results of these calculations *under*estimated the longitudinal cross section and *over*estimated the transverse cross section, both by about a factor of two. The calculation was brought into agreement by introducing modifications to the pion propagator (see Fig. 2, diagram (a)), in the case of the longitudinal cross section, and the width of Δ, in the case of the transverse cross section. Of particular relevance here is the modification of the pion propagator. It is such a modification that was suggested by Ericson and Thomas [4] to explain the EMC effect and gives rise to an excess of pions in the nucleus.

3.4 JLab E91003 Results

Although the focus of JLab experiment E91003 had been on quasifree pion electroproduction, it turns out that the coherent ^{3}He$(e, e'\pi^+)^3$H process was also in the experimental acceptance at the low $k = 0.2$ GeV/c setting (see Fig. 5). While the ^{3}H peak is clearly visible in the E91003 data, the missing mass resolution was not good enough to separate the coherent peak and quasifree final states event–by–event, but a pure ^{3}H sample could be obtained via a Monte Carlo background subtraction.

Although the E91003 data did not have the high kinematic precision of the Mainz data, it did benefit in other ways. In the E91003 data, the outgoing pion momentum was large (≈ 1.0 GeV), such that final state interactions were small. In the Mainz data, final state interaction effects were estimated to be about 30%. Also, the Mainz data was compared to a calculation that began with a model of the pion electroproduction cross section – this model is not constrained by any separated cross section data at the Mainz kinematics. The E91003 data, on the other hand, also included hydrogen targets so that a direct target comparison was possible.

The E91003 result is presented in the form the ratio of the coherent cross section to that from hydrogen. An estimate of this ratio, assuming no effects from nuclear pions, can be obtained from Eq. 4. The helium–3 form–factor (squared) can be written,

$$F^2(k) = \exp\left(-k^2/18a\right), \qquad (5)$$

where k is the momentum transfer to the nucleus and $a = 0.064$ fm. At E91003 kinematics, $k = 0.19$ GeV/c and the form–factor squared is $F^2(k) = 0.447$, while the kinematic factor, ρ, is 0.95. This gives a net ratio of $\sigma(^3He)/\sigma(H) \approx 0.42$. The measured ratios yield $0.50 \pm 0.04(\text{stat}) \pm 0.07(\text{sys})$ for the longitudinal cross sections and $0.24 \pm 0.04(\text{stat}) \pm 0.04(\text{sys})$ for the transverse.

Although the longitudinal ratio is consistent our simple estimate, the transverse ratio is smaller by about a factor of two. Thus by our simple calculation, it seems there is not strong evidence for an enhancement of the longitudinal cross section, although the results are consistent with the trend seen in the Mainz data that the transverse cross section is suppressed more than the longitudinal. Detailed calculations at the E91003 kinematics would be useful to help make a more quantitative statement regarding medium modifications to the longitudinal and transverse cross sections.

3.5 Pion Electroproduction Summary

We have seen that the JLab results of measurements of quasifree longitudinal pion electroproduction from nuclei yield a suppression at low virtual pion momentum, k and are basically consistent with no effect at high k where effects from the pion excess are expected. However, simple estimates of the effects from excess pions indicate that the expected enhancement of the cross section ratios is small enough that it cannot be discerned within the precision of the current data.

The results of the Mainz measurement of the coherent $^3He(e, e'\pi^+)^3H$ process hint at an enhancement of the longitudinal cross section due to medium modifications of the pion propagator. This would be a strong signal for the presence of nucelar pions. However, the E91003 data for the same process, which benefit from a direct comparison to the hydrogen cross section, do not see a significant enhancement. It is not obvious whether the inconsistency lies in the lack of a direct comparison to hydrogen in the Mainz data, or if the estimate of Eq. 4 is too simple to be meaningfully compared to the JLab data.

4 Where Do We Go from Here?

Clearly, the question reqgarding the presence or absence of nuclear pions is not settled by either the early DIS, Drell–Yan, and polarization transfer experiments or the pion electroproduction experiments described above. While the DIS and Drell–Yan experiments seem hampered by theoretical uncertainties in interpreting their results in terms of nuclear pions, the polarization transfer and pion electroproduction experiments suffer from experimental uncertainties. In the case of pion electroproduction, this could in part be overcome by optimizing kinematics and taking data on ^{4}He, where the expected effect should be about twice as large and the pion rescattering effects still controllable. Nonetheless, the predicted effect is still only about 25% in the best case. Fortunately, there may be another probe that gives a much larger effect.

It has been suggested that one may measure the inclusive longitudinal cross section from nuclei to look for effects from nuclear pions. G. Miller has developed a model of nuclei using light–front dynamics that gives results consistent with the Drell–Yan ratios discussed earlier, yet predicts significant effects in inclusive scattering [19]. In this model, the ratio of longitudinal cross sections between a nucleus and deuterium is given by,

$$\frac{\sigma_L(A)}{\sigma_L(D)} = 1 + x\frac{2}{3}f_\pi(\xi)\frac{\nu^2}{(Q^2+\nu^2)}\frac{F_\pi^2(Q^2)}{F_2^D R_D}(1+R_D), \qquad (6)$$

where $f_\pi(\xi)$ is the pion distribution function, ξ is the Nachtmann variable, F_π is the pion form–factor, F_2^D is the usual F_2 structure function in deuterium, and R_D is the ratio of longitudinal to transverse cross sections. Even ignoring Δ degrees of freedom in nuclei (which accounts for much of the pion excess in the model of Friman et al. [2]), Miller predicts large effects (75%) at low Q^2 ($\approx 0.3~(\text{GeV}/c)^2$). Such a measurement would be difficult since R_D is quite small in the region of interest (making a Rosenbluth separation technically challenging), but may be possible at JLab.

References

1. H. Yukawa, Proc. Phys. Math. Soc. Jap. **17**, 48 (1935).
2. B. L. Friman, V. R. Pandharipande and R. B. Wiringa, Phys. Rev. Lett. **51**, 763 (1983).
3. J. J. Aubert *et al.* [European Muon Collaboration], Phys. Lett. B **123**, 275 (1983).
4. M. Ericson and A. W. Thomas, Phys. Lett. B **128**, 112 (1983).
5. J. Ashman *et al.* [European Muon Collaboration], Z. Phys. C **57**, 211 (1993).
6. J. Gomez *et al.*, Phys. Rev. D **49**, 4348 (1994).
7. L. B. Rees *et al.*, Phys. Rev. C **34**, 627 (1986).
8. J. B. McClelland *et al.*, Phys. Rev. Lett. **69**, 582 (1992).
9. T. N. Taddeucci *et al.*, Phys. Rev. Lett. **73**, 3516 (1994).
10. D. M. Alde *et al.*, Phys. Rev. Lett. **64**, 2479 (1990).
11. D. S. Koltun, Phys. Rev. C **57**, 1210 (1998).
12. G. E. Brown, M. Buballa, Z. B. Li and J. Wambach, Nucl. Phys. A **593**, 295 (1995).
13. D. Gaskell *et al.*, Phys. Rev. Lett. **87**, 202301 (2001).
14. R. J. Loucks and V. R. Pandharipande, Phys. Rev. C **54**, 32 (1996).
15. K. I. Blomqvist *et al.*, Nucl. Phys. A **626**, 871 (1997).
16. M. Kohl *et al.* [A1 Collaboration], Phys. Lett. B **530**, 67 (2002).
17. D. Drechsel, O. Hanstein, S. S. Kamalov and L. Tiator, Nucl. Phys. A **645**, 145 (1999).
18. D. Gaskell *et al.*, Phys. Rev. C **65**, 011001 (2002).
19. G. A. Miller, Phys. Rev. C **64**, 022201 (2001).

POLARIZATION OBSERVABLES

R. GILMAN

Rutgers University, 136 Frelinghuysen Road, Piscataway, NJ 08854 USA
and
Jefferson Lab, 12000 Jefferson Avenue, Newport News, VA 23606 USA
E-mail: gilman@jlab.org

Polarization observables provide information that is inaccessible through cross section measurements. In some cases, experimental limitations do not allow information about smaller amplitudes to be extracted in practice, whereas in other cases, the information is not present in cross section measurements which average over spin states. I will introduce and review several recent topics investigated with polarization observables in Jefferson Lab experiments.

1 Introduction

In these notes I provide an introduction to and review of several topics in spin physics. In all cases, experiments have been carried out at Jefferson Lab, and polarization observables have been important in advancing our physical understanding. I will introduce both the theory and experimental techniques.

In some topics, the polarization observables are important because, although the information could in principle be extracted from cross section measurements, it is not possible in practice to measure cross sections precisely enough. This is particularly true for small amplitudes, which may contribute to the cross section at a level below the level of the experimental uncertainties. An example of this is the extraction of the two elastic electromagnetic proton form factors in $ep \to ep$ scattering.

Here is a simple demonstration of the enhanced sensitivity of the polarization observables to small amplitudes. Let us assume that there are two amplitudes, which physically have the values $M_{afi}^{physical} = 1$ and $M_{bfi}^{physical} = 0.1i$. The cross section is then proportional to $|M_{fi}|^2 = |1 + 0.1i|^2 = 1.01$. Measurements with 1% accuracy are extremely difficult, and thus it will be hard to constrain amplute b. A typical polarization observable might be $P \propto \Im[1 \times 0.1i]/|1 + 0.1i|^2 = 0.099$. Now let us assume a theorist calculates $M_{afi}^{calc} = 0.99$, and $M_{bfi}^{calc} = 0.2i$. The calculated cross section is $|M_{fi}|^2 = |0.99 + 0.2i|^2 = 1.02$. This is only off by 1%, so the theory appears to be in pretty good shape. The calculated polarization is $P \propto \Im[0.99 \times 0.2i]/|0.99 + 0.2i|^2 = 0.194$. The polarization is off by nearly a factor of two. This difference can be somehwat easily measured, and it will indicate a need to reexamine the theory.

More generally, unpolarized cross sections sum over all spin states, and the more detailed information present in the polarization observables is lost. Polarization measurements allow the determination of several independent combinations of the underlying reaction mechanism amplitudes, and thus provide the most stringent test of our theoretical understanding. Polarization data have often proven *too* difficult for detailed understanding – but before one starts thinking of blaming theorists, it should also be noted that experiments can be difficult and there are numerous cases of bad and / or inconsistent experimental data.

In the following sections, I will discuss the following: elastic *ep* scattering and the nucleon form factors, elastic *ed* scattering and the deuteron form factors, deuteron photodisintegration and the limits of meson-baryon theories, meson photoproduction and baryon resonances, and deep-inelastic scattering, nucleon spin structure, and sum rules. I have, perhaps unfairly, focussed on measurements with which I have been more involved over the past several years.

2 Nucleon form factors and elastic scattering

When an electron elastically scatters from a nucleon electromagnetically, the cross section is given by:

$$\frac{d\sigma}{d\Omega} = \left[\frac{d\sigma}{d\Omega}\right]_{point} \times F^2(\mathbf{q}). \tag{1}$$

The cross section is the same as that for a point particle, multiplied by the square of a function that we call the form factor. In a nonrelativistic interpretation, the form factor, $F(\mathbf{q})$, arises from the Fourier transform of the charge distribution:

$$F(\mathbf{q}) = \int \rho(\mathbf{x}) e^{i\mathbf{q}\cdot\mathbf{x}} d^3\mathbf{x}. \tag{2}$$

There are numerous cases which can be evaluated analytically. Of particular interest to nuclear and particle physicists is the exponential charge distribution leading to the dipole form factor:

$$\rho(x) = N e^{-\Lambda x} \iff F(q) \equiv G_D = \left(\frac{\Lambda^2}{\Lambda^2 + q^2}\right)^2 = \left(\frac{1}{1 + q^2/\Lambda^2}\right)^2 \tag{3}$$

For a potential that is limited in spatial extent, one generally expects the wave function to extend beyond the range of the potential, with an exponential tail to the wave function, and thus to the charge distribution. For a very short-range potential, like a δ function, the wave function and charge distribution will be purely exponential.

The discussion above was written as if there were only charge scattering. More generally, for a particle of spin J, there are $2J + 1$ electromagnetic form factors. There is no unique way to choose the $2J + 1$ form factors, since one can always transform to other combinations of two independent functions. A common choice for the spin-$\frac{1}{2}$ proton (spin-1 deuteron) are the electric and magnetic (electric, magnetic, and quadrupole) form factors, because they have a simple, obvious, intuitive description, at least in the nonrelativisitic limit.[a] The most common alternative choice is to use the Dirac and Pauli form factors, since the electromagnetic current of the free nucleon, $j^\mu = e\bar{u}\Gamma^\mu u$, with u the electron wave function, is more simply expressed with these form factors: $\Gamma^\mu = F_1\gamma^\mu + \frac{i\kappa}{2m}F_2\sigma^{\mu\nu}q_\nu$. Here F_1 (F_2) is the Dirac (Pauli) form factor, $q_\nu = (\omega, \vec{q})$ is the four-momentum transfered in ep scattering, γ^μ is the Dirac equation γ matrix, and $\sigma^{\mu\nu} = \frac{i}{2}(\gamma^\mu\gamma^\nu - \gamma^\nu\gamma^\mu)$. The sets of form factors are related by: $G_E \equiv F_1 - \frac{\kappa Q^2}{4m^2}F_2$ and $G_M \equiv F_1 + \kappa F_2$, with κ the anomalous magnetic moment of the nucleon, $\kappa = \mu - 1 = \frac{g-2}{2}$. The four-momentum transfer $Q^2 = -q^2 = 4EE'\sin^2\frac{\theta}{2}$ is defined to so that one uses a positive quantity, as $q^2 < 0$ for ep elastic scattering.[b] The functions are normalized at $Q^2 = 0$ so that $G_E(0) = 1$, $G_M(0) = \mu$, $F_1(0) = 1$, and $F_2(0) = 1$. It is common to use $\tau = \frac{Q^2}{4m^2}$ to simplify the appearance of these and other equations.

The dipole form factor works amazingly well at describing the nucleon elastic form factors. For many years, it has been believed that the nucleon elastic form factors[c] are, to within perhaps 10%, consistent with the dipole

---

[a] More correctly, these form factors provide the charge and magnetization distributions in the "Breit", or brick-wall frame, rather than in the nucleon rest frame. The Breit frame is reached by boosting the lab frame along the outgoing proton momentum, or the c.m. frame transverse to the momentum transfer, so that one reaches a frame in which the proton initial (final) momentum is $-\vec{q}/2$ ($\vec{q}/2$), and the energy transfer $\omega = 0$. The difference between the frames is small, for low Q^2, nonrelativistic kinematics. The relativistic interpretation is much more difficult. A recent general discussion concerning interpretation issues can be found in [1]. Although not discussed there, modern discussion often focusses on the interpretation of the neutron charge radius. A general concern is that one needs to worry about what the extracted radius means, unless the size of the probe is much smaller than the size of the structure being probed.

[b] E, E', and θ are not independent for elastic scattering, as $x = 1 = Q^2/2m\omega$ with $\omega = E - E'$ leads to $E'/E = (1 + \frac{2E}{m}\sin^2\frac{\theta}{2})^{-1}$.

[c] An exception is the neutron electric form factor, which is 0 at $Q^2 = 0$, and is discussed further below.

form factor over a large range of Q^2. The most noted contribution of Jefferson Lab so far is the determination of the proton elastic form factor at large Q^2, which shows very strong deviations from the dipole formula.

Traditionally, the nucleon form factors have been determined by measuring elastic ep scattering cross sections. The laboratory cross section is given by $\left.\frac{d\sigma}{d\Omega}\right|_{lab} = \left[\frac{\alpha^2}{4E^2 \sin^4 \frac{\theta}{2}}\right] \frac{E'}{E} \times \left[(F_1^2 + \tau\kappa^2 F_2^2)\cos^2 \frac{\theta}{2} + 2\tau(F_1 + \kappa F_2)^2 \sin^2 \frac{\theta}{2}\right]$.
The interference terms between F_1 and F_2 make this formula not so useful for determining the form factors. Instead, in terms of the electric and magnetic form factors, $\left.\frac{d\sigma}{d\Omega}\right|_{lab} = \left[\frac{\alpha^2}{4E^2 \sin^4 \frac{\theta}{2}}\right] \frac{E'}{E} \times \cos^2 \frac{\theta}{2} \times \left[\frac{G_E^2 + \tau G_M^2}{1+\tau} + 2\tau G_M^2 \tan^2 \frac{\theta}{2}\right]$.
Here, the Rosenbluth separation technique becomes clear. The cross section is measured for the same momentum transfer, but at multiple beam energies. At higher beam energies, the same momentum transfer is reached at smaller scattering angles. The different kinematic factors affecting G_E and G_M will allow them to be simply extracted.

It is common to make a definition like $\left.\frac{d\sigma}{d\Omega}\right|_{lab} \equiv \left.\frac{d\sigma}{d\Omega}\right|_0 \times \left.\frac{d\sigma}{d\Omega}\right|_{reduced}$ with $\left.\frac{d\sigma}{d\Omega}\right|_0 = \left[\frac{\alpha^2}{4E^2 \sin^4 \frac{\theta}{2}}\right] \frac{E'}{E} \times \cos^2 \frac{\theta}{2}$. The Rosenbluth separation can then be shown as a plot of $\left.\frac{d\sigma}{d\Omega}\right|_{reduced}$ vs. $\tan^2 \frac{\theta}{2}$. The plot will be linear, with a slope of $2\tau G_M^2$, and an intercept of $\frac{G_E^2 + \tau G_M^2}{1+\tau}$. The wide range of $\tan^2 \frac{\theta}{2}$ makes this plot visually unappealing, so it is is now standard to instead use the variable $\epsilon^{-1} = \left[1 + 2(1 + \tau)\tan^2 \frac{\theta}{2}\right]$, which is bounded from 0 to 1, and spaces the data points more evenly. The cross section is typically rewritten as $\epsilon\frac{1+\tau}{\tau}\left.\frac{d\sigma}{d\Omega}\right|_{reduced} = G_M^2 + \frac{\epsilon}{\tau}G_E^2$, and $\epsilon\frac{1+\tau}{\tau}\left.\frac{d\sigma}{d\Omega}\right|_{reduced}$ is plotted vs. ϵ. The intercept is G_M^2, and the slope is $\frac{G_E^2}{\tau}$.

The Rosenbluth separation is useful at moderate Q^2 kinematics, in which the electric and magnetic form factors contribute more or less equally to the cross section. It is not useful at small Q^2, for which the magnetic contributions are suppressed by a factor of τ, and the cross section is dominated by electric scattering. It is also not useful at large Q^2, for which the magnetic contributions are enhanced by a factor of τ, and the electric contributions are difficult to determine. In these extremes, polarization techniques are needed to enhance the smaller contribution. An extended description of the highest Q^2 Rosenbluth separation measurements on the proton, which showed $G_E \approx G_M/\mu \approx G_D$ can be found in [2].

Figure 1 shows an example of a Rosenbluth separation plot. Plotting the reduced cross section as a function of ϵ allows the points to all be seen and distinguished on a reasonable scale. The small differences that occur when G_E is modified, in this case by about 40%, can be seen to have only a small

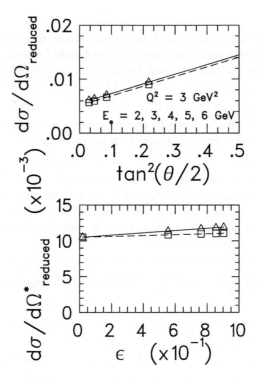

Figure 1. An example of a Rosenbluth separation for $Q^2 = 3$ GeV^2. The top plot shows the pseudo-points for various energies plotted as a function of $\tan^2 \theta/2$. The lowest energy, 2 GeV, point is well off scale in the top panel, at $\tan^2 \theta/2 \approx 15$. The bottom plot shows the reduced cross section $\epsilon \frac{1+\tau}{\tau} \left| \frac{d\sigma}{d\Omega} \right|_{reduced}$ plotted as a function of ϵ. Large $\tan^2 \theta/2$ corrsponds to small ϵ. The solid curve with triangular symbols uses the standard dipole form factors for both G_E and G_M. The dash curve with square symbols uses the same G_M, but a reduced G_E corresponding to the Jefferson Lab polarization results.

effect on the cross section, justifying the need for polarization observables, as discussed above.

Two techniques have been used, the scattering of polarized electrons on unpolarized targets, measuring the recoil nucleon polarization, and the asymmetry in scattering polarized electrons on polarized nucleons. In the former case, the polarization transfers[3] are given by

$$P^x = -2\sqrt{\tau(1+\tau)} \tan(\theta/2) G_E G_M / \left| \frac{d\sigma}{d\Omega} \right|_{reduced} \text{ and} \qquad (4)$$

$$P^z = \frac{k+k'}{m}\sqrt{\tau(1+\tau)}G_M^2 \tan^2(\theta/2)/\left|\frac{d\sigma}{d\Omega}\right|_{reduced}. \tag{5}$$

Here, the x and z components of the polarization are both in the scattering plane, transverse and parallel to the nucleon momentum. Since the reduced cross sections involves a combination of G_E^2 and G_M^2, it can be seen that each of the polarization components actually depends on the ratio of the form factors, $\frac{G_E}{G_M}$. The main point of these formulas is that the small G_E contribution comes in linearly to the polarizations, as opposed to quadratically to the cross sections, and thus it is easier to determine.

2.1 Recoil polarimetry

Measurement of proton polarization has been a standard technique, especially at hadronic facilities. At Jefferson Lab, and other electron machines, focal plane polarimeters (FPPs) are put in spectrometers to measure outgoing nucleon polarizations in nuclear reactions.

The basis for much proton polarimetry is the spin-orbit nuclear force, which is proportional to $L \cdot S$. For simplicity, let us consider a proton with its spin oriented vertically upward, which scatters from a nucleus in the horizontal plane. Depending on whether the proton approaches the left or right side of the nucleus, its orbital angular momentum will be oriented either vertically upward or downward, parallel or anti-parallel to the spin direction. Thus, protons approaching the nucleus on opposite sides will have, in addition to the large spin-independent central force, an increased or decreased interaction strength from the spin orbit force. The systematic difference between the two sides of the nucleus generates a left – right scattering asymmetry. The size of the asymmetry depends on the analyzer nucleus. Carbon is typically chosen, as it is inexpensive, safe, and can be easily obtained in a variety of shapes and sizes. Hydrogen has a larger analyzing power, but the low density and safety issues with cryogenic hydrogen generally preclude its use. Plastics such as CH_2 provide a nice compromise.

The measurement of the polarization then proceeds as follows. In some nuclear reaction, a distribution of protons is produced which will be transported from the target to the spectrometer focal plane. In general, the protons can have spin components in the x, y, and z directions.[d] As the protons are

---

[d]Indeed, the choice of coordinate system is an enduring problem in spin physics experiments, as, to a neutral observer, there are many different "standard" conventions, concerning how to label the axes, in what directions the axes point, and in what frame they are defined. There are also the additional related problems of what amplitudes to use and how to name the polarization observables. I admit to having personally made the morally repugnant

transported to the spectrometer focal plane, the spin will precess about the magnetic fields so that, in general, the focal plane and target spins are related by a unitary matrix,

$$\begin{pmatrix} S_{fp}^x \\ S_{fp}^y \\ S_{fp}^z \end{pmatrix} = \begin{pmatrix} S_{xx} & S_{xy} & S_{xz} \\ S_{yx} & S_{yy} & S_{yz} \\ S_{zx} & S_{zy} & S_{zz} \end{pmatrix} \begin{pmatrix} S_{tgt}^x \\ S_{tgt}^y \\ S_{tgt}^z \end{pmatrix}.$$

For these notes, we shall for simplicity consider only the simplest case: All protons scatter in a horizontal plane, in essentially the same direction. The magnetic focussing can be considered negligible, and the spectrometer has a purely horizontal magnetic field, bending the protons in the vertical direction. It is somewhat conventional to have the z axis along the proton momentum, and the x and y axes perpendicular to the momentum, in and perpendicular to the scattering plane.[e] With these simplifications, the spin transport becomes:

$$\begin{pmatrix} S_{fp}^x \\ S_{fp}^y \\ S_{fp}^z \end{pmatrix} = \begin{pmatrix} 1 & 0 & 0 \\ 0 & \cos\chi & \sin\chi \\ 0 & -\sin\chi & \cos\chi \end{pmatrix} \begin{pmatrix} S_{tgt}^x \\ S_{tgt}^y \\ S_{tgt}^z \end{pmatrix},$$

with the spin precession angle $\chi = \kappa\gamma\theta_B$, with κ the anomalous magnetic moment of the proton, $\gamma = E/m$ the standard relativistic factor, and θ_B the angle through which the proton trajectory bends in going through the spectrometer. The polarimeter then measures asymmetries related to the transverse spin components in the focal plane, $S_{fp}^x = S_{tgt}^x$ and $S_{fp}^y = \cos\chi S_{tgt}^y + \sin\chi S_{tgt}^z$. The left – right asymmetry discussed above becomes an azimuthal asymmetry, with the cross section for scattering in the analyzer proportional[f] to $\left|\frac{d\sigma}{d\Omega}\right|_0 \left[1 + A(S_{fp}^x \cos\phi + S_{fp}^y \sin\phi)\right]$.

The polarimeter analyzing power is calibrated with ep elastic scattering. In this reaction, a polarized electron beam transfers polarization to only two components of the recoil nucleon spin. There is no induced polarization,

---

choice of a left-handed coordinate system at times, a choice which could have been easily obscured by naming axes with an n, l and t convention instead of an x, y, and z, or x', y', and z' convention. I can provide no advice that is gauranteed to keep you out of trouble. My best suggestion is to try to use the same conventions that most others do, since it should ease the comparison of theory to data, and to avoid close collaboration with colleagues with overwhelming ideological biases on these issues.

[e] Most physics coordinate systems have y out of the reaction plane and thus generally vertical for many experiments. This is the coordinate system we use here. Spectrometer coordinate systems often use the transport convention, in which x is along the momentum dispersion direction, and thus vertical for most experiments with small acceptance spectrometers, which typically bend particles vertically.

[f] Here again, the choice of the origin and rotation direction of the azimuthal angle ϕ is arbitrary.

independent of the beam helicity, in the one-photon exchange approximation
– this is a good approximation since the electromagnetic coupling, $\alpha \approx 1/137$,
is small. With polarization transfers[9] P^x and P^z given in Eqs. 4 and 5 above,
and electron beam polarization h, the focal spin components are $S^x_{fp} = hP^x$
and $S^y_{fp} = h \sin \chi P^z_{tgt}$. The ratio of polarization components is largely free
of systematic uncertainties, other than the spin transport, since it does not
depend on either the analyzing power or the beam helicity – these do affect
the magnitude of the asymmetry however and thus the size of the uncertainty.
The ratio is given by:

$$\frac{P^x}{P^z} = -\frac{2m}{(E + E') \tan(\theta/2)} \frac{G_E}{G_M}$$

Thus, the ratio of the form factors leads to a phase shift in the azimuthal
scattering distribution from the FPP. The overall magnitude of the azimuthal
asymmetry depends on the product of beam polarization times analyzing
power, allowing the analyzing power to be determined if the beam polarization is independently known.

2.2 The Proton Form Factors at Large Q^2

We have now presented in a hopefully simple form the basis for the Jefferson
Lab Hall A recoil polarization measurements of the proton electric form factor,
reported in [4,5,6,7]. In Figure 2, we show a small sample of the measured
polarization transfer observables, to give an indication of the behavior of these
observables as a function of angle. The longitudinal polarization transfer P^z
is insensitive to the form factor ratio, which only appears in the denominator
in Eq. 4. The transverse polarization transfer P^x depends almost linearly
on the form factor ratio, which also appears as a factor in the numerator of
Eq. 5. The plot makes it clear that the polarization transfer measurements
are best taken at intermediate electron angles, so that the two polarization
components are moderately large, and the model sensitivity becomes large in
absolute terms as well. The data shown are a relatively low statistics, parasitic
measurements, taken as a background in a $\gamma p \to p\pi^0$ experiment[5]. The data
can be seen to agree slightly better with a faster decrease of G_E, as compared
to G_M.

Figure 3 shows the high statistics recoil polarization measurements from
Bates[8], Mainz[9], and Jefferson Lab[4,6,7]. These data much more clearly demon-

---

[9] Almost always P indicates an induced polarization, indpendent of the beam helicity. However, it is conventional in ep elastic scattering to also use P to indicate the polarization
transfers.

92

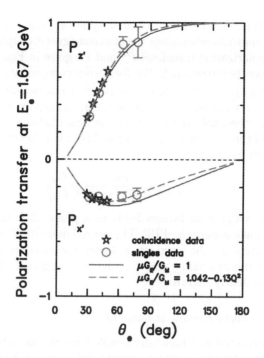

Figure 2. Sample polarization transfer data in *ep* elastic scattering. The curves show the expected results if both form factors follow the dipole formula, and if the electric form factor falls off faster with momentum transfer.

strate the fall off of the proton electric form factor. The fall off is nearly linear, and suggests that G_E changes sign at $Q^2 \approx 8$ Gev^2. The theoretical curves arise from several different approaches, and will be discussed further below.

We now consider the implications of these results for G_E/G_M on the ratio of the Pauli and Dirac form factor, F_2/F_1. This ratio is given by

$$\frac{F_2}{F_1} = \frac{1 - G_E/G_M}{\kappa(\tau + G_E/G_M)}.$$

The ratio is of particular interest because it is generally believed that the form factors at some large, but as yet unknown, momentum transfer can be explained through perturbative QCD, which predicts that $Q^2 F_2/F_1 \to$ constant. Generally, pQCD is taken to imply that hadronic helicity[h] conserving processes dominate, and helicity nonconserving processes are suppressed by

[h] Helicity is the component of the spin along the momentum direction.

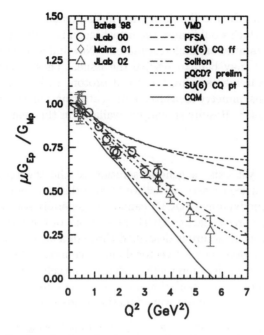

Figure 3. Polarization transfer data for the proton form factor ratio, compared to several theoretical calculations and fits.

factors of the momentum transfer squared, Q^2. Thus, the helicity nonconserving Pauli F_2 form factor decreases faster by a power of Q^2 than the helicity conserving Dirac F_1 form factor. While the Rosenbluth separation measurements had indicated that indeed this was the case, the Jefferson Lab data instead show that this ratio is still increasing, but that QF_2/F_1 is constant.[i] Ralston, Jain and colleagues[10] in particular have emphasized that the constancy of QF_2/F_1 is an indication of the importance or quark orbital angular momentum in the nucleon, a viewpoint they have championed in explaining several puzzles in a number of other reactions.

As is the case for many disagreeing physics measurements, there is as yet no clear resolution to the disagreement between the Rosenbluth separation

---

[i]For $\tau >> G_E/G_M$, $G_E/G_M \to$ constant $\Rightarrow Q^2F_2/F_1 \to$ constant, and the pQCD limit can be seen to hold. For $\mu G_E/G_M \approx 1.04 - 0.13\, Q^2$, as in the Jefferson Lab data, the constancy of QF_2/F_1 is not generally true. It holds only over a limited range of Q^2 from about 2 to 10 GeV^2. Thus, precision measurements at higher Q^2 should be able to distinguish between QF_2/F_1 constant and G_e/G_M linear.

measurements and the recoil polarization measurements. There are no known physical effects which would cause the measurements to disagree, so it appears that some uncertainties are underestimated.[j] New independent measurements are planned that may help to resolve the issue. These include Rosenbluth separation measurements using the recoil proton, rather than the scattered electron[11], and asymmetry measurements scattering polarized electrons from a polarized target[12]. Results should be available in the near future.

2.3 Models

There does not yet exist a general solution to the theoretical problem of describing the nucleon elastic structure at an arbitrary momentum transfer. Quantum Chromodynamics (QCD) cannot be generally solved, and there are not yet extensive precise lattice QCD calculations of the form factors. Thus, the nucleon charge structure is described through a variety of models.

Vector meson dominance / vector dominance models (VMD / VDM) are based on the photonuclear interaction occuring largely through the hadronic content of the photon, its oscillation into vector mesons of the same quantum numbers [k] That is, the photon wave function is given by

$$|\gamma_{physical}>= a_\gamma|\gamma_{bare}> +a_\rho|\rho> +a_\omega|\omega> +a_\phi|\phi> + \ldots$$

This model leads to the proton form factor being a sum of monopoles,

$$F = \sum_i \left(\frac{a_i}{1 + Q^2/m_i^2}\right) + 1 - \sum_i a_i,$$

with the m_i's ideally being identified as the masses of known vectors mesons, and the a_i's related to the strength of the vector meson component of the wave function and its coupling to the nucleon. The last terms serve to preserve the correct normalization ($F = 1$) at $Q^2 = 0$. In practice, the m_i's and a_i's are treated as parameters to be fit, and for better fits at least some terms cannot be identified with known vector mesons.

Early fits[13,14] used only the sum of monopoles. More recently, it has become common[15,16,17] to build in an approach to perturbative QCD behavior at large momentum transfer. The generally accepted pQCD prediction has

---

[j] As a polarimeter person, with a biased viewpoint, I note that if the 1.6 GeV spectrometer data[2] had *not* been renormalized, the SLAC Rosenbluth separation measurements would have been in much better agreement with the Jefferson Lab results.

[k] Of course, there is an ambiguity as to whether a meson in the Feynman diagram should be considered part of the proton's structure, the reaction mechanism, or the photon's structure. Some cases have a clear resolution, e.g., a pion is not a vector meson and cannot be part of the photon, but others do not.

been that the leading helicity conserving Dirac form factor should vary like $(Q^2)^{1-n}$, where n is the number of valence quarks in the nucleon (or other object). Helicity nonconserving form factors decrease faster by an extra power of Q^2. This leads to a dominance of the nucleon magnetic form factor (but the electric form factor for objects of integral spin, such as the deuteron).

The quark meson coupling model treats the nucleon as a quark core, that is much smaller than the physical nucleon, surrounded by a meson, mostly π, cloud. The meson cloud is much more extended than the bare quark core, and accounts for most of the nucleon size. The virtual γ can couple either to one of the quarks or to the meson cloud. This model is one particular formulation of the general idea that the proton wave function includes meson cloud components. In hadronic terms,

$$|p_{physical}> = a_p|p_{bare}> + a_{\pi^0}|\pi^0 p> + a_{\pi^+}|\pi^+ n_{bare}> + a_{K^+}|K^+ \Lambda^0> + \ldots$$

Constituent quark models which fit the electromagnetic form factors are also possible. Schlumpf[18] noted that harmonic oscillator wave functions fit the form factors at low Q^2, but produced too little strength at large Q^2. He instead chose $\phi(M) \propto (M^2 + \alpha^2)^{-3.5}$ as an orbital wave function to describe the 3-quark bound state. The model has two parameters, a quark mass and a confinement scale. The relativistic approach results in enhanced form factors at larger Q^2 that fit the experimental data. Miller[19] has recently reinvigorated this approach, emphasizing how the key feature that allows for the reduction of G_E/G_M is the orbital angular moomentum inherent in the quark wave function in the nucleon. While the large Q^2 behavior of the form factors is predicted from the 3-quark core of the nucleon, the low Q^2 behavior is best understood as arising from the mson cloud of the nucleon. A detailed description showing how to calculate the form factors is given in [20].

The modern generalized parton distribution (GPD) approach takes the point of view that the photon has a hard interaction with a single quark in the nucleon; the momentum is shared with other quarks through the wave function, described through the nucleon GPDs. At this point the GPDs that describe the quark distributions in the nucleon are not well understood. From this perspective, the elastic form factors, being an integral of the GPDs over x, the quark momentum fraction, can be used to constrain model parameters. Afanasev[21] showed that the proton form factor ratio can be approximately reproduced with several functional forms for the GPDs, and further that the orbital angular content of the quarks in the nucleon is significant and only weakly dependent on the functional form chosen. Ralston and colleagues[10] suggest that in the GPD approach, the Pauli form factor arises from an interference of $L = 0$ and $L = 1$ components to the wave function. The near unity of QF_2/F_1 indicates that these two components of the proton wave function are nearly equal in size. Furthermore, the photon probe breaks rotational symmetry in probing the proton, so that the proton is deformed, a result also emphasized by Miller within the framework of relativistic constituent quark models.

2.4 More Form Factors

Form factors are an old topic, but one which has generated much excitement recently with the surprising measurements of the high Q^2 behavior of the proton electric form factor. There is insufficient space to do justice to the future possibilities or to recent measurements of the neutron form factors, so we shall simply summarize the recent experimental progress and likely future.

The measurements of the proton form factor will continue up to $Q^2 \approx$ 9 GeV^2 with a new experiment in Jefferson Lab Hall C[22]. The magnetic form factor has been measured up to $Q^2 \approx 31$ GeV^2 with simple cross sections, and assuming that the electric form factor does not become large[23]. Mesurements of the proton form factors at low Q^2 will be improved by internal polarized proton target asymmetry measurements with the MIT Bates BLAST detector[24].

Until a recent series of polarization measurements, neutron form factors were largely extracted from electron scattering from the deuterium nucleus. The most extensive determination of G_E^n was done with elastic ed scattering[25], for $Q^2 \approx 0.2 - 1$ GeV^2. Care must be taken in extracting G_E^n from elastic ed scattering. Meson-exchange current and relativistic corrections are needed to extract an isoscaler electric form factor, and then G_E^p is needed to extract the much smaller G_E^n. Circular reasoning is quite possible, as discussed in [26]. The neutron electric form factor has now been extracted in a series of polarization measurements. The most recent results measure asymmetries in the polarized electron – polarized deuteron $d(e, e'n)$ reaction[27]. The Galster parameterization[28] continues to work amazingly well. These measurements can be extended to much higher Q^2 using a polarized ^{3}He target in Hall A[29].

The neutron magnetic form factor has been extracted[30] up to $Q^2 = 4$ GeV^2 from quasifree $ed \rightarrow eX$. Jefferson Lab Hall B measurements[31] comparing quasifree $d(e, e'n)p$ and $d(e, e'p)n$ can be used to extract the neutron magnetic form factor precisely at higher Q^2. At low Q^2, precise cross section measurements for quasifree $d(e, e'n)p$ have also been used, along with polarization measurements, but recent experiments have had deviations exceeding their claimed uncertainties. The most recent Jefferson Lab polarization results[32] agree well with the Mainz cross section measurements[33,34], suggesting problems with much of the other data.

3 Deuteron Structure Probed with *ed* Elastic Scattering

The structure of the deuteron should be of interest to any nuclear physicist. As the only two-nucleon nucleus, it is the preeminent test case for our ability to construct a microscopic description of nuclear structure. Given a model for the NN force and for the γN interaction, the deuteron properties as investigated through photonuclear reactions are calculable with excellent precision, and free from the many-body ambiguities inherent in heavier nuclei. Furthermore, the deuteron is the nucleus most amenable to relativistic calculations.

But this argument assumes that the deuteron can be described with hadronic degrees of freedom. Should not the deuteron instead be described with quark-gluon degrees of freedom? It would appear to be intuitive that, at least at high momentum transfers, corresponding to short distance scales, the underlying quark-gluon degrees of freedom will become evident, as hadronic theories begin to break down. Even at low momentum transfers, if quarks are exchanged between nucleons, the individual nucleons lose their identity, and hadronic degrees of freedom may have trouble representing the resulting deuteron structure.

But this argument misses the point of hadronic theories. Hadrons are the particles that are manipulated and detected in our scattering experiments. One can always generate an effective[l] theory using hadronic degrees of freedom that will describe data, since any experimental observable ultimately results from a matrix element connecting initial and final hadronic states. The problem then is not whether in priniciple hadronic theories can describe the data, it is whether one can construct an efficient and consistent hadronic description of large amounts of data. From this viewpoint, the evidence for the reality of quarks and QCD is that one can provide a simple systematic explanation of large amounts of data from a theory with only a few parameters.

Our viewpoint is that in principle one can describe the deuteron with either hadronic or quark-gluon degrees of freedom. In practice, neither, one, or both of these approaches might work. What works will depend on the kinematic region. One of the main goals of intermediate energy nuclear physics is to investigate what are the limits on the practical application of hadronic theories, and when quark-gluon based theories become relevant for describing experiments. Investigating this question on the deuteron is particularly

---

[l] as opposed to fundamental

important, as the theory has the fewest ambiguities, and thus the least uncertainty, in this case.

Thus, the deuteron's electromagnetic structure is of great interest,
prompting three recent reviews[35,36,37]. As the reviews cover many issues in
great detail, here I shall present a more introductory description, focussing
towards polarization details.[m]

3.1 Elastic Scattering Cross Sections and Form Factors

Elastic scattering from the deuteron is described by three form factors, usually chosen to be G_C, G_M, and G_Q, but elastic ed scattering involves only
two independent functions. Thus a polarization measurement is required in
principle to understand the deuteron structure, as opposed to just being convenient in practice, the case for the proton. The elastic scattering formulas
are however very similar to those for the proton. The scattering cross sections
is given by

$$\frac{d\sigma}{d\Omega} = \frac{d\sigma}{d\Omega}\bigg|_{NS}\left[A(Q) + B(Q)\tan^2(\theta/2)\right] \equiv \frac{d\sigma}{d\Omega}\bigg|_{NS} S_d(Q,\theta) \qquad (6)$$

where

$$\frac{d\sigma}{d\Omega}\bigg|_{NS} = \frac{\alpha^2 E' \cos^2(\theta/2)}{4E^3 \sin^4(\theta/2)} = \sigma_M \frac{E'}{E} = \sigma_M \left(1 + \frac{2E}{m_d}\sin^2\frac{\theta}{2}\right)^{-1} \qquad (7)$$

The structure functions $A(Q)$ and $B(Q)$[n] depend on the three electromagnetic
form factors as

$$A(Q) = G_C^2(Q) + \frac{8}{9}\eta^2 G_Q^2(Q) + \frac{2}{3}\eta G_M^2(Q)$$

$$B(Q) = \frac{4}{3}\eta(1+\eta)G_M^2(Q), \qquad (8)$$

with $\eta = Q^2/4m_d^2$. A Rosenbluth separation determines A and B, and thus
G_M. At very low Q^2, calculations show G_C dominates A, but a polarization
observable is needed to actually separate G_C and G_Q.

If the deuteron were made only of nucleons, then the form factors could
be calculated from the deuteron wave function and the isoscalar form factors,
$G_E^{IS} = G_E^p + G_E^n$ and $G_M^{IS} = G_M^p + G_M^n$. In a nonrelativistic theory, G_C

---

[m] I have a perhaps understandable fondness for [35]. The reader is referred to that and the
other reviews for a more expanded and advanced discussion of the issues addressed here.

[n] I use here $Q = \sqrt{Q^2}$, but one must be careful since in the literature Q might be used to
indicate a three momentum vector, rather than the magnitude of a four momentum.

and G_Q arise from G_E^{IS}, while G_M has contributions from both G_E^{IS} and G_M^{IS}. Actual calculations include also meson exchange currents and relativistic corrections, within a variety of theoretical approaches. Within a hadronic model, one should also consistently include other non-nucleonic components to the deuteron wave function, such as $\Delta\Delta$ configurations, but these are far off shell and appear to be small.

3.2 Polarizations

Hundreds of cross section measurements have been made for *ed* elastic scattering over nearly 50 years. In contrast, 20 years of efforts have yielded about 40 polarization measurements, with half the data being for t_{20}, defined below.

The deuteron, being spin 1, can have both vector and tensor polarizations. The physics for the vector polarizations is similar to that for the spin-$\frac{1}{2}$ proton. A polarized electron beam will transfer a z component of polarization to the recoil proton, that is proportional to G_M^2/S_d, and an x component of polarization that is proportional to $G_M(G_C + \frac{\eta G_Q}{3})/S_d$. One could, similar to the proton polarization transfer experiments, take a ratio of polarization components to determine $(G_C + \frac{\eta G_Q}{3})/G_M$, or similar to the polarized target experiments one could measure asymmetries in polarized electron scattering from polarized targets oriented in the x and z directions, relative to the momentum transfer. While vector polarimetry of deuterons is similar to proton polarimetry, and polarized deuteron targets exist, no such measurement has ever been made. The reasons for this include that deuteron vector polarimetry is not as common as proton polarimetry, and that the predicted absolute magnitude of the polarizations tend to be small, typically a few percent rather than the larger proton polarizations, e.g., shown in Figure 2.

All of the *ed* polarization measurements have concerned tensor polarizations. Tensor polarizations can more simply be illustrated with a polarized target. A polarized target will have will have probabilities n_+, n_0, and n_- of being in the $S_z = +1$, 0, and -1 substates, respectively, with $n_+ + n_0 + n_- = 1$. The target vector polarization is given by $(n_+ - n_-)/(n_+ + n_-)$. The interesting tensor polarization of the target is given by $t_{20} = 1 - 3n_0$. Given a vector polarization for the target along with t_{20} (and the normalization condition), the spin state is completely specified.

There are three tensor polarization observables, each of which depends on a combination of form factors divided by S_d: t_{20} involves all the form factors, t_{21} depends on G_M and G_Q, and t_{22} depends only on G_M^2 and provides no new information beyond what has been learned from B. The most measured of these is t_{20}, with

$$-\sqrt{2}S\,t_{20} = \tfrac{8}{3}\,\eta\,G_C G_Q + \tfrac{8}{9}\,\eta^2 G_Q^2 + \tfrac{1}{3}\eta\left[1 + 2(1+\eta)\tan^2\frac{\theta}{2}\right]G_M^2.$$

An important point concerning t_{20} is that it is insensitive to the nucleon form factors. Instead, it largely reflects the deuteron structure. This can be recognized from the formulas above, with the additional knowledge that the deuteron magnetic form factor contributions to S and to t_{20} are small. Also, recall that in a purely nucleon model of the deuteron, both G_C and G_Q are proportional to G_E^{IS}.

The tensor polarization observables are single spin measurements, in which one measures asymmetries in scattering from polarized targets or recoil polarization in unpolarized ed elastic scattering. Consider for example the recent Jefferson Lab t_{20} measurements[38]. When the scattering produces the deuteron with some value of t_{20} the deuteron is polarized along its momentum direction. The $m = 0$ and $m = \pm 1$ substates are different in whether the deuteron density is distributed primarily along the momentum direction, or transversely to the momentum direction. In a simple classical model, a larger transverse distribution implies a wider deuteron that is more likely to interact with an analyzer target atom. Thus, the absolute scattering cross section in the analyzer is changed. For the t_{20} experiment, the analyzing reaction was $dp \to pp + n$, using a cryogenic hydrogen analyzer, and kinematically constraining the neutron to (likely) be a spectator. The polarimeter was calibrated with measurements at Saturne, before begin brought to Jefferson Lab for the ed measurements.

A summary of the world polarization data[o] is presented in Table 1. In the Bates and JLab experiments, a conventional electron beam of tens of microamps struck a cryogenic target, and recoiling deuterons were analyzed in a polarimeter. In the Bonn experiment, a low luminosity beam struck a polarized target and asymmetries were measured, but with poor precision. The Novosibirsk and NIKHEF experiments used electron rings, which leads to currents up to hundreds of mA, with internal polarized targets. The key features of the target system include cryogenic nozzles, producing a large density of slow atomic gas, intense sextupole magnets and RF transition units, to prepare the spin state of the deuterons, and a storage cell to build up the deuterium gas density.

One now has the problem of extracting form factors from the numerous cross sections, and few polarizations, which are not measured in the same

---

[o]Note the convention that a lower case t refers to a recoil polarization measurement, while an upper case T refers to a polarized target asymmetry.

Table 1. World data for tensor polarization observables.

Experiment	Type	Q (GeV)	Observables	# of points	Year and Reference
Bates	polarimeter	0.34, 0.40	t_{20}	2	1984 [39]
Novosibirsk VEPP-2	atomic beam	0.17, 0.23	T_{20}	2	1985 [40,41]
Novosibirsk VEPP-3	storage cell	0.49, 0.58	T_{20}	2	1990 [42]
Bonn	polarized target	0.71	T_{20}	1	1991 [43]
Bates	polarimeter	0.75 - 0.91	t_{20}, t_{21}, t_{22}	3	1991 [44,45]
Novosibirsk VEPP-3	storage cell	0.71	T_{20}	1	1994 [46]
NIKHEF	storage cell	0.31	T_{20}, T_{22}	1	1996 [47]
NIKHEF	storage cell	0.40 - 0.55	T_{20}	3	1999 [48]
JLab Hall C 94-018	polarimeter	0.81 - 1.31	t_{20}, t_{21}, t_{22}	6	2000 [38]
Novosibirsk VEPP-3	storage cell	0.63 - 0.77	T_{20}	5	2001 [49]

kinematics. One possibility is to fit the more numerous A and B functions from the world cross section data with a set of smooth functions, removing some of the statistical variation of the data, and generating a smooth and continuous distribution. The charge and quadrupole form factors can then be extracted at each of the few polarization data points. A recent set of fits performed by the Jefferson Lab t_{20} collaboration is presented in [50].

3.3 The Deuteron Data vs. Hadronic Theories

A summary of the experimental data is shown in Figure 4. The data are compared to several, mostly relativistic hadronic calculations, and it becomes clear that, at least approximately, the electromagnetic structure of the deuteron can be explained by hadronic theories. There is no necessity for quark models.

The A structure function is generally well reproduced as it drops 8 orders of magnitude. An important point is that the deuteron wave function, and nucleon form factors, are sufficiently well understood that it becomes clear that a nonrelativistic nucleon theory does not describe the data. Since the meson exchange currents are of relativistic origin, it then is clear that in hadronic models relativity is needed to understand the deuteron. The large drop in A hides significant problems, such as 10% disagreements between overlapping experiments that each claim significantly better than 10% precision. Further, a detailed study of the low Q data even indicates that the sign of the relativistic corrections at low Q is not certain, although that might be where one expects the correction to be clearest.

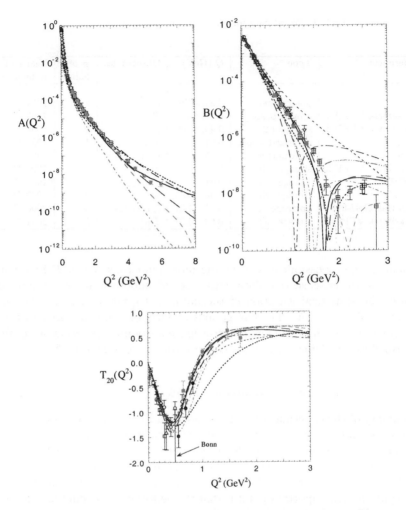

Figure 4. The structure functions A, B, and t_{20} for ed elastic scattering.

The B structure function is not so well reproduced, but note what is happening. The magnetic form factor goes through 0 and changes sign near $Q^2 \approx 2\,\mathrm{GeV}^2$, leading to a minimum in B. The exact position of the minimum relies on the delicate cancellation of various terms in the theories, and one should not be too critical if the cancellation is slightly off. Finally, t_{20}, which is insensitive to the nucleon form factors, is generally well reproduced by most of the theories.

While we see that the deuteron data are well produced, the discussion here has hidden an important issue in these calculations. Earlier, I presented in the discussion of the electromagnetic current of the proton the two nucleon form factors. The constraint that there are only two form factors arises from the algebra / symmetries of the field theory. In the usual derivation, it comes from having the same proton mass in initial and final states in the Dirac equation. When one couples an electron to a nucleon in the nucleus, the nucleon is off shell, and its mass will not be the same in initial and final states. There will be additional terms, with additional form factors, in the current. These form factors are in principle unconstrained, aside from a typical convention like $F_3(Q^2 = 0) = 1$. The Q^2 dependence is generally *assumed* to follow G_D, simply because the nucleon elastic form factors roughly follow G_D. There are in addition poorly constrained meson-exchange currents, such as the "famous" $\rho\pi\gamma$ current. Without a more fundamental theory to predict these poorly constrained ingredients, one has to wonder what physics has been effectively included in them. Relativistic theories have been most extensively developed for elastic *ed* scattering, and their extension to photo- and electro-disintegration, along with other nuclei, could illuminate this issue.

3.4 The Deuteron Data vs. Quark Theories

Can any of the deuteron data also be explained by quark theories, or, more generally, by QCD inspired theories? Several such theories have been applied to the deuteron. The ones most discussed are perturbative QCD, at high momentum transfer, and effective field / chiral perturbation theory, at low momentum transfer.

Perturbative QCD

Let us first consider whether perturbative QCD (pQCD) is in agreement with the high Q behavior of the structure functions or of t_{20}. When the first t_{20} polarization data came out, it was still believed that pQCD might apply to the deuteron structure at modest Q^2. Dominance of helicity conserving form factors is generally considered to reesult from pQCD – but note the discussion of orbital angular momentum in the proton. The single helicity-flip form factor falls faster by Q^2 than the favored helicity nonflip form factor, while the helicity double-flip form factor falls faster by Q^4. Rewriting t_{20} in terms of helicity non-flip, single-flip, and double-flip form factors then allows a simple prediction[p] to be made:

$$\tilde{t}_{20}(Q^2) \to -\sqrt{2}\,\{1 + \mathcal{O}(\Lambda/Q)\}\ . \tag{9}$$

Since the lowest Q data approach the minimum allowed value, $-\sqrt{2}$, predicted in Equation 9 there was much excitement. Now we see that $t_{20} \approx 0$ at the highest measured Q^2, near 2 GeV^2, and does not provide support for this prediction. The form factor analysis[50] further shows that the helicity flip form factors are not suppressed compared to the helicity nonlip one.

At slightly higher Q^2, the minimum in the B structure function is also taken as evidence that pQCD is not applicable, since one generally expects just a smooth falloff in the form factors rather than a minimum. While this is the case if the helicity nonflip form factor is very dominant, the behavior in B can be reproduced if the helicity nonflip form factor is only starting to dominate.

In pQCD, the deuteron form factor arises from 5 hard gluon exchanges that share the photon momentum among the six valence quarks of the deuteron. The gluon propagator of $1/Q^2$ then leads to an amplitude that varies as $1/Q^{10}$, and a form factor / A structure function that varies as $1/Q^{20}$. The last few data points in the A structure function are more or less consistent with this behavior, but this assessment depends on how critically one looks. The absolute magnitude of the form factor has only been evaluated once, by summing more than 300,000 diagrams that contribute to the hard scattering[51]. This leading twist pQCD estimate is $10^3 - 10^4$ times smaller than the measured deuteron form factor, implying large soft contributions, and suggesting that pQCD should not be used as an explanation for the form factor.

What should one conclude from this discussion? There appears to be little predictive power in the pQCD approach for explaining the deuteron form factor data. It appears that there are important soft physics contributions.

In an attempt to include some of the soft physics, and to extend the region of applicability of pQCD down to lower momentum transfers, Brodsky and Hiller introduced reduced nuclear amplitudes[52] (RNA). In this model, for *ed* elastic scattering, the gluon exchanges within a nucleon give the kinematic dependences of experimental nucleon form factors. The result is that the

---

[p]There is only a slight angle dependence to t_{20} at fixed Q^2, which arises from the small magnetic term. Since one does not wish to entirely ignore it, there are two conventions for adjusting data to make them more comparable. First, correct the magnetic contribution to adjst the data to a scattering angle of 70°, generating $t_{20}(70°)$. Second, remove the effect of the magnetic term entirely, generating $\tilde{t}_{20}$.

deuteron form factor has the form

$$F(Q^2) = F_n(Q^2/4)F_p(Q^2/4)f_d(Q^2),$$

where $f_d(Q^2)$ is the reduced nuclear form factor, of two point nucleons in the deuteron. It would thus be expected to have a monopole form, $f_d(Q^2) \approx (1 + Q^2/m^2)^{-1}$, with m some appropriate scale. This formula reproduces the A structure function amazingly well, even down to Q^2 of 1 – 2 GeV^2, once m and the overall normalization are fit.[q]

Chiral Perturbation Theory

Pionless effective field theory ($\not\pi$EFT) / chiral perturbation theory (χPT) attempt to divide physics up in low energy, long range physics which is treated exactly, and high energy, short range physics which either has no significant influence, or can be integrated over and represented by an adjusted parameter. The theories are expected to hold up to an energy scale of the pion / rho meson masses. They have been applied with good success to various processes, including nucleon nucleon scattering and the deuteron structure. It provides both an alternative to, and a justification for, conventional meson-baryon theories.

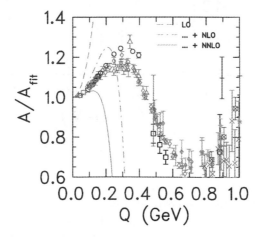

Figure 5. The A structure function compared to $\not\pi$EFT. Data and theory have been divided by an arbitrary Q^2-dependent fit function to remove much of the Q^2 dependence, allowing differences to be seen.

---

[q] However, recall that the nucleon form factors largely follow the dipole formula, but are not generally believed to result from pQCD.

The effective range theory of Bethe[53] may be considered an early version of $\not\pi$EFT. Modern χPT started with Weinberg's study of NN scattering[54]. Much has been learned, particular concerning the difficulty of a perturbative treatment of the strong tensor part of one pion exchange, and difficulties with reproducing the deuteron quadrupole moment, which appears to involve short range physics. A good recent (and well written) $\not\pi$EFT calculation[55] provides a nice description of the lowest momentum transfer A data – Figure 5 has been generated from the published formulas. Comparisons with the lowest Q^2 form factor data can be found in the article, but there is only an overlap with the first few t_{20} data points.

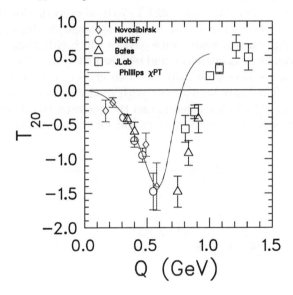

Figure 6. Measured data for t_{20} compared to a χPT calculation.

One can instead include pions, using χPT to extend the validity of the theory over a larger range. An extended description may be found in previous lecture notes[56]. Calculations[56,57,58] provide a generally nice description of the deuteron structure functions A and B and form factors up to $Q \approx 0.5$ GeV. As we are focussing here on polarization observables, we reproduce the calculation of [56] in Figure 6, as an example of what is possible.

QCD and the Deuteron

What should you conclude from this discussion of QCD-based theories and the deuteron structure? In elastic ed scattering, measurements have been carried out to large Q^2, indicating a short distance scale, but in elastic scattering the energy scale is fixed to $W = m_d$, due to the matching of the energy and momentum transfer. In these kinematics, the baryon resonances and perhaps "free" quark configurations probed in, e.g., (e, e') deep inelastic scattering, are far off shell. Effective theories with nucleon degrees of freedom work well, whether conventional relativistic theories or QCD-inspired χPT. The limited range of the latter leaves a continuing role for conventional relativistic theory at large Q^2. There is no clear need for explicit quarks, and apparently little predictive power for pQCD. Thus, in practice, elastic scattering reveals a deuteron made of nucleons and mesons.

4   Deuteron Photodisintegration

The interpretation I have presented for elastic ed scattering leads nicely into deuteron photodisintegration, since in this reaction the deuteron is probed at large momentum transfer and large center of mass energy.[r] High energy photodisintegration moves the kinematics well into and above the resonance region. As beam energies rise to 4 GeV, for example, the two nucleons share 4.3 GeV of energy in the c.m., as opposed to a constant $W = m_d$ in the elastic scattering case. In a hadronic picture, calculating the amplitudes would then involve summing over hundreds of combinations of baryon resonances, which would be sensible to do by using quark degrees of freedom.

Momentum transfers to each nucleon are also large. For $\theta_{\text{c.m.}} = 90°$, the four momentum transfer to the nucleon is in excess of 1 GeV^2 starting at photon energies near 1 GeV. The transverse momentum transfer also reaches 1 GeV for photon energies near 1 GeV. Thus, deuteron photodisintegration provides both the large energy and momentum transfer that are viewed as making a quark based explanation likely.

The large energy in photodisintegration can be emphasized also in comparison to nucleon nucleon scattering. To reach the same c.m. energy, nucleon nucleon scattering requires that the incident nucleon have $T_N \approx 2E_\gamma$. Photodisintegration at 4 GeV corresponds to NN scattering at 8 GeV kinetic energy. This is a high energy region in which it is common to think of NN scattering in quark terms. Indeed, the NN interaction at these energies appears to be simpler to think about as involving quark exchange, rather than meson exchange. Most conventional NN potentials, especially modern high-precision ones, are only fit to data up to about 0.3 GeV kinetic energy, the π

---

[r]The discussion of this section is also related to the recent review of [35], which presents a more expanded and advanced discussion of the issues addressed here. Here I will strive for a simpler presentation.

production threshold. There are a only few cases in which NN potentials have been constructed to fit data up to ≈ 1 GeV. This observation in itself makes it clear that it will be difficult to construct a satisfactory hadronic theory for photodisintegration.

4.1 Photodisintegration Models

Hadronic Models

Low-energy photodisintegration has been of interest for many years, and there are now a variety of good calculations, some nicely incorporating relativistic effects. The recent calculations of Schwamb and Arenhövel[59] are probably the most detailed ones to date. Input to the calculations includes NN data, fit up to 800 MeV, and γN data. The interactions are determined with π, N, and Δ degrees of freedom. Effects such as meson retardation, meson exchange currents, and the meson dressing of the nucleon lines required by unitarity are included. Although no parameters are adjusted in calculating photodisintegration, these calculations provide a generally good description of cross sections and polarizations for photon energies up to about 400 MeV – but see the discussion of the induced polarization below.

In contrast, calculations that have been attempted at higher energies are relatively crude, as each calculation has tended to focus on a single aspect of the problem of a hadronic description of high-energy photodisintegration. Coupled channel calculations[60] using N, Δ and the $P_{11}(1440)$ (Roper) resonances suggest that final-state interactions significantly enhance the cross section for photon energies from $1 - 2$ GeV. An Bonn calculation [61] included pole diagrams involving π, ρ, η, and ω exchange with 17 well-established nucleon and Δ resonances. Nagornyi and collaborators[62,63] introduced a model based on the sum of 4 Feynman diagrams: three pole diagrams coming from the coupling of the photon to the three external legs of a covariant dnp vertex, plus a contact interaction designed to maintain gauge invariance. The dnp vertex has "soft" and "hard" parts, with the hard part designed to reproduce the pQCD counting rules, and with its strength determined by a fit to the data at 1 GeV. A relativistic Born approximation calculation using the Bethe-Salpeter formalism[64,65] found a cross section too small by a factor of 2 to 10, similar to previous Born approximation estimates.

Perhaps the main result to take from this discussion is that detailed hadornic calculations are difficult to formulate. Indeed, it is difficult to even imagine a calculation being formulated that takes into account microscopically the numerous resonance channels, as resonance-resonance scattering is very poorly understood.

Perturbative Quark Models and Implication of Hadronic Helicity Conservation
From pQCD, one expects the consitutent counting rules to also hold for deuteorn photodisintegration. For this reaction, $n = 13$ from 6 quarks in initial and in final states plus one photon, and the cross section should vary like

$$\frac{d\sigma}{dt} \propto s^{-11}.$$

Numerous photoreactions have been previously shown to follow the counting rules, and deuteron photodisintegration is no exception, as we shall show below.

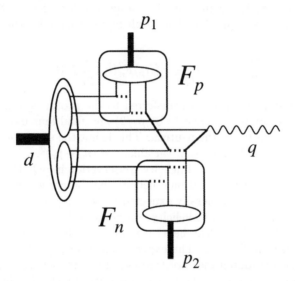

Figure 7. A sample Feynman diagrams for deuteron photodisintegration with quark-gluon degrees of freedom.

Figure 7 shows a sample Feynman diagram for deuteron photodisintegration. In pQCD, one has to consider all possible variations of the photon coupling to the quarks, the gluon couplings between the quarks, and the quark colors, leading to of order 10^6 diagrams to be summed. The reduced nuclear amplitudes approach has also been applied to deuteron photodisintegration. In the RNA approach, the gluon couplings within each nucleon, indicated by

Table 2. Helicity amplitudes for deuteron photodisintegration. The photon helicity may be assumed to be $\lambda_\gamma = 1$.

$\langle \lambda_p\, \lambda_n\, |O|\, \lambda_d \rangle$	Amplitude[66]
$\langle \pm\frac{1}{2} \pm \frac{1}{2} \lvert J \cdot \epsilon_+ \rvert 1 \rangle$	$F_{1\pm}/2m$
$\langle \pm\frac{1}{2} \pm \frac{1}{2} \lvert J \cdot \epsilon_+ \rvert 0 \rangle$	$F_{2\pm}/2m$
$\langle \pm\frac{1}{2} \pm \frac{1}{2} \lvert J \cdot \epsilon_+ \rvert -1 \rangle$	$F_{3\pm}/2m$
$\langle \pm\frac{1}{2} \mp \frac{1}{2} \lvert J \cdot \epsilon_+ \rvert 1 \rangle$	$F_{4\pm}/2m$
$\langle \pm\frac{1}{2} \mp \frac{1}{2} \lvert J \cdot \epsilon_+ \rvert 0 \rangle$	$F_{5\pm}/2m$
$\langle \pm\frac{1}{2} \mp \frac{1}{2} \lvert J \cdot \epsilon_+ \rvert -1 \rangle$	$F_{6\pm}/2m$

the boxes in Figure 7, are absorbed into a nucleon form factor, thus incorporating some of the soft physics. The resulting expression for the cross section is then

$$\frac{d\sigma}{dt} = \frac{1}{(s - m_d^2)^2}\, F_p^2(\hat{t}_p) F_n^2(\hat{t}_n) \frac{1}{p_T^2} f^2(\theta_{\rm cm}). \tag{10}$$

Here $\hat{t}_p$ is the four momentum transferred to the proton, given, e.g., by the difference in its initial (assumed to be $p_d/2$) and final four momenta, and $\hat{t}_n$ is the similar quantity for the neutron. This is not the same four momentum transfer as the usual Mandelstam variable, $t = (p_\gamma - p_p)^2$.

We now turn to the issue of helicity conservation in pQCD. At some high energy, it is expected that hadronic helicity will be conserved. At Jefferson Lab energies, this is not generally expected to be true, but one might see the start of a smooth approach to helicity conservation.[s]

There are 12 independent helicity amplitudes for deuteron photodisintegration, shown in Table 2. The expectation from pQCD is that hadronic helicity conserving amplitudes are leading, and each helicity flip suppresses the amplitude by a power of the four-momentum transfer t. Thus, we find:

$$\begin{array}{ll}
F_{1+},\ F_{3-},\ F_{5\pm} & \text{leading} \\
F_{2\pm},\ F_{4\pm},\ F_{6\pm} & \text{suppressed by } t^{-1} \\
F_{1-},\ F_{3+} & \text{suppressed by } t^{-2}
\end{array} \tag{11}$$

We now consider what happens to polarization observables in deuteron photodisintegration. The polarization observables generally involve sums of

---

[s] Of course, the discussion above concerning orbital angular momentum in the proton must be recalled, and it has been called into question whether one should expect helicity conservation as a prediction of pQCD. We shall assume that this is the case, although it is not supported by existing data.

Table 3. Expressions for polarization observables in terms of the helicity amplitudes. Each observable is to be divided by $f(\theta) = \sum_{i=1}^{6} \sum_{\pm} |F_{i\pm}|^2$.

Observable	Helicity amplitude combination
Σ	$2\text{Re} \sum_{i=1}^{3} (-)^i \left[-F_{i+} \, F^*_{(4-i)-} + F_{(3+i)+} \, F^*_{(7-i)-} \right]$
p_y	$2\text{Im} \sum_{i=1}^{3} \left[F^*_{i+} \, F_{(i+3)-} + F_{i-} \, F^*_{(i+3)+} \right]$
$C_{x'}$	$2\text{Re} \sum_{i=1}^{3} \left[F^*_{i+} \, F_{(i+3)-} + F_{i-} \, F^*_{(i+3)+} \right]$
$C_{z'}$	$\sum_{i=1}^{6} \left\{ |F_{i+}|^2 - |F_{i-}|^2 \right\}$

products of two amplitudes. For any observables, there is no term involving the product of two helicity conserving amplitudes, so the observable ends up approaching 0 as t^{-1} or t^{-2}, depending on the details of the amplitude. At high energies, the observables that have been measured include the Σ asymmetry, and the recoil proton polarizations. Recoil proton polarization measurements have been discussed earlier, but the Σ asymmetry is a new quantity, as is the use of photon beams.

Unpolarized high energy bremsstrahlung photon beams are produced by putting a high energy electron beam into a convenient high Z material. Copper, for example, is fine do to its good heat conductivity and high melting point. Bremsstrahlung is a continuous radiation, with energies of the photons ranging from essentially 0 up to the end point at the beam's kinetic energy. If the electron beam is longitudinally polarized, the bremsstrahlung will be circularly polarized. Why is this? For beams like electrons and protons, the polarization is used to indicate the direction of the spin. For photon beams, the polarization indicates instead the directions of the electric and magnetic fields. When the electron kinetic energy is largely converted into a photon, the photon will also carry the electron's spin, and the longitudinal spin of the photon corresponds to circular polarization with transverse rotating E and B fields. The degree of polarization is calculated with a simple formula,

$$\frac{P_\gamma}{P_e} = \frac{x(4-x)}{4 - 4x + 3x^2}$$

with $x = E_\gamma / E_e$. As long as the photons are close to the end point, the photon beam polarization is essentially equal to the incident electron beam polarization.

The Σ asymmetry comes from an azimuthal asymmetry that occurs when the beam photons are linearly polarized. Linear polarized photons can be produced by a variety of techniques, include Compton backscattering of linearly polarized laser photons from high-energy electrons, bremsstrahlung of electrons from an oriented crystal radiator, or by selection of a subset of normal bremsstrahlung photons with a particular azimuthal angle, by detecting the scattered electron to break the normal rotational invariance of the scattering. The crytal radiator technique is probably most common, often with a diamond crystal. The crystal is oriented both in, and perpendicular to the reaction plane. The cross section asymmetry, divided by the beam linear polarization, then gives the Σ asymmetry.

These relations between these observables and the helicity amplitudes are given in Table 3. Examing the formulas, it becomes apparent that we expect[t] that p_y and $C_{x'}$ both approach 0 as $1/t$. For $\theta_{c.m.} = 90°$, one can see from their similarity that $F_{1+} \approx F_{3-}$, and $F_{5+} \approx F_{5-}$. Thus, near $90°$, $C_{z'}$ goes to 0 as $1/t^2$, while Σ approaches -1 as $1/t$. These limits do not apply at other angles, as $C_{z'}$ is not generally required to vanish, and Σ is generaly required to vanish at $0°$ and $180°$, where there is no distinction between parallel and perpendicular.

Nonperturbative Quark Models

We have already introduced the RNA extension of pQCD, which attempts to incorporate some soft physics to extend the region of validity to lower momentum trasnfer. We now introduce three other nonpreturbative quark models.

Radyushkin[73] proposed a different analysis of the diagram of Figure 7. The photon is absorbed on two quarks being exchanged between the nucleons. This is a short distance, hard process, which can be viewed as having only a slight kinematic dependence, compared to the soft physics through which the momentum is shared between the quarks in the nucleons. The quarks are different flavors, since if they are the same flavor cancellations reduce the cross section.[u] Phase space factors are also needed, so that the cross section

---

[t] An important point here is in what frame these observables are measured. The helicity amplitudes relations given define the observables in the c.m. frame. The observables Σ and p_y are identical in the lab and c.m. frames, since the spin directions are transverse to the boosts, but the observables $C_{x'}$ and $C_{z'}$ are mixed by the boost from the lab to the c.m. The key point, for those unfamiliar with relativistic effects on spin, is that the spin may be treated as a four vector, subject to the usual Lorentz boosts, with the 0th component equal to 0 in the proton rest frame. See [67] and references therein, or advanced textbooks.

[u] This may be viewed as being similar to the dominance of charged meson exchange currents over neutral meson exchange currents.

can be written as:

$$\frac{d\sigma}{dt} = \frac{1}{(s - m_d^2)^2} \, F_{p \to n}^2(\hat{t}_p) F_{n \to p}^2(\hat{t}_n) f^2(\theta_{cm}).$$ (12)

Here, f represents the short distance process in which the photon is absorbed on the exchanged quarks, and the $p \leftrightarrow n$ transition form factors may be assumed to have a dipole form, similar to the elastic form factors. The basic difference between Eq. 12 and Eq. 10 is the factor of $1/p_T^2$, a difference that allows a good fit to the data.

Frankfurt, Miller, Strikman, and Sargsian[74,75] proposed yet another analysis for the diagram of Figure 7. They consider the photon to be absorbed by a quark in the nucleon, and the nucleons to share momentum through a hard-rescattering final-state interaction. The hard rescattering model (HRM) takes as input the deuteron wave function, most importantly at low momenta, the photo-quark coupling, and the NN scattering amplitude, taken from experimental data, by interpolating / extrapolating the measured data to the same s and $\hat{t}_p$ as the photodisintegration. This model provides an absolute calculation of the $90°$ cross sections, but tends to overpredict the forward angle data unless a simple scaling function is used.

The quark gluon string (QGS) model[76,77] instead considers the reaction from the viewpoint of large N_c QCD, and the dominance of planar diagrams. Photodisintegration then proceeds by 3 quark exchange, with an arbitrary number of gluon exchanges to share momentum. The cross section is evaluated by Regge theory techniques.[v] The most recent work[77] empahsizes the importance of nonlinear Regge trajectories, in achieving a generally good fit to the entire cross section data set. Polarization calculations are also underway.

4.2 Cross Sections

Figure 8 shows the high-energy photodisintegration cross sections, multiplied by s^{11} to remove their rapid energy dependence. From pQCD, we expect the cross sections to become flat in this plot at sufficiently high energy, as generally appears to happen. The onset of this behavior varies from about 1 GeV at $\theta_{c.m.} = 90°$ to 4 GeV at $\theta_{c.m.} = 37°$. It corresponds to a nearly constant transverse momentum transfer of $p_T \approx 1.3$ GeV. The dotted line, from Eq. 10, is the RNA extension of pQCD. It is interesting that the data are closer to the asymptotic s dependence than to the extension that incorporates some soft physics. The product of phase space times form factors, the long – short dashed curve from Eq. 12, is actually quite similar to s^{-11}. It is the

---

[v] Regge theory will be discussed in the next section, on meson photoproduction.

114

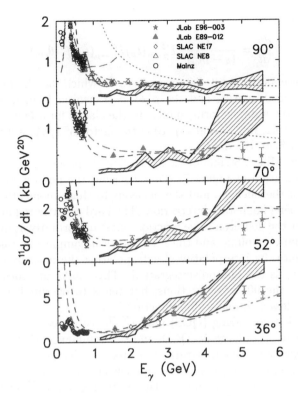

Figure 8. Published data for high-energy deuteron photodisintegration, compared to several models.

factor of $1/p_T^2$ in the RNA approach that causes the significant differences with s^{-11} above 1 GeV photon energy. The QGS model is shown as the dash line. It generally fits the data well from just below 1 GeV until 3 to 4 GeV. At the higher energies, it tends to overpredict the small angle cross sections while underpredicting the large angle cross sections. The hatched region is the HRM prediction. It too agrees reasonably with data, within about a factor of two, for energies above about 1.5 or 2 GeV.

4.3 Polarization Observables

There are over 1100 polarization data taken for deuteron photodisintegration since the first measurements about forty years ago. We will only be concerned with a few recent high energy measurements in this discussion.

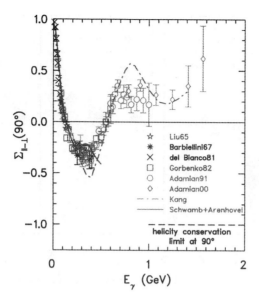

Figure 9. The Σ asymmetry in deuteron photodisintegration, compared to two hadronic calculations and the pQCD limit.

Figure 9 compares the Σ asymmetry to two meson baryon calculations and to the pQCD limit. The Σ asymmetry starts at unity,[w] goes negative, peaks near the Δ resonance, and again becomes positive at high energies. Of particular note are the recent Yerevan data[68] above 1 Gev, which give the impression of trending towards 1 again. With the large uncertainties, the data are also consistent with being about constant. In any case, they are far from the pQCD limit and give no indication of approaching it. The data are reasonably reproduced by both calculations, the low energy calculation of Schwamb and Arenhövel[59] – which does appear to have a slight problem at the Δ resonance – and the more extended calculation of Kang et al.[61]

Figure 10 shows the high energy recoil polarization measurements from Jefferson Lab[69]. The behavior of the induced polarization p_y is particularly striking. Near the Δ resonance, the induced polarization is large, much larger then expected in any of the hadronic theories. The explanation of this be-

---

[w]One must be careful, as some authors use the convention $|| - \perp$ as we do here, while others use $\perp - ||$. We emphasize that the conventions here are consistent, and the data are almost opposite the pQCD limit.

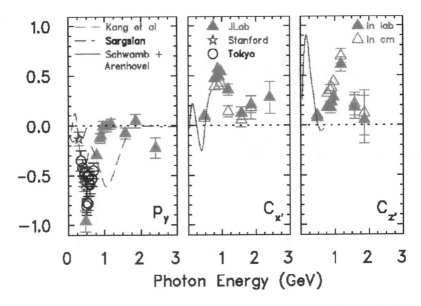

Figure 10. Recoil proton polarization in deuteron photodisintegration. See text for details.

havior is an unsolved problem that has existed for over 20 years.[x] In the calculations of [59], for example, the induced polarization angular distribution has a behavior that is qualitatively like $\sin 3\theta$, with a minimum near 90°, whereas the data appear to be more like $\sin\theta$, with a maximum near 90°. At higher energies, the induced polarization rapidly becomes smaller, and is consistent with vanishing for all energies above about 1 GeV, and thus consistent with pQCD expectations. In contrast, the hadronic calculations of [61] predict additional structure in the induced polarization, particularly a peak near 1 GeV due to the D_{13} and F_{15} resonances interfering with the Born amplitude. A calculation of the recoil polarization in the HRM, shown as the dashed line slightly below $P_y = 0$ above 2 GeV , also indicates that the induced polarization is likely to be small.

Figure 10 also shows the polarization transfer components. These are the

---

[x]The observation many years ago prompted much discussion about the contribution of dibaryon resonances to deuteron photodisintegration, and there were many calculations that added Breit-Wigner dibaryon resonances to the photodisintegration amplitudes. Such explanations have for good reason largely fallen out of favor, but it is clear from the induced polarization that some significant aspect of the physics on the Δ resonance is being missed.

first measurements of these observables. The low energy points are in good agreement with the calculation of [59]. These polarization transfers have to be small near 500 MeV, since p_y is large, close to -1. As p_y goes to 0, these observables both increase in magnitude, and apparently peak near 1 GeV, before decreasing again at the highest energies. While they become small, the highest energy points for $C_{x'}$ are all positive, inconsistent with vanishing, and thus inconsistent with helicity conservation. A smooth approach to helicity conservation would imply a $1/t$ approach to vanishing for $C_{x'}$. While the data are not inconsistent with this behavior, they do not favor it. Although no HRM prediction is shown, an estimate has been made. The longitudinal polarization transfer $C_{z'}$ is expected to vanish, while the transverse polarization transfer $C_{x'}$ should be small and positive. The data is consistent with both of these estimates.

Figure 10 shows the polarization transfer components in both the lab and c.m. frames. For these data, the boost effects are small and do not significantly affect the interpretation, but this observation is partly due to the polarizations being small and thus not well determined, in percentage terms. Because of spin transport in the spectrometer, $C_{z'}$ has large uncertainties at 2.4 GeV and must be considered unknown. Thus, one cannot determine a value for $C_{x'}$ in the c.m. system. This observation indicates that although the amplitudes are usually defined in the c.m. system, theory may be more constrained by a comparison to data in the lab system. While the theory is equally well determined in either frame, the data may not be.

4.4 Understanding Deuteron Photodisintegration, Future Tests

From the discussion given above, we can draw several conclusions about how to understand high energy deuteron photodisintegration.

First, from the comparison of the data to various models, it is apparent that the energy dependence has told us little so far. All of the models tend to more or less agree with each other and with the data. Perhaps this is not so surprising, since over the measured region there is not much difference between s^{-11} and the product of phase space time form factors. As long as models are not required to give absolute normalizations, this situation is unlikely to change.

Second, we are unlikely to ever be able to understand the data with a traditional detailed hadronic calculation. By this we mean, for example, a coupled channels calculation that includes all contributing baryon resonances and their interactions. The needed data to formulate such a model does not exist, and there are simply too many possible channels to expect any numerical accuracy. It is possible that a simpler, but reasonable model will be formulated and acceptable, in particular as the polarizations do not strongly indicate resonance contributions. It probably makes statistical sense that if there are numerous resonance contributions, their combined effect will largely average out and the induced polarization will vanish. It is surprising that it occurs so quickly in deuteron photodisintegration.

Third, with the current best understanding of pQCD, it does not appear that the absolute magnitude of the photodisintegration cross section can be reproduced. The approximate success of the constituent counting rules in this and other reactions, although predicted by pQCD, does not appear in fact to arise out of pQCD.

Fourth, it appears that there are some nonperturbative quark models that more or less account for the data, may be correct, and are good enough to require further testing. This will occur through more extensive cross section measurements[70], extending the polarization measurements[71], and possibly through measurements of γpp photodisintegration[72].

5 Meson Photoproduction

Elastic scattering provides only a limited view of the nucleon structure. Further insight is gained through inelastic scattering, leading to the excited states of the nucleon, the baryon resonances, and beyond to the deep inelastic scattering (DIS) regime.

The major difficulty in baryon resonance determinations is that there are numerous, overlapping, broad resonances. This makes the separation of the various resonances and parameterization of their masses, widths, and quantum numbers challenging. Polarization observables are important due to their usefulness in constraining the reaction amplitudes, and in increasing sensitivity to small amplitudes. Here, we consider polarization in meson photoproduction reactions, in particular a set of our measurements on π^0 photoproduction from the proton.[67] These measurements were not focussed on extracting resonances. Rather, they focussed on covering a broad range of energy to see if we could get above the resonance region, and use the polarizations to give indications about the reaction mechanism.

5.1 Physical Models

In simple hadronic models, single meson photoproduction is viewed as arising from t, u, and s channel processes. In the t channel, the nucleon emits a virtual meson, while remaining on shell. The meson absorbs the full energy of the photon, knocking it on shell and forward, at small t. In the u channel, the nucleon emits an on-shell meson, so the nucleon becomes virtual. The nucleon then absorbs the full energy of the photon, knocking it on shell and forward. The meson then is kinematically constained to go backward, at small u. In the s channel, the nucleon absorbs a photon, producing an excited system, a far off shell nucleon or possibly a nearly on shell baryon resonance. The resonance then decays with some probability to $p\pi^0$. This process is neither strongly forward peaked, as is the t-channel process, nor strongly backward peaked, as is the u-channel process.

Since resonances are a main focus of lower energy photoproduction measurements, it is common to analyze these data in terms of partial wave analyses. In the simplest form, such an analysis would use, for example, a polynomial series to describe "background" processes plus a series of Breit-Wigenr functions to describe resonances. The polynomials and Breit-Wigners would exist in each partial wave, and the polynomial coefficients and Breit-Wigner masses and widths would be determined by fitting data. There are also various resonance photoproduction models that try incorporate the s, t, and u channel processes described above with more of a physics calculation, and less of a fit, though still fitting parameters to the data. Two large analysis projects that attempt to understand the entire data base are the SAID[78] and MAID[79] models. SAID in particular tries to understand NN, γN, and πN, which is necessary since, e.g., any extracted resonance parameters in photoproduction can be modified by a πN final state rescattering. There are at present a large number of photoproduction data, over 15000 points (not all of them good). For the polarization data in particular, the data are scarce for energies much above 1 GeV.

A general problem related to these analyses is how much data must be taken to uniquely determine the amplitudes. There are four helicity amplitudes in pseudoscalar meson photoproduction. Since the overall phase is arbitrary, this leads to seven independent parameters. Chiang and Tabakin[80] recently showed that the amplitudes can be uniquely determined by sets of eight properly selected measurements – there were problems in several previous analyses. One needs to measure the cross section $d\sigma/d\Omega$, the three single polarization observables, Σ, p_y, and the polarized target asymmetry T, and four properly selected double polarization observables. Polarized beams (B), targets (T), and recoil (R) measurements are all needed, but it is possible to have two measurements each of the types BT and BR, for example. This is good since TR measurements tend to be much more difficult. Triple polarizations have no new information. Of course, measuring all observables is preferable as it gives redundant information and consistency checks. Such a measurement is even feasible, in the case of $K^+\Lambda^0$ photoproduction, since the $\Lambda^0 \to p\pi^-$ is common, self-analyzing decay channel.[86] However, at present, the world database consists of cross sections and largely single polarization measurements. Double polarization measurements, such as we describe here, are in their infancy.

The reaction helicity amplitudes are defined as $A = \langle\ \lambda_p\ \lambda_\pi|T|\lambda_\gamma\ \lambda_p\rangle$, with

$$N = \langle\ 1/2\ 0|T|1\ -1/2\rangle,$$

$$S_1 = \langle\ 1/2\ 0|T|1\ 1/2\rangle,$$

$$S_2 = \langle -1/2\ 0|T|1\ -1/2\rangle,$$

and

$$D = \langle -1/2\ 0|T|1\ 1/2\rangle.$$

Here N, S, and D refer to non, single, and double *spin* flip, the change in total angular momentum along the c.m. momentum directions between initial to final states. The hadronic helicity conserving amplitudes are S_1 and S_2, since one neglects here the photon helicity. The amplitudes N and D are single hadronic helicity flip. With these amplitudes, the recoil polarization observables are $p_y d\sigma/dt = 2\Im(S_2 N^* + S_1^* D)$, $C_x d\sigma/dt = -2\Re(S_2 N^* + S_1^* D)$, and $C_z d\sigma/dt = S_2^2 - S_1^2 - N^2 + D^2$.

At higher energies, the t channel production of a single meson could arise with the intermediate meson being any of a number of mesons m_{int}, such that the $m_{int}pp$ and $\gamma m_{int}\pi^0$ couplings are allowed.[y] The theory has to sum over all possible intermediate states. Regge theory provides a technique for doing this summation. Figure 11 shows Regge trajectories for three families of mesons, in the J vs. m^2 plane. The trajectories are all straight lines with similar slopes, but different intercepts. The basic technique in the theory is to replace the meson propagator, typically of the form $(t - m_\pi^2)^{-1}$ by a Regge propagator, that has the effect of summing all the mesons. Better fits to data are often achieved by allowing the trajectory to be nonlinear,[81] although this brings up the question as to what is the physical interpretation of the procedure. In [81], the trajectory also is made to saturate, so that the calculated cross sections approach pQCD behavior at large momentum transfer. The Regge techniques can also be applied to u channel exchange.

---

[y]This is also the case at low energies, except that at low energies the π^0 is more nearly on shell, while the heavier mesons are far off shell, with a much smaller contribution to the reaction amplitude.

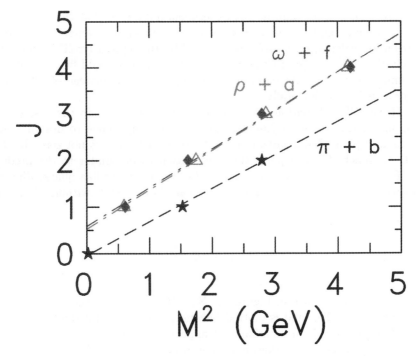

Figure 11. Some Regge trajectories. See text for details.

At very high energies, one has to consider whether pQCD or some other quark model applies. The implications of pQCD should by now be obvious. For the $\gamma p \to p\pi^0$ reaction, the cross sections will vary like

$$\frac{d\sigma}{dt} \propto s^{-7}, \tag{13}$$

and the angular distribution will become independent of energy, once this factor is taken into account. It has long been known[82] that this approximately holds for beam energies of several GeV. If there is an approach to helicity conservation at high momentum transfer, the spin observables also approach simple limits. From the expressions given above, one finds that p_y and $C_{x'}$ will aproach 0 like $1/t$, but $C_{z'} \to (S_2^2 - S_1^2)/(S_2^2 + S_1^2)$. $C_{z'}$ is not required to vanish, and can be calculated in various models.

A pQCD calculation[83] was done for the amplitudes for many meson photo-production channels. The calculated cross sections are typically 1 – 2 orders of magnitude below the data. Perhaps surprisingly, the calculated cross sections decrease with t faster than do the experimental data, so that the agreement is best at small angles. Similar results were obtained in a generalized parton distribution calculation by Afanasev[21], and by Huang and Kroll[85], who note merely that π^0 photoproduction in the GPD approach would be expected to be similar in size to real Compton scattering, $\gamma p \to \gamma' p'$, instead of 100 times larger as it is experimentally.

The pQCD calculation also gives predictions for $C_{z'}$. In the case of π^0 photoproduction, using asymptotic distribution amplitudes to describe the proton and pion, The polarization tends to be small, but is not zero. In the GPD approach,[84] one expects the polarization to be similar to the prediction for polarizations transfer in scattering from a point-like quark, diluted somewhat by wave function effects. The calculated result, without dilution, is:

$$C_{z' \; cm} = \frac{s^2 - u^2}{s^2 + u^2} \tag{14}$$

The prediction is shown in Figure 12. Note that the prediction is only valid for large s, t, and u, and is plotted beyond the limits of its validity. While the GPD approach is in its infancy, it appears at present that there is additional soft physics that it does not incorporate. Thus, we do not at present understand how in quark models to generate the large photoproduction cross sections.

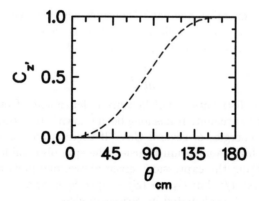

Figure 12. Calculation of polarization transfer to a quark.

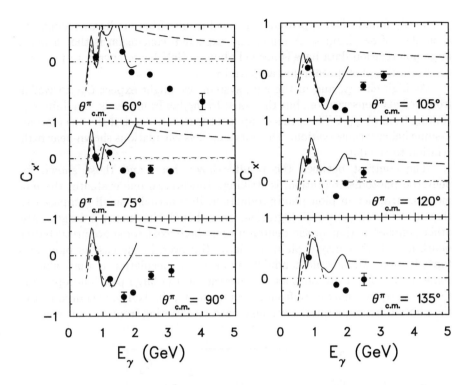

Figure 13. Transverse polarization transfer as a function of energy for the $\gamma p \rightarrow p\pi^0$ reaction. Data are shown in the lab frame. The solid (dashed) lines at low energy are from SAID (MAID). The long (short) dashed lines at high energy are from the polarization transfer on a quark (pQCD) calculation).

5.2 Data vs. Calculations

A sample of our pion photoproduction data is shown in Figure 13. We generally find that, even though the amplitudes were not uniquely constrained by previous data, and even though the polarization transfers are a new, never before measured combination of amplitudes, the SAID and MAID analyses tend to reproduce the data moderately well. The data and calculations tend to diverge with increasing energy, and SAID tends to reproduce the data slightly better than MAID. A new fit including this data has been performed.[87] It reproduces our measurements well, with little change to resonance amplitudes previously found. The improved fit instead results from moderate changes to a number of amplitudes. A main, and expected, conclusion is that a much large polarization data base is needed above 1 GeV beam energy to properly constrain the photoproduction amplitudes.

At high energy, and in the c.m. system, we might expect $C_{x'}$ to vanish from helicity conservation, but the boost to the lab frame leads to a finite $C_{x'}$ coming from $C_{z'\text{c.m.}}$. Thus, this component is nonzero even if the calculations assume helicity conservation. The quark models calculations shown bear little relation to the data.

The contrast between the results shown here, and those shown for deuteron photodisintegration, is striking. However, if one evaluates the momentum transfer in pion photoproduction, it is actually small compared to that in deuteron photodisintegration, even though the energy is larger. Another contrast is that in the deuteron case there exist some nonpreturbative quark models that may explain the data. However, in the case of pion photoproduction, we have no candidate models. A speculation is that, akin to some models for deuteron photodisintegration, the large pion photoproduction cross section may arise from a vector meson, the hadronic component of the photon, having a quark exchange hard rescattering from the nucleon.

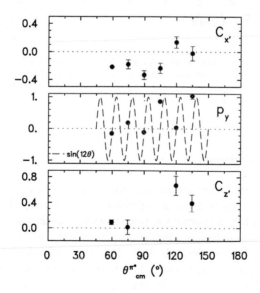

Figure 14. Recoil polarizations as a function of angle at 2.5 GeV or the $\gamma p \rightarrow p\pi^0$ reaction.

Figure 14 shows the 2.5 GeV angular distributions for the three recoil polarization observables. Particularly striking is the induced polarization, which rapidly oscillates and approaches the limit of unity at two angles. This is amazing in that it implies the dominace of two amplitudes, perhaps a single resonance and the background Born amplitude, of nearly equal size and nearly 90° out of phase, so that the interference is nearly purely imaginary. Since 2.5 GeV corresponds to $W > 2$ GeV, above the nominal resonance region, it is perhaps further suprising that in the single meson photoproduction channel there appears to be a single large and dominant resonance, rather than many overlapping resonances. The $\sin 12\theta$ curve has no theoretical significance. It merely indicates that the oscillations are rapid, indicative of a high spin resonance. Note that $C_{z'}$ could not be measured at intermediate angles due to poor spin transport.

5.3 Conclusions about Meson Photproduction

What have we learned from meson photoproduction? At low energies, it is clear that one needs a much larger set of polarization data to better constrain the reaction amplitudes. The experiment was a step in that direction, and further such measurements are needed. No new resonances were found, but one should hardly expect that in a simple survey.

As the energy increased to be above the nominal resonance region, we saw no clear indication of an approach to a smooth behavior, with no indication of resonances. Instead, there are strong oscillations and large induced polarizations at 2.5 GeV, which are echoed in the 3.1 GeV data. Perhaps here a more meaningful resonance determination can be made. Any attempts to go beyond the resonance region require a new experiment at higher energies, preferrable with a calorimeter in coincidence with the proton arm, to improve the data by removing backgrounds.

Acknowledgments

I would like to thank the HUGS organizers for giving me the opportunity to present the lectures and draft this manuscript. I thank the students for their attention and questions, which tended to be much more interesting than those at the usual seminar. My research is funded by the National Science Foundation, grant PHY-00-98642 to Rutgers University. I am indebted to experimental colleagues too numerous to name. I also thank Andrei Afanasev, Franz Gross, Anatoly Radyushkin, Misak Sargsian, and Igor Strakovski for their great patience.

References

1. J.P. Ralston and P. Jain, hep-ph/0207129.
2. L. Andivahis *et al.*, Phys. Rev. D **50**, 5491 (1994).
3. We use here the conventions of R.E. Arnold, C.E. Carlson, and F. Gross, Phys. Rev. C **23**, 363 (1981). The problem of polarized lepton – polarized hadron scattering appears to have first been addressed by J. Scofield, Phys. Rev. **113**, 1599 (1959), with an extension to recoil polarization in J. Scofield, Phys. Rev. **141**, 1352 (1966). N. Dombey, Rev. Mod. Phys. **41**, 236 (1969) provides a detailed derivation with modern notation. See also A.I. Akhiezer and M.P. Rekalo, Sov. J. Part. Nucl. **4**, 277 (1974).
4. M. Jones *et al.*, Phys. Rev. Lett. **84**, 1398 (2000).
5. O. Gayou *et al.*, Phys. Rev. C **56**, 038202 (2001).
6. O. Gayou *et al.*, Phys. Rev. Lett. **88**, 092301 (2002).
7. V. Punjabi *et al.*, to be published.
8. B. Milbrath *et al.*, Phys. Rev. Lett. **80**, 452 (1998), Phys. Rev. Lett. **82**, 2221(E) (1999).
9. Th. Pospischil *et al.*, Eur. Phys. J. A **12**, 125 (2001).
10. J.P. Ralston, R.V. Buniy, and P. Jain, hep-ph/0206063; J.P. Ralston, P. Jain, and R.V. Buniy, hep-ph/0206074; T. Gousset, B. Pire, and J.P. Ralston, Phys. Rev. D **53**, 1202 (1996).
11. J. Arrington, R. Segel, *et al.*, Jefferson Lab E01-001.
12. R. Madey, *et al.*, Jefferson Lab Hall C E93-038.
13. T. Janssens *et al.*, Phys. Rev. **142**, 922 (1966).
14. G. Höhler *et al.*, Nucl. Phys. B **114**, 505 (1976).
15. M.F. Gari and W. Krümpelmann, Phys. Lett. B **274**, 159 (1992).
16. P. Mergell, Ulf-G. Meißner, D. Drechsel, Nucl. Phys. A **596**, 367 (1996).
17. E. Lomon, Phys. Rev. C **64**, 035204 (2001).
18. F. Schlumpf, J. Phys. G **20**, 237 (1994).
19. G.A. Miller, Phys. Rev. C **66**, 032201 (2002), nucl-th/0207007; G.A. Miller, nucl-th/0206027.
20. G.A. Miller and M.R. Frank, Phys. Rev. C **65**, 065205 (2002), nucl-th/0201021.
21. A. Afanasev, hep-ph/9910565.
22. C. Perdrisat *et al.*, Jefferson Lab Hall C E01-109.
23. A.F. Sill, Phys. Rev. D **48**, 29 (1993).
24. H. Gao, J.R. Calarco, H. Kolster and the Bates BLAST Collaboration.
25. S. Platchkov *et al.*, Nucl. Phys. A **510**, 740 (1990).
26. R. Gilman *et al.*, Jefferson Lab Hall A E02-004.

27. H. Zhu *et al.*,Phys. Rev. Lett. **87**, 081801 (2001).
28. S. Galster, Nucl. Phys. B **32**, 221 (1971).
29. B. Wojtsekhowski *et al.*, Jefferson Lab Hall A E02-013.
30. A. Lung *et al.*, Phys. Rev. Lett. **70**, 718 (1993).
31. W. Brooks *et al.*, Jefferson Lab Hall B E94-017.
32. W. Xu *et al.*, submitted to Phys. Rev. C; W. Xu *et al.*, Phys. Rev. Lett. **85**, 2900 (2000).
33. H. Anklin Phys. Lett. B **428**, 248 (1998).
34. G. Kubon Phys. Lett. B **524**, 26 (2002).
35. R. Gilman and F. Gross, J. Phys. G **28**, R37 (2002).
36. M. Garçon and J.W. Van Orden, *Advances in Nucl. Phys.* **26**, 293 (2001).
37. I. Sick, Prog. Part. Nucl. Phys. **47**, 245 (2001).
38. D. Abbott *et al.*, Phys. Rev. Lett. **84**, 5053 (2000).
39. M.E. Schulze *et al.*, Phys. Rev. Lett. **52**, 597 (1984).
40. V.F. Dmitriev *et al.*, Phys. Lett. B **157**, 143 (1985).
41. B.B. Wojtsekhowski *et al.*, Pis'ma Zh. Eksp. Teor. Fiz. **43**, 567 (1986) [JETP Lett. **43**, 733 (1986)].
42. R. Gilman *et al.*, Phys. Rev. Lett. **65**, 1733 (1990).
43. B. Boden *et al.*, Z. Phys. C **49**, 175 (1991).
44. I. The *et al.*, Phys. Rev. Lett. **67**, 173 (1991).
45. M. Garçon *et al.*, Phys. Rev. C **49**, 2516 (1994).
46. S.G. Popov *et al.*, *Proceedings of the 8th International Symposium on Polarization Phenomena in Nuclear Physics, Bloomington, Indiana, 1994*, AIP Conf. Proc. No. 339 (AIP, New York, 1995).
47. M. Ferro-Luzzi *et al.*, Phys. Rev. Lett. **77**, 2630 (1996).
48. M. Bouwhuis *et al.*, Phys. Rev. Lett. **82**, 3755 (1999).
49. D.M. Nikolenko *et al.*, Nucl. Phys. A **684**, 525c (2001); D.M. Nikolenko *et al.*, accepted by Phys. Rev. Lett.
50. D. Abbott *et al.*, Eur. Phys. J. A **7**, 421 (2000)
51. G.R. Farrar, K. Huleihel and H. Zhang, Phys. Rev. Lett. **74**, 650 (1995).
52. S.J. Brodsky and J.R. Hiller, Phys. Rev. C **28**, 475 (1983) and Phys. Rev. C **30**, 412(E) (1984).
53. H.A. Bethe, Phys. Rev. **76**, 38 (1949).
54. S. Weinberg, PLB **251**, 288 (1990).
55. D.R. Phillips, R. Rupak, and M.J. Savage, PLB **473**, 209 (2000).
56. D.R. Phillips, nucl-th/0203040 and private communication.
57. D.R. Phillips and T.D. Cohen Nucl. Phys. A **668**, 45 (2000).
58. M. Walzl and Ulf G. Meissner, Phys. Lett. B **513**, 37 (2001).

59. M. Schwamb and H. Arenhövel, Nucl. Phys. A **696**, 556 (2001); M. Schwamb and H. Arenhövel, Nucl. Phys. A **690**, 682 (2001); M. Schwamb and H. Arenhövel, Nucl. Phys. A **690**, 647 (2001); M. Schwamb, H. Arenhövel, P. Wilhelm, and Th. Wilbois, Phys. Lett. B **420**, 255 (1998).

60. T-S.H. Lee, *Few Body Syst. Supplement* **6**, 526 (1992); T-S.H. Lee, *Argonne National Laboratory Preprint* PHY-6886-TH-91 (1991); T-S.H. Lee, *Argonne National Laboratory Preprint* PHY-6843-TH-91 (1991).

61. Y. Kang, P. Erbs, W. Pfeil, and H. Rollnik *Abstracts of the Particle and Nuclear Intersections Conference*, (MIT, Cambridge, MA, 1990); Y. Kang, Ph.D. thesis Bonn (1993).

62. S.I. Nagornyi, Yu A. Kasatkin, and I.K. Kirichenko, Yad. Fiz. **55**, 345 (1992) [Sov. J. Nucl. Phys. **55**, 189 (1992)].

63. A.E.L. Dieperink and S.I. Nagornyi, Phys. Lett. B **456**, 9 (1999).

64. K.Yu Kazakov and D.V. Shulga *Preprint* nucl-th/0101059.

65. K.Yu Kazakov and S.Eh. Shimovsky, Phys. Rev. C **63**, 014002 (2000).

66. V.P. Barannik *et al.*, Nucl. Phys. A **451**, 751 (1986).

67. K. Wijesooriya *et al.*, Phys. Rev. C **66**, 034614 (2002).

68. F. Adamian *et al.*, Eur. Phys. J. A **8**, 423 (2000).

69. K. Wijesooriya *et al.*, Phys. Rev. Lett. **86**, 2975 (2001).

70. P. Rossi *et al.*, Jefferson Lab Hall B E93-017.

71. R. Gilman *et al.*, Jefferson Lab Hall A E00-007, E00-107.

72. R. Gilman *et al.*, Jefferson Lab proposal E03-005.

73. A. Radyushkin, private communication.

74. L.L. Frankfurt, G.A. Miller, M.M. Sargsian, and M.I. Strikman, Phys. Rev. Lett. **84**, 3045 (2000).

75. L.L. Frankfurt, G.A. Miller, M.M. Sargsian, and M.I. Strikman, Nucl. Phys. A **663**, 349 (2000).

76. L.A. Kondratyuk *et al.*, Phys. Rev. C **48**, 2491 (1993).

77. V.Yu Grishina *et al.*, Eur. Phys. J. A **10**, 355 (2001).

78. R.A. Arndt *et al.*, Phys. Rev. C **53**, 430 (1996) and Phys. Rev. C **56**, 577 (1997). See also http://gwdac.phys.gwu.edu/.

79. D. Drechsel *et al.*, Nucl. Phys. A **645**, 145 (1999). See also http://www.kph.uni-mainz.de/MAID/maid98/maid98.html.

80. W.-T. Chiang and F. Tabakin, Phys. Rev. C **55**, 2054 (1997).

81. M. Guidal, J.-M. Laget, and M. Vanderhaeghen, Nucl. Phys. A **627**, 645 (1997).

82. R.L. Anderson *et al.*, Phys. Rev. D **14**, 679 (1976).

83. G.R. Farrar, K. Huleihel, and H.-Y. Zhang, Nucl. Phys. B **349**, 655 (1991).
84. A. Afanasev, unpublished.
85. H.W. Huang and P. Kroll, Eur. Phys. J. C **17**, 423 (2000).
86. F. Klein *et al.*, Jefferson Lab Hall B E02-112.
87. R.A. Arndt, I.I. Strakovsky, and R.L. Workman, preprint nucl-th/0212055.
88. M. Amarian *et al.*, Phys. Rev. Lett. **89**, 242301 (2002).
89. E. Schulte *et al.*, Phys. Rev. C **66**, 042201R (2002).
90. E. Schulte *et al.*, Phys. Rev. Lett. **87**, 102302 (2001).
91. C Jarlskog in *CP Violation*, ed. C Jarlskog (World Scientific, Singapore, 1988).
92. J.D. Bjorken and I. Dunietz, Phys. Rev. D **36**, 2109 (1987).
93. C.D. Buchanan *et al*, Phys. Rev. D **45**, 4088 (1992).

QUARK-HADRON DUALITY: A PEDAGOGICAL INTRODUCTION

MARTIN A. DEWITT

North Carolina State University, Physics Department, Raleigh, NC 27695 USA
E-mail: madewitt@unity.ncsu.edu

SABINE JESCHONNEK

The Ohio State University, Physics Department, Lima, OH 45804 USA
E-mail: jeschonnek.1@osu.edu

Quark-hadron duality and its potential applications are discussed. We focus on theoretical efforts to model duality. These lectures were given at HUGS 2002 by Sabine Jeschonnek, and were written up by Martin DeWitt.

1 Introduction

In its broadest sense, the term duality means that there exist two different languages to describe the same thing. Both of these languages are correct. However, one of the languages may be more convenient in a particular situation. In hadronic physics, reactions may be described either in terms of hadrons or in terms of quarks and gluons. This has been termed "degrees of freedom" duality[1]. In principle, we can describe any hadronic reaction in terms of quarks and gluons by solving Quantum Chromodynamics (QCD). However, in the vast majority of cases, a full QCD solution is not possible. Unfortunately, in many of these same situations, performing a complete hadronic calculation is equally impractical.

However, another more practical form of duality has been found to exist in many reactions including $e^+e^- \rightarrow hadrons$[2], semileptonic decays of heavy quarks[3,4], dilepton production in heavy ion reactions[5], hadronic decays of the τ lepton[6], and inclusive inelastic electron scattering. This special type of duality states that, in a certain kinematic regime, properly averaged hadronic observables can be described by perturbative QCD (pQCD). Since pQCD calculations can be performed, this version of duality has a much wider range of applicability than the "degrees of freedom" duality described above.

As one example of the usefulness of duality, consider the case of inclusive electron–nucleon scattering. In this situation, duality establishes a connection between the resonance region and the deep–inelastic region. It is well known that measurements in the deep–inelastic region have extremely low count rates compared with measurements taken in the resonance region. Because of this, it may be difficult if not impossible to measure certain observables at high four-momentum transfer. However, the connection supplied by duality could allow one to use easily attainable resonance region data to calculate observables in the deep-inelastic region.

Electron scattering provides one of the best illustrations of duality, and it will be the main focus of this paper. As such, we will begin in Sec. 2 with a basic review of inclusive electron-nucleon scattering. Following this, in Sec. 3, we will examine the phenomenon of duality in detail and present some recent experimental evidence. We will also present some specific examples of how duality has been applied.

Even though the existence of duality has been verified experimentally, our theoretical understanding of it is still quite limited. Being able to use duality to reliably extract information requires knowing exactly where it holds and how accurate it is. Understanding the origins of duality is of vital importance for this. In Sec. 4 we will discuss some of the theoretical efforts which have been made toward understanding how duality arises. One of the main topics of this discussion will be a particular model which has been quite successful in qualitatively reproducing all of the experimentally observed characteristics of duality.

2 Inclusive Electron–Nucleon Scattering — A Review

2.1 Kinematics

We begin with a review of inclusive electron–nucleon scattering which contains material taken from Refs.[7,8,9]. Figure 1 shows the diagram for the inelastic

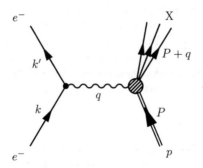

Figure 1. Diagram for the reaction $ep \to eX$

reaction $ep \rightarrow eX$. Here, the final state, X, may consist of several particles (*e.g.* proton and pions) or simply one excited particle (*e.g.* $\Delta(1232)$). The invariant mass of the final state is denoted by W, and, in the rest frame of the nucleon, is given by

$$W^2 = (q + P)^2 = q^2 + P^2 + 2q \cdot P$$
$$= M^2 - Q^2 + 2M\nu, \tag{1}$$

where M is the nucleon mass, $\nu = q^0$ is the energy transfer, and $Q^2 = \vec{q}^{\,2} - \nu^2$. In the case of elastic scattering where the final state is the same as the incoming nucleon, $W = M$, and Eq. (1) becomes

$$Q^2 = 2M\nu \Rightarrow \nu = \frac{Q^2}{2M}. \tag{2}$$

Equation (2) demonstrates that for elastic scattering there is a fixed relationship between the energy and three–momentum transfer. In inelastic scattering (Eq. (1)), the invariant mass of the final state may be anything, and no fixed relation between transferred energy and three–momentum exists.

Figure 2 shows a typical cross section for electron–proton scattering. There are three main regions which can be seen here: the elastic region, the resonance region, and the deep inelastic region. The large spike in the cross section at $W = M_p = 938\ MeV$ indicates elastic scattering from the proton. Here the energy of the exchanged virtual photon is so low (*i.e.* it has such a long wavelength) that the electron only "sees" the proton as a single structureless particle with charge e and magnetic moment $\mu = 2.79e/(2M_p)$. However, as the energy of the virtual photon increases, the invariant mass also increases and the cross section goes through a series of bumps or resonances. These resonances imply the existence of a substructure inside the proton. At high energy, a short wavelength virtual photon is necessary in order for this substructure to be seen. As the invariant mass increases further, the resonances disappear. The region above $W \approx 2\ GeV$ is known as the deep–inelastic region. Here the energy of the virtual photon is so high that it is able to resolve the quarks themselves. We now turn to a review of form factors and structure functions in these three regions.

2.2   Elastic Form Factors

For elastic electron–nucleon scattering, the cross section is calculated beginning with a leptonic and a hadronic current. The leptonic current is simply $j_\ell^\mu = \bar{u}\gamma^\mu u$. However, since the nucleon has an extended structure the current associated with it must be written as $j_h^\mu = \bar{u}\Gamma^\mu u$, where Γ^μ is the most general

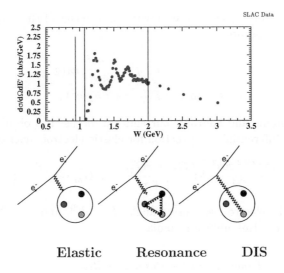

Elastic   Resonance   DIS

Figure 2. Cross section for electron–proton scattering as a function of the invariant mass of the final state, W. Figure reproduced with the kind permission of Ioana Niculescu.

four–vector that can be constructed from k, k', q, and γ^μ; more explicitly

$$\Gamma^\mu = \gamma^\mu B_1 + i\sigma^{\mu\nu}(k' - k)_\nu B_2 + i\sigma^{\mu\nu}(k' + k)_\nu B_3 \\ + (k' - k)^\mu B_4 + (k' + k)^\mu B_5 \, , \tag{3}$$

where each of the coefficients, B_i, is a function of the only independent scalar, q^2. The number of terms can be reduced by employing the Gordon decomposition. Once this is accomplished, current conservation (*i.e.* $q_\mu j^\mu = 0$) implies that

$$j_h^\mu = \overline{u}[F_1(q^2)\gamma^\mu + \frac{\kappa}{2M}F_2(q^2)i\sigma^{\mu\nu}q_\nu]u \, . \tag{4}$$

If the form factors F_1 and F_2 are rewritten in terms of the Sachs electric and magnetic form factors,

$$G_E(q^2) = F_1(q^2) + \frac{\kappa q^2}{4M^2} F_2(q^2)$$
$$G_M(q^2) = F_1(q^2) + \kappa F_2(q^2), \tag{5}$$

then the leptonic and hadronic currents can be used to obtain the following form of the differential cross section for elastic electron–nucleon scattering,

$$\frac{d\sigma}{dE'd\Omega} = \frac{4\alpha^2 E'^2 \cos^2 \frac{\theta}{2}}{Q^4} \left(\frac{G_E^2 + \tau G_M^2}{1 + \tau} + 2\tau G_M^2 \tan^2 \frac{\theta}{2} \right) \delta \left(\nu - \frac{Q^2}{2M} \right), \tag{6}$$

where θ is the electron scattering angle, $E' = k'^0$, and $\tau = Q^2/(4M^2)$. This is known as the Rosenbluth formula.

2.3   Structure Functions in the Resonance Region

For inelastic electron–nucleon scattering, the final hadronic state, that may contain many particles, cannot be represented by an ordinary Dirac spinor. Therefore, the cross section cannot be parameterized on the level of the current. The cross section can instead be considered to be proportional to the product of a leptonic tensor and a hadronic tensor. One then parameterizes on the level of the tensor. The leptonic tensor is completely known, $L^{\mu\nu} = 2[k^\mu k'^\nu + k^\nu k'^\mu + g^{\mu\nu}(m_e^2 - k \cdot k')]$. The hadronic tensor is written as the most general tensor that can be constructed from k, q, and $g^{\mu\nu}$; more specifically

$$W^{\mu\nu} = -W_1 g^{\mu\nu} + \frac{W_2}{M^2} p^\mu p^\nu + \frac{W_4}{M^2} q^\mu q^\nu + \frac{W_5}{M^2} (p^\mu q^\nu + q^\mu p^\nu), \tag{7}$$

where the W_i are functions of two independent variables often chosen to be Q^2 and ν, or Q^2 and the combination $Q^2/(2p \cdot q)$ (or $Q^2/(2M\nu)$ in the nucleon rest frame), known as Bjorken x (or x_{Bj}). Technically this is not the most general tensor since it only contains terms (symmetric in μ and ν) which will survive after contraction with the completely symmetric leptonic tensor. Once again, applying current conservation (i.e. $q_\mu W^{\mu\nu} = 0$), simplifies the expression to

$$W^{\mu\nu} = W_1(x_{Bj}, Q^2) \left(-g^{\mu\nu} + \frac{q^\mu q^\nu}{q^2} \right)$$
$$+ \frac{W_2(x_{Bj}, Q^2)}{M^2} \left(p^\mu - \frac{q \cdot p}{q^2} q^\mu \right) \left(p^\nu - \frac{q \cdot p}{q^2} q^\nu \right). \tag{8}$$

W_1 and W_2 are called structure functions. Using this form of the hadronic tensor, the cross section is

$$\frac{d\sigma}{dE'd\Omega} = \frac{4\alpha^2 E'^2 \cos^2\frac{\theta}{2}}{Q^4}\left(W_2(x_{Bj},Q^2) + 2W_1(x_{Bj},Q^2)\tan^2\frac{\theta}{2}\right). \quad (9)$$

If the form of Eq. (9) is compared with that of Eq. (6), the elastic contributions to the structure functions are found to be

$$W_1^{el} = \tau G_M^2 \frac{1}{\nu}\delta(1 - x_{Bj})$$

$$W_2^{el} = \frac{G_E^2 + \tau G_M^2}{1+\tau}\frac{1}{\nu}\delta(1 - x_{Bj}). \quad (10)$$

It is clear then that elastic scattering occurs when the variable Bjorken x is exactly equal to 1. This condition is equivalent to that given in Eq. (2).

2.4 Structure Functions in the Deep Inelastic Region

As the energy transfer of the virtual photon gets large, the electron is able to resolve the individual quarks inside the nucleon. Inelastic scattering from the nucleon then becomes elastic scattering from a pointlike quark. The structure functions for a pointlike fermion with mass, m, are

$$2mW_1^{point} = \frac{Q^2}{2m\nu}\delta\left(1 - \frac{Q^2}{2m\nu}\right)$$

$$\nu W_2^{point} = \delta\left(1 - \frac{Q^2}{2m\nu}\right). \quad (11)$$

These structure functions depend on only one variable, $Q^2/(2m\nu)$. To reflect the fact that these dimensionless structure functions go from being a function of two independent variables to one independent variable, they are usually redefined as $MW_1^{point} = F_1$ and $\nu W_2^{point} = F_2$. Summing over all of the partons with charges e_i and momentum distributions $f_i(x)$ gives

$$F_1(x_{Bj}) = \frac{1}{2x_{Bj}}\sum_i e_i^2 x_{Bj} f_i(x_{Bj})$$

$$F_2(x_{Bj}) = \sum_i e_i^2 x_{Bj} f_i(x_{Bj}). \quad (12)$$

The structure functions, which were originally functions of both x_{Bj} and Q^2, become independent of Q^2 at some point in the deep inelastic region. This behavior is known as scaling, and the section of the deep–inelastic region in which scaling occurs is termed the scaling region. In like manner, the structure functions $F_{1,2}$ are sometimes referred to as scaling functions, and the independent variable, x_{Bj}, is called a scaling variable.

3   What is Duality?

3.1   Bloom–Gilman Duality

Two different types of duality were mentioned in the introduction: a (1)"degrees of freedom" duality in which the quark–gluon degrees of freedom are equivalent to the hadronic degrees of freedom, and (2)a special form in which properly averaged hadronic observables can be described by pQCD. The latter of these two types of duality was first observed by Bloom and Gilman in 1970[10]. In their work on inclusive electron–proton scattering, Bloom and Gilman observed a surprising correspondence between low Q^2 resonance region data, and high Q^2 scaling region data. In order to delve into this finding in more detail, we must first examine the dependence of the proton structure functions on the scaling variable x_{Bj}.

It turns out that x_{Bj} is a very useful quantity which allows us to compare data taken in different kinematic regions. For example, in inclusive electron–proton scattering, the dimensionless structure function νW_2 is a function of both Q^2 and x_{Bj}. If we were to plot νW_2 solely as a function of x_{Bj}, for fixed values of Q^2, this would look like Fig. 3.

First, examine panel (a). The elastic peak is shown at $x_{Bj} = 1$, and as x_{Bj} decreases (i.e. as the energy transfer, ν, increases), the characteristic "bumps" of the resonance region can clearly be seen. As x_{Bj} decreases further, we see the transition between the resonance region and the deep inelastic region. At each value of Q^2, the same resonances appear (compare panels (a), (b) and (c)). However, as Q^2 increases, the resonances get shifted to higher and higher values of x_{Bj}. This can be seen quantitatively by starting with the expression for the invariant mass of the final state,

$$W^2 = M^2 + 2M\nu - Q^2 = M^2 + Q^2 \frac{(1 - x_{Bj})}{x_{Bj}} , \qquad (13)$$

and noting that if we have a resonance in the final state with an invariant mass, W_{res}, the value of x_{Bj} at which the resonance will occur is given by

$$x_{Bj_{res}} = \frac{Q^2}{W_{res}^2 - M^2 + Q^2} . \qquad (14)$$

As Q^2 increases, the value of x_{Bj} at which a given resonance occurs also increases. In the limit as Q^2 goes to infinity, all of the resonances get pushed to $x_{Bj} = 1$.

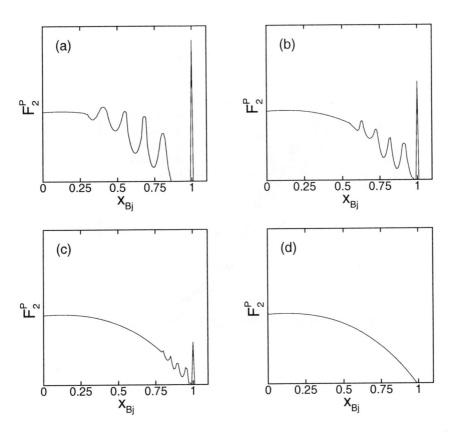

Figure 3. F_2^p plotted as a function of x_{Bj} for (a) low Q^2, (b) medium Q^2, (c) high Q^2, (d) very high Q^2. Please note that this is not actual data. The plots are for purposes of illustration only.

Notice that as Q^2 increases, the range of x_{Bj} which contains the resonance region shrinks. At the same time, the range of x_{Bj} which contains the deep inelastic region broadens. If Q^2 is very large (panel (d)), then the resonance region would be pushed right up against $x_{Bj} = 1$, and the entire graph would be showing just the deep inelastic behavior. This curve, obtained from high Q^2 (scaling region) data is called a "scaling curve." The scaling curve can be compared with the other curves at lower Q^2. This allows us to compare high Q^2 data from the deep inelastic region at certain values of the scaling variable, x_{Bj}, with low Q^2 resonance region data at the same values of x_{Bj}.

138

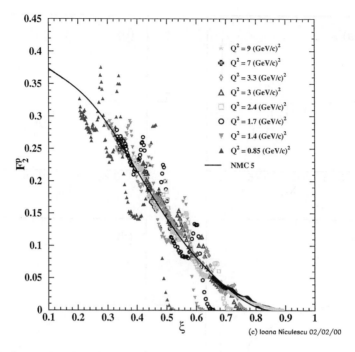

Figure 4. Recent data from Jefferson Lab plotted against Nachtmann ξ. This figure shows data published in [11].

If we take the low Q^2 resonance region data (say in panel (a)) and overlay it onto the scaling curve (panel (d)), a striking similarity is revealed. The overall shape of the resonance curve seems to follow the shape of the scaling curve. If we properly average the oscillations of the F_2 structure function in the resonance region, then this averaged curve will closely match the scaling curve. This is the essence of Bloom and Gilman's discovery and is sometimes referred to as "Bloom–Gilman duality".

More recent studies of the F_2 structure function of the proton have verified local duality. Figure 4 shows new data obtained at Jefferson Lab[11]. The resonance data here are plotted versus a different scaling variable, Nachtmann ξ, defined by

$$\xi = \frac{2x_{Bj}}{1 + \sqrt{1 + \frac{4M^2 x_{Bj}^2}{Q^2}}} \,. \tag{15}$$

The origin of Nachtmann ξ will be discussed in Sec. 4. For now, it is sufficient to note that in the Jefferson Lab data, duality is shown to hold all the way down to about $Q^2 = 0.5\,GeV^2$.

3.2 Application: Extracting the Proton's Magnetic Form Factor

It has been shown that Bloom–Gilman duality connects the resonance region with the deep inelastic region. This version of duality tends to indicate that each of the resonances individually averages to the scaling curve. This idea has been termed "local" duality. If we take local duality seriously, then we can obtain information about a single resonance from structure function data in the scaling region or vice versa. One such calculation has been carried out by Ent, Keppel, and Niculescu (EKN)[12]. This group used the idea of local duality to extract the proton elastic form factor, G_M, from deep inelastic data. A brief introduction to moments is essential for understanding this calculation.

Moments are a useful tool in analyzing structure function data. The basic definition is,

$$M_n(Q^2) = \int dx\, x^{n-2} F(x, Q^2)\,, \tag{16}$$

where x represents a scaling variable and F represents some structure function. For certain types of moments, the structure function is modified, e.g. to take target mass effects into account. If $x = x_{Bj}$ the moments are called Cornwall–Norton Moments, and another version of the moments,

$$M_n(Q^2) = \int dx \frac{\xi^{n+1}}{x^3} \frac{3 + 3(n+1)r + n(n+2)r^2}{(n+2)(n+3)} F_2(x, Q^2) \tag{17}$$

is referred to as Nachtmann moments[13], with $r = \sqrt{1 + \frac{4M^2 x^2}{Q^2}}$. Cornwall–Norton and Nachtmann Moments of F_2^p extracted from the world's electron–proton data are shown in Fig. 5[14]. Note how the moments flatten out as Q^2 increases. This indicates the onset of a type of duality that is slightly different from local duality. The range of integration in Eq. (16) is over the entire range of the scaling variable x. This means that in the resonance region (low to moderate Q^2), all of the resonances are being integrated over to obtain the moment. In the scaling region (high Q^2), the moment is simply the integral of the scaling curve. The value of Q^2 at which the moments flatten out is the point at which the integral over all of the resonances becomes equal to the integral over the scaling curve. This is known as "global" duality to

140

distinguish it from "local" duality where individual resonances are compared with the scaling curve.

To see how these moments allow one to calculate G_M from deep inelastic data, let us begin by writing down the moment for the proton structure function, following [15]:

$$M_n^P(Q^2) = \int_0^{\xi_{el}} d\xi \, \nu W_2(\xi, Q^2) \xi^{n-2} . \tag{18}$$

Here, we use Nachtmann's variable ξ, but an analogous derivation can be obtained using x_{Bj} based moments. The position of the elastic peak is indicated by ξ_{el}; note that $\xi_{el} \to \infty$ for $Q^2 \to \infty$. Now, if we ignore global duality and make the "very strong" assumption that duality is valid "very locally" (i.e.

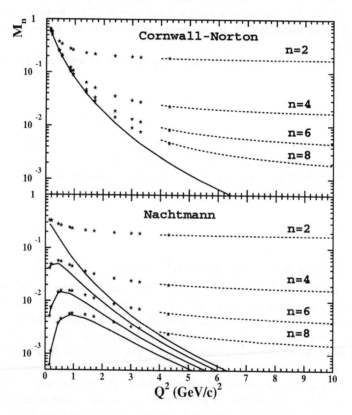

Figure 5. Moments of the proton structure function. This figure is reproduced from [14].

just in the elastic region), then we can write,

$$\int_{\xi_{thr}}^{\xi_{el}} d\xi \, \nu W_2(\xi, Q^2) \xi^{n-2} = \int_{\xi_{thr}}^{\xi_{el}} d\xi \, F_2^{DIS}(\xi) \xi^{n-2} , \qquad (19)$$

where F_2^{DIS} represents the scaling curve. Note that the integration runs over just the elastic region with ξ_{thr} being the value of ξ at the pion production threshold (i.e. where $x_{Bj} = Q^2/[(M + m_\pi)^2 - M^2 + Q^2]$). As Q^2 increases, the value of ξ_{thr} gets closer to 1 and the range over which the curves are integrated decreases.

The elastic contribution to the structure function is given by Eq. (10). This can be used in Eq. (19) to evaluate the left-hand side. However, since we will be integrating over ξ, it is useful to first rewrite the delta function in terms of ξ. Using the fact that $\xi = (1/M)\sqrt{(\nu^2 + Q^2)} - \nu$ we find that

$$\frac{1}{\nu} \delta(1 - x_{Bj}) = \delta\left(\nu - \frac{Q^2}{2M}\right)$$
$$= \delta\left(\frac{Q^2(1 - \xi) - M^2\xi^2}{2M\xi}\right). \qquad (20)$$

After some careful maniputlation, this becomes

$$\frac{2M\xi^2}{M^2\xi^2 + Q^2} \delta(\xi - \xi_{el}) , \qquad (21)$$

where $\xi_{el} = (2\tau)/(\tau + \sqrt{\tau(1+\tau)})$. Inserting this back into Eq. (19) and performing the integration we obtain:

$$(Q^2 - M^2\xi^2) \frac{G_E^2 + \tau G_M^2}{1 + \tau} \frac{1}{M^2\xi^2 + Q^2} \xi_{el}^{n-1} = \int_{\xi_{thr}}^{\xi_{el}} d\xi \, F_2^{DIS}(\xi) \xi^{n-2} . \qquad (22)$$

Using the condition from the delta function that $Q^2(1 - \xi^2) - M^2\xi^2 = 0$, this can be simplified to

$$\frac{\xi_{el}^n}{2 - \xi_{el}} \frac{G_E^2 + \tau G_M^2}{1 + \tau} = \int_{\xi_{thr}}^{\xi_{el}} d\xi \, F_2^{DIS}(\xi) \xi^{n-2} . \qquad (23)$$

Note that this depends on both the electric and magnetic form factors. In order to extract G_M, some assumption about the relationship between G_E and G_M must be made. If one makes the popular assumption of a dipole form (i.e. $\mu = G_M/G_E$), and uses only the lowest moment ($n = 2$) then G_M is given by

$$G_M^2(Q^2) = \frac{1 + \tau}{\frac{1}{\mu^2} + \tau} \frac{2 - \xi_{el}}{\xi^2} \int_{\xi_{thr}}^{\xi_{el}} d\xi \, F_2^{DIS}(\xi) . \qquad (24)$$

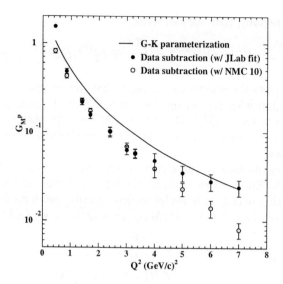

Figure 6. G_M extracted from the scaling curve. This figure is reproduced from [12].

Two different scaling curves are used by EKN in Eq. (24) to extract the proton's magnetic form factor. The first of these is an NMC parameterization of deep inelastic data taken at $Q^2 = 10\ GeV^2$, and the second (JLab fit) is generated by averaging the world's electron–proton resonance excitation data[16]. The resulting values of G_M for each scaling curve are shown in Fig. 6. EKN extracted the form factor by integrating the scaling curve over ξ from $\xi = 1$ to the deep inelastic region, and subtracting the resonance data, integrated up to ξ_{thr}. The extracted values agree reasonably well with the data which is shown using the model parameterization of Ref.[17]; maximum deviations are only around 30%.

There are factors that tend to limit the range of validity of the above analysis. At high Q^2, high ξ data are scarce. This brings into question the accuracy of the NMC parameterization in the high ξ region. Also, as Simula[18] has pointed out, the resonance region data used for the JLab fit extends only to about $\xi = 0.86$, corresponding to the $\Delta(1232)$ at $Q^2 = 8\ GeV^2$. In light of these difficulties, the extractions done with both of these scaling curves should be constrained to low Q^2. In this way, the contribution from integrating over high values of ξ will be a small fraction of the total moment. For values of $Q^2 < 4\,GeV^2$ these high ξ contributions remain smaller than about 30%[19]. Therefore, we should conservatively say that EKN's analysis only verifies the existence of this highly local duality up to about $Q^2 = 4\,GeV^2$.

Another difficulty which has only recently been brought to light is that the dipole assumption ($\mu = G_M/G_E$) used in the derivation of Eq. (24) may be incorrect. Most of the world's data seem consistent with the assumption that $\mu G_E/G_M$ is constant and equal to one. However, new Hall A data from Jefferson Lab[20] clearly shows that this ratio dramatically decreases with increasing Q^2. It is unclear which of these two data sets is to be believed, and the only way to circumvent this inconsistency is to find a way to extract G_M which does not involve G_E. This can be done by using the structure function $W_1(\xi, Q^2)$ instead of $W_2(\xi, Q^2)$. In a similar fashion to the above derivation, it can be shown that

$$G_M^2 = 2\frac{2 - \xi_{el}}{\xi_{el}^n} \int_{\xi_{thr}}^1 d\xi \, \xi^{n-2} F_1^{DIS}(\xi, Q^2) \,. \tag{25}$$

The derivation of this equation does not involve G_E at any point, and so does not require any assumption about the relationship between G_E and G_M. New W_1 data will be available from Jefferson Lab shortly which will allow an integral over the resonance region to be generated. This integral, together with a parametrization of DIS data, can then be used in Eq. (25) to extract the magnetic form factor.

3.3   Application: Extracting the Nucleon Polarization Asymmetry

In the preceding section, it was demonstrated how to use local duality to obtain elastic form factors from deep–inelastic data. It is also possible to carry out the converse of such an analysis (*i.e.* to obtain information on the deep–inelastic region using elastic form factor data). The polarization asymmetry of the nucleon in deep inelastic electron scattering is one example of a quantity that can feasibly be extracted in this manner. The nucleon polarization asymmetry, A_1^N, is defined as

$$A_1^N = \frac{\sigma_{1/2} - \sigma_{3/2}}{\sigma_{1/2} + \sigma_{3/2}} \,, \tag{26}$$

where the subscripts $1/2$ and $3/2$ indicate that the polarizations of the electron beam and target nucleon are such that the final state has $J_z = 1/2$ and $J_z = 3/2$ respectively. Predictions of the polarization asymmetry have been made using both simple quark models and pQCD calculations. These predictions can be compared with the values for A_1^N extracted from data using duality. Here, we mainly follow [8].

Looking at electron–nucleon scattering in terms of a parton model, it is fairly simple to determine the polarization asymmetry. In this process, the incoming electron emits a virtual photon which then collides with the nucleon. If one assumes that the nucleon has $J_z = +1/2$ and the virtual photon is transversely polarized with $J_z = \pm 1$, then the cross sections correspond to:

$$\sigma_{3/2} \to \gamma^*(J_z = +1) + N(J_z = +1/2)$$
$$\sigma_{1/2} \to \gamma^*(J_z = -1) + N(J_z = +1/2) \,. \tag{27}$$

Let us consider the nucleon to be made up of three valence quarks with $L_z = 0$. Assuming that the quarks inside the nucleon have no transverse momentum so that the photon–quark collision is collinear, then the only allowed transitions are:

$$\gamma^*(J_z = -1) + q(J_z = +1/2) \to q(J_z = -1/2)$$
$$\gamma^*(J_z = +1) + q(J_z = -1/2) \to q(J_z = +1/2) \,. \tag{28}$$

The overall photon–nucleon cross section can be built up from the photon–quark cross section. Combining the information in Eqs. (27) and (28) we can obtain:

$$\sigma_{3/2} \propto [\gamma^*(J_z = +1) + N(J_z = +1/2)] \propto \sum_i e_i^2 q_i(J_z = -1/2)$$
$$\sigma_{1/2} \propto [\gamma^*(J_z = -1) + N(J_z = +1/2)] \propto \sum_i e_i^2 q_i(J_z = +1/2) \,, \tag{29}$$

where the index i denotes the quark flavor. Using this, the polarization asymmetry can now be written in terms of quark distribution amplitudes,

$$A_1^N = \frac{\sigma_{1/2} - \sigma_{3/2}}{\sigma_{1/2} + \sigma_{3/2}} = \frac{\sum_i e_i^2 [q_i \uparrow - q_i \downarrow]}{\sum_i e_i^2 [q_i \uparrow + q_i \downarrow]} \,, \tag{30}$$

where $q_i \uparrow (\downarrow)$ is the probability distribution to find a quark of flavor i with spin parallel (anti-parallel) to the nucleon spin. Examining a naive, SU(6) symmetric, quark model wavefunction for the proton and neutron such as can be found in Refs.[8,21] these amplitudes are:

$$Proton: \quad u\uparrow = 5/9 \quad u\downarrow = 1/9 \quad d\uparrow = 1/9 \quad d\downarrow = 2/9$$
$$Neutron: \quad u\uparrow = 1/9 \quad u\downarrow = 2/9 \quad d\uparrow = 5/9 \quad d\downarrow = 1/9 \,. \tag{31}$$

Substituting these into Eq. (30) yields $A_1^p = 5/9$ and $A_1^n = 0$.

Using pQCD to calculate the amplitudes for the process $ep \to eX$, it has been shown[22,23] that at the elastic point, $x_{Bj} = 1$, the polarization asymmetry of the nucleon (proton and neutron alike) is exactly equal to 1. This is in stark contrast to the predictions of unbroken spin–flavor SU(6) derived in the preceding paragraph. Unfortunately, no precise data, or no data at all for the polarization asymmetry of the nucleon exist, especially at high x_{Bj}. However, if duality is used, it is possible to extract the polarization asymmetry from existing nucleon elastic form factor data.

That structure function behavior in the scaling region can be calculated using elastic form factor data was first shown by Bloom and Gilman in 1971[24]. Following their procedure, Melnitchouk[25] has used elastic form factor data to calculate the polarization asymmetry for the proton and neutron in the deep–inelastic region. We present his method here.

First, recall that the most general hadronic tensor contains both spin–independent (F_1,F_2) and spin–dependent (g_1,g_2) structure functions[26]:

$$
W_{\mu\nu} = F_1\left(-g_{\mu\nu} + \frac{q_\mu q_\nu}{q^2}\right) + \frac{F_2}{p\cdot q}\left(p_\mu - \frac{p\cdot q\, q_\mu}{q^2}\right)\left(p_\nu - \frac{p\cdot q\, q_\nu}{q^2}\right)
$$
$$
+ \frac{ig_1}{p\cdot q}\epsilon_{\mu\nu\lambda\sigma}q^\lambda s^\sigma + \frac{ig_2}{(p\cdot q)^2}\epsilon_{\mu\nu\lambda\sigma}q^\lambda((p\cdot q)\, s^\sigma - (s\cdot q)\, p^\sigma). \quad (32)
$$

The relevant cross sections are given in terms of both the spin–independent and spin–dependent structure functions. The polarization asymmetry becomes

$$
A_1^N = \frac{\frac{4\alpha^2 E'}{M\nu Q^2 E}[(E + E'\cos(\theta))g_1^N - \frac{Q^2}{\nu}g_2^N]}{2\sigma_{Mott}[\frac{1}{\nu}F_2^N + \frac{2}{M}F_1^N\tan^2(\theta/2)]}. \quad (33)
$$

In order to calculate this ratio, the deep inelastic structure functions must first be related to the elastic form factors.

In Ref.[24], Bloom and Gilman derive a finite–energy sum rule for the structure function $\nu W_2(x, Q^2)$. Making the assumption that duality is valid very locally in the elastic region, their relation takes on a form similar to

$$
\int_{x_{thr}}^1 dx_{Bj}\, F_2(x_{Bj}) = \frac{2M}{Q^2}\int d\nu\, F_2^{el}(x_{Bj}, Q^2), \quad (34)
$$

where x_{thr} is the value of x_{Bj} at the pion production threshold. It is well known that the elastic contributions to the structure functions are:

$$F_1^{el} = M\tau G_M^2 \, \delta\left(\nu - \frac{Q^2}{2M}\right)$$

$$F_2^{el} = \frac{2M\tau}{1+\tau}(G_E^2 + \tau G_M^2)\, \delta\left(\nu - \frac{Q^2}{2M}\right)$$

$$g_1^{el} = \frac{M\tau}{1+\tau}G_M(G_E + \tau G_M)\, \delta\left(\nu - \frac{Q^2}{2M}\right)$$

$$g_2^{el} = \frac{M\tau^2}{1+\tau}G_M(G_E - G_M)\, \delta\left(\nu - \frac{Q^2}{2M}\right). \tag{35}$$

Substituting F_2^{el} into Eq. (34), performing the integration on the right-hand side, and then differentiating both sides with respect to Q^2 gives

$$F_2(x_{Bj} = x_{thr}) = \beta\left[\frac{G_M^2 - G_E^2}{2M^2(1+\tau)^2} + \frac{2}{1+\tau}\left(\frac{dG_E^2}{dQ^2} + \tau\frac{dG_M^2}{dQ^2}\right)\right], \tag{36}$$

where $\beta = (Q^4/M^2)(\xi_{el}^2/\xi^3)(2x - \xi)/(4 - 2\xi_{el})$. This gives the structure function, F_2, evaluated at $x_{Bj} = x_{thr}$, expressed in terms of elastic form factors. Expressions for F_1, g_1, and g_2 evaluated at x_{thr} can be obtained in a similar fashion:

$$F_1(x_{thr}) = \beta\frac{dG_M^2}{dQ^2}$$

$$g_1(x_{thr}) = \beta\left[\frac{G_M(G_M - G_E)}{4M^2(1+\tau)^2} + \frac{1}{1+\tau}\left(\frac{d(G_E G_M)}{dQ^2} + \tau\frac{dG_M^2}{dQ^2}\right)\right]$$

$$g_2(x_{thr}) = \beta\left[\frac{G_M(G_M - G_E)}{4M^2(1+\tau)^2} + \frac{\tau}{1+\tau}\left(\frac{d(G_E G_M)}{dQ^2} + \frac{dG_M^2}{dQ^2}\right)\right]. \tag{37}$$

These expressions can now be substituted back into Eq. (33) to calculate the polarization asymmetry for the proton and neutron. Melnitchouk's results are shown in Fig. 7.

It is important to keep in mind that since $x = x_{thr}$, each value of x corresponds to a different Q^2. The point $x = 0.8$ corresponds to $Q^2 \approx 1\,GeV^2$, and as $x \to 1$, $Q^2 \to \infty$. The solid lines show the values extracted from data, while the dashed lines are simply extrapolations. In the region $0.6 < x < 0.9$, $A_1^{(p,n)}$ seems to be roughly consistent with the quark model predictions of 5/9 and 0 respectively. However, as x increases, both polarization asymmetries show a strong tendency to move towards one, the prediction of pQCD. The predictions of duality should be much more reliable at large Q^2 than at very small Q^2. Therefore, this particular extraction would tend to favor the pQCD

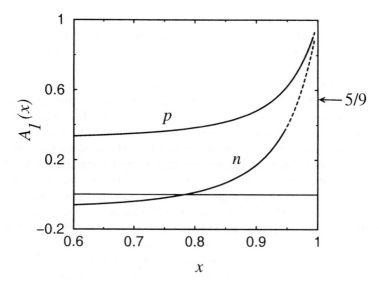

Figure 7. Polarization Asymmetry for the Proton and Neutron. This figure is reproduced from [25].

prediction over the naive, SU(6) symmetric quark model prediction. More high–x data is certainly needed in order to verify the $x \rightarrow 1$ behavior of the polarization asymmetry. However, Melnitchouk's analysis is an excellent example of how duality can be used to obtain information in regions which are not easily accessible to direct experimentation.

4 Exploring the Origins of Duality

4.1 The Operator Product Expansion—A First Explanation

One of the first attempts to explain why duality exists involved Wilson's Operator Product Expansion (OPE). In 1973, Nachtmann[13] was the first to use the OPE to analyze moments of structure functions. This connection was further developed by DeRújula, Georgi, and Politzer (DGP)[27,15]. These analyses showed that the OPE can be used to relate moments of structure functions to certain operators of definite spin. This can be expressed as

$$\int_0^1 d\xi \, \xi^{n-2} F(\xi, Q^2) = A_n(Q^2) + \sum_{k=1}^{\infty} \left(n \frac{M_0^2}{Q^2} \right)^k B_{nk}(Q^2) \,. \qquad (38)$$

The term $A_n(Q^2)$, known as the "leading twist" term, includes radiative corrections and is therefore the perturbative QCD result. However, it is not purely perturbative, as it still contains coefficients A_n, which are numbers, that contain the strong interaction dynamics. One may think of this term as arising as the "soft" input of a parton distribution function is evolved perturbatively. Note that the situation in inclusive electron scattering is more complicated than in $e^+e^- \to hadrons$, where a pure pQCD calculation without any non-perturbative input describes the situation. The remaining "higher twist" terms, $B_{nk}(Q^2)$, involve initial and final state interactions of partons. In the limit as Q^2 goes to infinity the higher twist terms vanish leaving only the perturbative result, $A_n(Q^2)$. However, in the resonance region where interactions are strong, the higher twist terms become more important.

The authors of Ref.[27] show that $A_n(Q^2)$ can be written as the moment of a quark distribution function. This function can be identified with the average function $< \nu W_2(x, Q^2) >$ which arises in Bloom–Gilman local duality[28]. Following Ref.[29], we will denote this average function by $F_2^{TM}(\xi, Q^2)$, and refer to it as the "dual" structure function. Therefore we have

$$A_n(Q^2) = \int_0^1 d\xi \, \xi^{n-2} F_2^{TM}(\xi, Q^2) \,. \tag{39}$$

Furthermore, DGP show that

$$
\begin{aligned}
F_2^{TM}(\xi, Q^2) &= \frac{x_{Bj}^2}{\left(1 + \frac{4M^2 x_{Bj}^2}{Q^2}\right)^{3/2}} \frac{F_2^S(\xi, Q^2)}{\xi^2} \\
&\quad + 6\frac{M^2}{Q^2} \frac{x_{Bj}^3}{\left(1 + \frac{4M^2 x_{Bj}^2}{Q^2}\right)^2} \int_\xi^1 d\xi' \frac{F_2^S(\xi', Q^2)}{\xi'^2} \\
&\quad + 12\frac{M^4}{Q^4} \frac{x_{Bj}^4}{\left(1 + \frac{4M^2 x_{Bj}^2}{Q^2}\right)^{5/2}} \int_\xi^1 d\xi' \int_{\xi'}^1 d\xi'' \frac{F_2^S(\xi'', Q^2)}{\xi''^2} \,.
\end{aligned} \tag{40}
$$

Here, $F_2^S(\xi, Q^2)$ represents the scaling structure function fitted to high Q^2 data and evolved down to low Q^2 by the Altarelli–Parisi equations. The variable, Nachtmann ξ, and the integrals in Eq. (40) arise naturally out of the OPE analysis, and represent target–mass corrections to the structure function (hence the superscript on F_2^{TM}).

The ξ dependence of $F_2^S(\xi, Q^2)$ and $F_2^{TM}(\xi, Q^2)$ is shown in Fig. 8 for various values of four-momentum transfer squared[29]. It can be seen that the pQCD evolution introduces a significant Q^2 dependence into $F_2^S(\xi, Q^2)$. However, once target–mass corrections have been taken into account, the dual

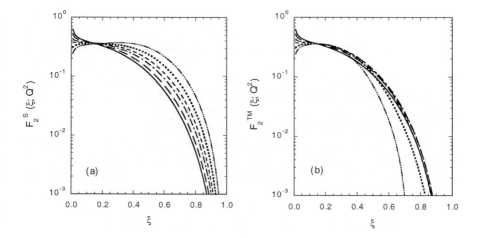

Figure 8. Q^2 dependence of F_2^S and F_2^{TM}. This figure is reproduced from [29].

structure function, $F_2^{TM}(\xi, Q^2)$, exhibits only a slight Q^2 dependence for $Q^2 > 1.5\,GeV^2$. Moreover, the author of Ref.[29] points out that the dual structure function at $Q^2 > 1.5\,GeV^2$ is essentially equal to the function F_2^S at high Q^2:

$$F_2^{TM}(\xi, Q^2) \sim F_2^S(\xi, Q_{High}^2) \,. \tag{41}$$

If we take this approximate relationship, combine it with Eq. (39), and substitute the result back into Eq. (38) we see that

$$\int_0^1 d\xi\, \xi^{n-2} F_2(\xi, Q^2) = \int_0^1 d\xi\, \xi^{n-2} F_2^S(\xi, Q_{High}^2)$$
$$+ \sum_{k=1}^{\infty} \left(n\frac{M_0^2}{Q^2} \right)^k B_{nk}(Q^2) \,. \tag{42}$$

If duality holds, then it is clear that the moment of the F_2 structure function is approximately equal to the moment of the scaling curve (*i.e.* the l.h.s. is approximately equal to the first term on the r.h.s.). Therefore, the contribution from all of the higher–twist terms, $B_{nk}(Q^2)$, must be quite small. This could mean either that every single higher–twist term is small, or that they almost cancel each other out. Either way, this result is quite surprising. While it was expected that the higher–twist terms would vanish at high Q^2, they were expected to be more significant at low Q^2. However, given that recent measurements show that duality can hold down to relatively small values of Q^2, it seems evident that the higher–twist contribution remains small in this region as well. We can rephrase this idea and say that, in order for duality to exist, the contribution of all the higher–twist terms must be very small. The problem of explaining how duality arises then becomes a matter of explaining why the higher–twist terms behave this way. Unfortunately, a clear explanation of why this is so has not been forthcoming.

Today, there remain many unanswered questions regarding the origins of duality. How does the transition occur between the region where duality holds and where it does not? How is it possible to see precocious scaling in a region where interactions are strong? For which observables in which kinematic regimes can we apply duality, and how precise are our results going to be? Current efforts aimed at answering these questions fall into two main areas: refining the analysis and modeling. The former includes trying to determine an appropriate scaling variable, debating Cornwall–Norton vs. Nachtmann moments, working with target mass corrections, and developing new averaging schemes. The latter involves developing a simple model for hadrons which contains the essential features necessary to reproduce the phenomenon of duality. For the remainder of this paper we will focus on a particular model, presented in Refs.[1,30,31], which has been quite successful in this regard.

4.2 Introduction to a Simple Relativistic Model

When developing a model for duality it is desirable to keep things as simple as possible by first attempting to gain only a qualitative understanding of the phenomenon. This allows one to make strong simplifying assumptions, and keep only the most essential physical features. The general approach is to choose a fully solvable model for hadrons, calculate the relevant observables, and compare these results to the — hypothetical — free quark results. At this point, all models assume that after the excitation from the ground state to an excited level, N, the quark will remain in its excited state (bound–bound transition). In other words, the resonance that is produced is assumed not to decay. The results obtained for the bound–bound transition are summed over and compared to the case where in the final state, the binding potential is switched off, and the quark is "free" (bound–free transition). The bound–free transition corresponds to the pQCD situation. Any model for duality must fulfill the following criteria which have been derived from experimental observation:

1. Scaling must be reproduced. Moreover, the scaling curve for the bound–

bound transition must be the same as for the bound–free transition.

2. The resonance region results must oscillate around the scaling curve (*i.e.* local duality).

3. The moments of the structure functions must flatten out at large Q^2 (*i.e.* global duality).

The particular approach used in Refs.[1,30,31] incorporates confinement and relativity into a valence quark model. It is assumed that spin is not necessary in order to observe duality, so the quarks are treated as scalar particles. Furthermore, it is assumed that only one of the quarks in a baryon will carry charge and therefore interact with an incoming photon. The other two quarks are combined into one spectator object which can be thought of either as a diquark or an antiquark. In the limit as the spectator mass goes to infinity, the Bethe–Salpeter equation for the two-body system reduces to a one-body Klein–Gordon equation:

$$\left(\frac{\partial^2}{\partial t^2} - \nabla^2 + m^2 + V^2 \right) \Phi(x) = 0 \,, \tag{43}$$

where $\Phi(x) = \Phi(\vec{r})exp(-iEt)$. At this point, a relativistic harmonic oscillator confining potential, $V^2(\vec{r}) = \alpha r^2$, is chosen. With a little foresight, the Klein–Gordon equation can be written in the form

$$\left[-\frac{1}{2m}\nabla^2 + \frac{1}{2}\left(\frac{\alpha}{m}\right)r^2 - \left(\frac{E^2 - m^2}{2m}\right) \right] \Phi(\vec{r}) = 0 \,. \tag{44}$$

If this result is compared with the Schrödinger equation for the harmonic oscillator,

$$\left[-\frac{1}{2m}\nabla^2 + \frac{1}{2}\kappa r^2 - E \right] \Psi(\vec{r}) = 0 \,, \tag{45}$$

which has the solution

$$E_N = \sqrt{\frac{\kappa}{m}}(N + \frac{3}{2})$$

$$\Psi(x) = \sqrt{\frac{\beta}{2^{n_x}n_x!\sqrt{\pi}}} H_{n_x}(\beta x)exp\left(-\frac{1}{2}x^2\beta^2 \right) \,, \tag{46}$$

then we can see that $\kappa \leftrightarrow \alpha/m$ and $E \leftrightarrow (E^2 - m^2)/(2m)$. Substituting these expressions into the eigenvalue in Eq. (46) and solving for the energy gives

$$E_N^{rel} = \sqrt{2\sqrt{\alpha}\left(N + \frac{3}{2} \right) + m^2} \,. \tag{47}$$

The eigenfunction for the relativistic case will be the same as in Eq. (46) with $\beta \leftrightarrow \alpha^{1/4}$. Notice that in the nonrelativistic case $E_N \propto N$, while in the relativistic case $E_N^{rel} \propto \sqrt{N}$. This results in a much higher density of excited states in the relativistic case. We will see later that this high density of states is crucial in order for the model to exhibit duality.

Now that the hadron spectrum has been established, the results can be used to calculate excitation form factors and structure functions related to bound–bound transitions. Interestingly enough, duality is found to be manifested in the form factors, as well as the structure functions, when compared with the free quark results. We will examine the form factors first.

4.3 Duality in the Model Form Factors

The form factor for a free, scalar particle is defined by the conserved electromagnetic current,

$$j_{em}^\mu = ie(p + p')F(Q^2) \,. \tag{48}$$

If the particle is a free, pointlike quark then the form factor is just equal to 1. However, in the hadronic case, the quark is bound and the scattering is inclusive. Therefore, a form factor must be calculated for each resonance, and then the resonances must be summed over incoherently. In order to check for duality, this result is then compared with the free quark result of 1.

The conserved electromagnetic current for scalar bound states is given explicitly by

$$j_{em}^\mu = i \int d^4x \, e^{-iq \cdot x} [\Phi_f^*(x)\partial_\mu \Phi_i(x) - (\partial_\mu \Phi_f^*(x))\Phi_i(x)] \,, \tag{49}$$

where the indicies i and f represent the initial and final states respectively. Considering only the 0–component of this current,

$$j_{em}^0 = \int dt \, e^{-i\nu t} e^{iE_f t} e^{-iE_i t} \int d^3\vec{r} \, e^{i\vec{q}\cdot\vec{r}} (E_i + E_f)\Phi_f^*(\vec{r})\Phi_i(\vec{r})$$

$$= \delta(E_f - E_i - \nu)(E_i + E_f) \int d^3\vec{r} \, e^{i\vec{q}\cdot\vec{r}} \Phi_f^*(\vec{r})\Phi_i(\vec{r}) \,, \tag{50}$$

the form factor is given by

$$F(\vec{q}^{\,2}) = \int d^3\vec{r} \, e^{i\vec{q}\cdot\vec{r}} \Phi_f^*(\vec{r})\Phi_i(\vec{r}) \,. \tag{51}$$

The excitation form factor for a model hadron going from the ground state to an excited state, N, can now be calculated by substituting the wavefunctions

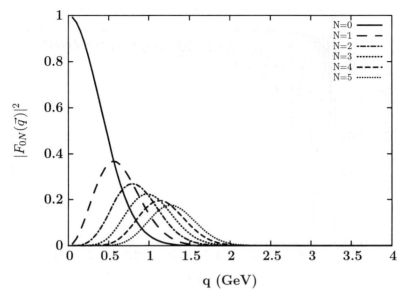

Figure 9. The excitation form factor squared for the excitations N=0 to N=5 with $\beta = 0.4 GeV$.

from Eq. (46) into Eq. (51). The square of the resulting form factor is

$$|F_{0N}(\vec{q})|^2 = \frac{1}{N!}\left(\frac{\vec{q}^{\,2}}{2\beta^2}\right)^N e^{-\frac{\vec{q}^{\,2}}{2\beta^2}}.$$ (52)

The square of each individual form factor is needed for the incoherent sum over resonances. A plot of the form factor squared versus three-momentum transfer for various excitation levels is shown in Fig. 9. Note that as N increases the curves shift to higher values of three-momentum. Additionally, as N increases the peaks tend to broaden and decrease in height.

Summing over resonances from $N = 0$ to some maximum value, N_{max}, gives

$$\sum_{N=0}^{N_{max}} |F_{0N}(\vec{q})|^2 = \exp\left(-\frac{\vec{q}^{\,2}}{2\beta^2}\right) \sum_{N=0}^{N_{max}} \frac{1}{N!}\left(\frac{\vec{q}^{\,2}}{2\beta^2}\right)^N.$$ (53)

It is clear that as $N_{max} \to \infty$, the sum on the r.h.s. of Eq. (53) becomes $\exp(\vec{q}^{\,2}/2\beta^2)$. In this limit, the sum of the excitation form factors squared is equal to 1, the same as the free quark result. However, in actuality the value of N_{max} is limited by kinematics. Electron scattering is a spacelike reaction

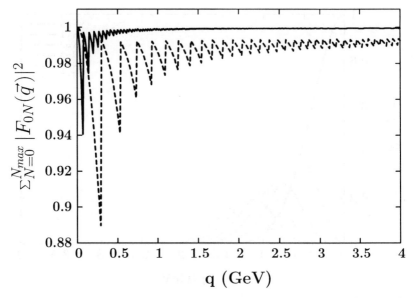

Figure 10. The sum of the excitation form factors squared for $\beta = 0.2\,GeV$ (solid line) and $\beta = 0.6\,GeV$ (dashed line).

(*i.e.* $q^2 < 0$). Therefore, it must be true that

$$|\vec{q}| > \nu = E_N - E_0 \,. \tag{54}$$

The value of $|\vec{q}|$ places an upper limit on the value of E_N, and consequently fixes N_{max} at some finite value. The sum of the form factors squared for two values of the binding strength, β, is shown in Fig. 10. The jagged peaks are a consequence of the fact that only certain resonances are allowed to contribute to the sum at a particular value of $|\vec{q}|$. At each threshold where N_{max} increases, the contribution of the newly allowed form factor causes the curve to jump upward. As $|\vec{q}|$ increases, the jumps decrease in size because the peak of the form factor squared decreases as N increases (refer back to Fig. 9). At high enough three-momentum transfer the curve approaches some limiting value. For the case in which $\beta = 0.2\,GeV$, the curve approaches a value of 0.9994, and therefore duality is only violated by about 0.06%. As the binding energy increases with $\beta = 0.6\,GeV$, it can be seen that duality does not work as well, but it does still hold to within 1%. That duality works best for low binding energies is to be expected since, as the binding energy increases, the situation becomes more and more non-perturbative. Also, it is

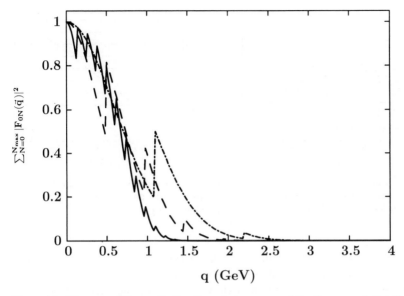

Figure 11. The sum of the excitation form factors squared for the non-relativistic case for $\beta = 0.2\,GeV$ (solid line) and $\beta = 0.6\,GeV$ (dashed line).

obvious that the degree to which duality is fulfilled is closely related to the number of resonances that are available to contribute to the sum. At low $|\vec{q}|$ where few resonances contribute, duality is violated significantly. However, as more and more states become available, the results more closely match those of the free quark.

Earlier it was mentioned that the high density of states generated by relativity plays a role in the manifestation of duality. Figure 11 shows the sum of the form factors for the non-relativistic case. Instead of approaching the free quark result as the three-momentum transfer increases, the curves fall off to zero. This is a direct result of the widely spaced energy eigenvalues of the non-relativistic solution (see Fig. 12). In the non-relativistic case, as each new threshold is reached, the sum picks up the contribution of a new form factor. However, in contrast to the relativistic case where the threshold is a the front or middle of the form factor curve (refer to Fig. 9), the non-relativistic thresholds fall on the backside of the new form factor curves. The tails of these curves contribute very little to the sum which eventually dies out. Again, we see that duality fails when there is not a sufficient number of available resonances.

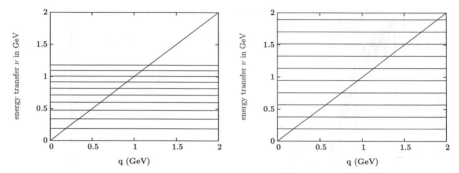

Figure 12. Energy transfers, ν, for the relativistic case (left panel) and for the non-relativistic case (right panel). The diagonal line represents the photopoint, *i.e.* $|\vec{q}| = \nu$. All energy transfers below this are allowed in the spacelike region.

Duality is fulfilled in the model form factors at about the 1–2% level. The realization of duality in the excitation form factors is closely related to completeness. The more resonances that contribute, the more basis states (in this case the Hermite polynomials in Eq. (46)) there are available with which to mathematically reproduce the free quark result. Only a relativistic description can provide a sufficient number of resonances for this purpose. However, even with a dense phase space, violations of duality can still occur, and these violations are directly related to the strength of the binding energy. We now turn to the calculation of structure functions.

4.4   Duality in the Model Structure Functions

Although we have explored duality in some sense using the model form factors, we have not yet seen how the model satisfies the three conditions for duality set forth on pages 21–22. In order to see this we must examine the behavior of the structure functions. In the case of a scalar electron scattering off of a scalar quark, the structure function is defined in terms of the cross section as

$$\frac{d\sigma}{dE' d\Omega} = \frac{g^4}{16\pi^2} \frac{E'}{E} \frac{1}{Q^4} W_{scalar} \,. \tag{55}$$

It is the quantity, W_{scalar}, that must meet the three criteria outlined above.

The first criterion states that scaling must be reproduced. However, one must first determine which quantity it is that is supposed to scale (*i.e.* which scaling variable to use). In Sec. 2.4 we saw how the nucleon structure function, W_2, could be rewritten as a scaling function, $F_2 = \nu W_2$. As $Q^2 \to \infty$, the scaling function F_2 became a function of x_{Bj} only. Therefore, one may first think to try x_{Bj} as the scaling variable. In fact, it is possible to derive the scalar analog of F_2 by starting with the cross section for a scalar electron scattering off of a free, scalar quark,

$$\frac{d\sigma}{dE'd\Omega} = \frac{g^4}{4\pi^2} \frac{E'}{8EE_{p_i}} \frac{1}{Q^4} \delta(p_f^2 - m^2) , \qquad (56)$$

where m, p_i, and p_f represent the mass, initial momentum, and final momentum of the free quark respectively. Comparing this with Eq. (55), it is obvious that the corresponding structure function is given by

$$W_{scalar}^{free} = \frac{1}{2E_{p_i}} \delta(p_f^2 - m^2) . \qquad (57)$$

If we evaluate this for the infinite–momentum frame where the momentum of the nucleon is $P = (\sqrt{P^2 + M^2}, 0, 0, P)$, and the quark is assumed to carry a fraction, x, of the nucleon momentum (*i.e.* $p_i = xP$), then the delta function in Eq. (57) becomes, $\delta(x^2 M^2 - Q^2 + 2xM\nu - m^2)$. Invoking the Bjorken limit (*i.e.* $Q^2 \to \infty$), the quark and target masses become negligible compared with Q^2, and the structure function can be written as

$$W_{scalar}^{free,Bjorken} = \frac{1}{E_{p_i}} \frac{1}{2M\nu} \delta(x - x_{Bj}) . \qquad (58)$$

Notice that if we multiply the structure function by the energy transfer, ν, then we obtain a function which depends only on one variable, x_{Bj}:

$$S_{free}^{Bj}(x_{Bj}) = \nu W_{scalar}^{free,Bjorken} . \qquad (59)$$

This is the Bjorken scaling function, and x_{Bj} is the corresponding scaling variable. The form of this scaling function is the same as that for real electron–nucleon scattering (*i.e.* νW).

Unfortunately, the Bjorken scaling variable and function do not work. In the derivation above, it was assumed that all masses were negligible compared to Q^2. However, duality is known to exist down to fairly low values of Q^2. In this region, the masses of the model baryons and constituent quarks cannot be neglected. Therefore, a truly appropriate scaling variable will still contain all of these masses. Nachtmann ξ is a slight improvement because it includes target–mass corrections, however it still neglects quark mass effects. The developers of this particular model chose to use a scaling variable derived by Barbieri *et al.*[32] for scattering from a free, on-shell quark with a momentum distribution:

$$x_{cq} = \frac{1}{2M} \left(\sqrt{\nu^2 + Q^2} - \nu \right) \left(1 + \sqrt{1 + \frac{4m^2}{Q^2}} \right) . \qquad (60)$$

The scaling function that corresponds to this scaling variable is given by

$$S_{cq} \equiv |\vec{q}|W = \sqrt{\nu^2 + Q^2}\,W \,. \qquad (61)$$

Since the model target has an infinite mass, it is necessary to rescale x_{cq} in the following way:

$$u \equiv \frac{M}{m} x_{cq} \,. \qquad (62)$$

Note that for a given Q^2, the new scaling variable, u, has a maximum value given by

$$u_{max} = \frac{Q}{2m}\left(1 + \sqrt{1 + \frac{4m^2}{Q^2}}\right). \qquad (63)$$

The structure function for the bound–bound transition is given by summing over the excitation form factors squared,

$$W^{b.b.}_{scalar}(\nu, \vec{q}^{\,2}) = \sum_{N=0}^{N_{max}} \frac{1}{4E_0 E_N} |F_{0N}(\vec{q})|^2\, \delta(\nu - E_N + E_0)\,. \qquad (64)$$

In order to evaluate this expression, a smoothing procedure is needed. This is done either by replacing the delta function with a Breit–Wigner shape,

$$\delta(\nu - E_N + E_0) \rightarrow \frac{\Gamma}{2\pi}\frac{f}{(\nu - E_N + E_0)^2 + 0.25\Gamma^2}\,, \qquad (65)$$

where the factor $f = \pi/[\frac{\pi}{2} + \arctan\frac{2(E_N - E_0)}{\Gamma}]$ ensures that the integral over the δ-function is identical to that over the Breit-Wigner shape, or by continuizing the sum over N to an integral. Panel A of Fig. 13 shows the results for the scaling function, $S^{b.b.}_{cq} = |\vec{q}|W^{b.b.}_{scalar}$, plotted as a function of u. It is easy to see that this function does in fact scale. For sufficiently high values of four–momentum transfer, the Q^2 dependence vanishes. An analytic expression for $S^{b.b.}_{cq}$ in the scaling region (*i.e.* the scaling curve) can be derived by taking the continuum limit for the energy and applying Stirling's formula. This scaling curve for the bound–bound transition reads:

$$S^{b.b.}_{cq}(Q^2_{high}) = \frac{1}{4\sqrt{\pi}\beta E_0}\exp\left(-\frac{(E_0 - mu_{Bj})^2}{\beta^2}\right), \qquad (66)$$

where $u_{Bj} = (M/m)x_{Bj}$. The scaling curve is plotted together with the scaling function in panel B of Fig. 13. As Γ decreases, the scaling function approaches the scaling curve.

In order to fully satisfy the first criterion, it is still necessary to show the scaling curve for the bound–bound transition coincides with the scaling

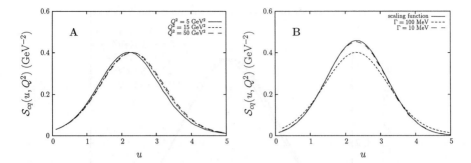

Figure 13. The scaling behavior of S_{cq} for the bound–bound transition as a function of u for various values of Q^2 (left panel). In the limit as $\Gamma \to 0$, the structure function approaches the scaling curve (right panel).

curve for the bound–free transition. The structure function for the bound–free transition is given by

$$
W^{b.f.}_{scalar}(\nu, \vec{q}\,^2) = \frac{1}{4\sqrt{\pi}\beta E_0 |\vec{q}|}
$$
$$
\times \left[\exp\left(-\frac{1}{\beta^2}\left(|\vec{q}| - \sqrt{(\nu + E_0)^2 - m^2}\right)^2 \right) \right.
$$
$$
\left. - \exp\left(-\frac{1}{\beta^2}\left(|\vec{q}| + \sqrt{(\nu + E_0)^2 - m^2}\right)^2 \right) \right]. \quad (67)
$$

The scaling function for the bound–free transition, $S^{b.f.}_{cq}$, is plotted in Fig. 14, and it is clear that scaling occurs for this case as well. Both the bound–free and bound–bound cases are shown in Fig. 15. As $\Gamma \to 0$, the scaling curve for the bound–bound transition nicely overlaps the scaling curve for the bound–free transition.

The second criterion dictates that the resonances inherent in the structure functions at relatively low Q^2 should oscillate around the scaling curve. Figure 16 shows the scaling function for the bound–bound case plotted for both low and high values of four–momentum transfer. The resonances seen here at low Q^2 do, at least qualitatively, seem to oscillate about the high Q^2 scaling curve. In order to illustrate how successful the scaling function, S_{cq}, is at reproducing Bloom–Gilman duality, compare these results with those of Fig. 17, in which the same information is plotted in terms of the Bjorken scaling function. Recall that x_{Bj} and S_{Bj} contain no information about the target or constituent quark masses. This fact manifests itself in a very poor duality

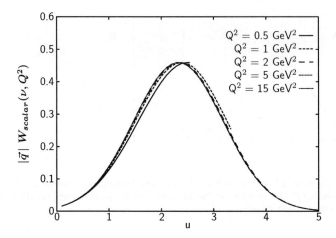

Figure 14. The scaling behavior of S_{cq} for the bound-free transition as a function of u for various values of Q^2

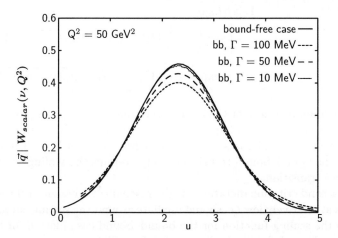

Figure 15. The scaling curve for the bound-free case is plotted here along with the scaling function for the bound-bound case at various values of the width, Γ.

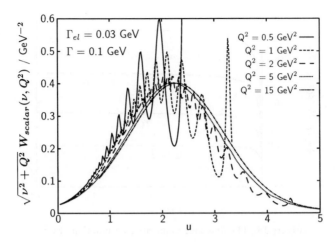

Figure 16. The low Q^2 behavior of S_{cq} as a function of u for $Q^2 = 0.5\,GeV^2$ (solid), $Q^2 = 1\,GeV^2$ (short-dashed), $Q^2 = 2\,GeV^2$ (long dashed), and $Q^2 = 5\,GeV^2$ (dotted).

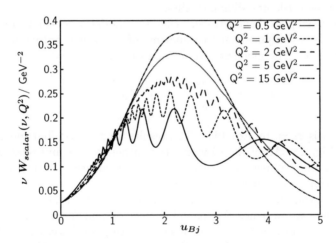

Figure 17. Low Q^2 behavior of S_{Bj} as a function of u_{Bj} for $Q^2 = 0.5\,GeV^2$ (solid), $Q^2 = 1\,GeV^2$ (short-dashed), $Q^2 = 2\,GeV^2$ (long dashed), and $Q^2 = 5\,GeV^2$ (dotted).

162

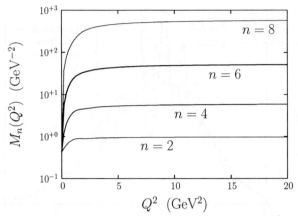

Figure 18. The first four moments as a function of Q^2

between the high and low Q^2 regions, even though the same model results are plotted in both cases. The choice of a suitable scaling function is crucial in order to observe Bloom-Gilman duality.

The last step in demonstrating that the model successfully reproduces duality involves verifying that the moments of the structure functions flatten out as $Q^2 \to \infty$. The moments are defined as

$$M_n(Q^2) = \int_0^{u_{max}} du \, u^{n-2} S_{cq}(u, Q^2) \, , \tag{68}$$

where u_{max} is given by Eq. (63). The results for the first four moments are shown in Fig. 18. All of the moments rise rather rapidly and tend to flatten out at high four-momentum transfer. Each of them rise to within about 10% of their asymptotic values by $Q^2 = 1 \, GeV^2$.

Going beyond the case of all scalars, the originators of the model have also analyzed the case in which the incoming electron and the interacting virtual photon have the proper spin (electromagnetic case). In this case, the structure functions can be determined from the response functions. The longitudinal response function is given by,

$$R_L(\vec{q}, \nu) = \sum_{N=0}^{\infty} \frac{1}{4E_0 E_N} |F_{0N}(\vec{q})|^2 \Big[(E_0 + E_N)^2 \, \delta(\nu - E_N + E_0)$$
$$- (E_0 - E_N)^2 \, \delta(\nu + E_N + E_0) \Big] \, , \tag{69}$$

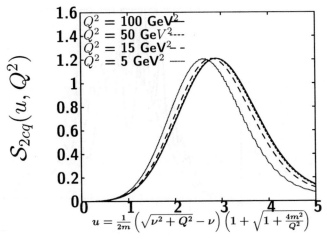

Figure 19. The high energy scaling behavior of S_{2cq} as a function of u for various values of Q^2

and the transverse response function is given by,

$$R_T(\vec{q}, \nu) = 8\frac{\alpha}{\vec{q}^2} \sum_{N=0}^{\infty} \frac{N}{4E_0 E_N} |F_{0N}(\vec{q})|^2 \Big[\delta(\nu - E_N + E_0)$$
$$-\delta(\nu + E_N + E_0)\Big]. \tag{70}$$

In terms of these quantities the structure functions are defined as

$$W_1(\nu, Q^2) = \frac{1}{2}R_T$$
$$W_2(\nu, Q^2) = \frac{Q^4}{(Q^2 + \nu^2)^2}R_L + \frac{Q^2}{2(Q^2 + \nu^2)}R_T. \tag{71}$$

The same scaling function and variable are used to analyze the behavior of the structure functions. The scaling function for the bound–bound transition, $S_{2cq}^{b.b.}$ (where the 2 indicates that we are examining the W_2 structure function), is plotted in Fig. 19. It can clearly be seen that scaling does still occur. By continuizing the sums in the response functions to integrals and using Stirling's formula, an analytic expression for the electromagnetic bound–bound scaling

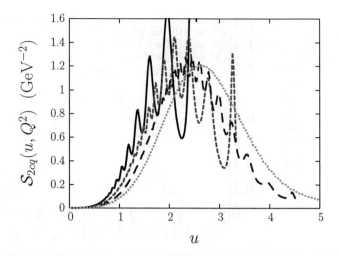

Figure 20. The low Q^2 behavior of S_{2cq} as a function of u for $Q^2 = 0.5\,GeV^2$ (solid), $Q^2 = 1\,GeV^2$ (short-dashed), $Q^2 = 2\,GeV^2$ (long dashed), and $Q^2 = 5\,GeV^2$ (dotted).

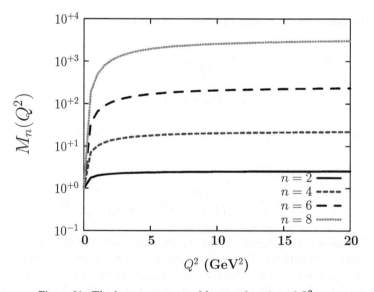

Figure 21. The lowest moments, M_n, as a function of Q^2

curve can be obtained,

$$S_{2cq}^{b.b.}(Q_{high}^2) = \frac{m^2 u_{Bj}^2}{\sqrt{\pi}\beta E_0} \exp\left(-\frac{(E_0 - mu_{Bj})^2}{\beta^2}\right), \tag{72}$$

which is very similar to the form of Eq. (66). Just as for the all scalar case, in the electromagnetic case it is possible to prove analytically that the scaling curves for the bound–bound and bound–free transitions are the same. By calculating the response functions for the bound–free case and then letting $Q^2 \rightarrow \infty$, it can be shown that the bound–free scaling curve is exactly equal to Eq. (72).

The electromagnetic case has very similar success to the all scalar case in demonstrating both local and global duality. Figure 20 shows the behavior of the bound-bound structure function at low Q^2. Just as in the all scalar case, the resonance oscillations do qualitatively tend to follow the scaling curve. The first four moments are shown if Fig. 21. It can clearly be seen that all of the moments flatten out, even though they do not all reach their asymptotic values at the highest Q^2 shown in this case.

5 Summary and Outlook

We have shown that duality is experimentally very well established and has many interesting and useful applications. While we have focused here on electron–nucleon scattering, duality appears in many different reactions and has the potential to be used to solve a broad range of problems.

Unfortunately, our current theoretical understanding of duality is limited. A deeper understanding of the origins of duality, where it holds, and how accurate it is, is necessary in order to be able to apply it successfully. In the past, theorists have attempted to explain duality using the operator product expansion. While the OPE shows that duality holds because the contribution of higher-twist corrections is small, it does not offer any explanation for why the higher-twist terms behave this way. More recently, a deeper understanding of where and how duality arises has been sought through the use of modeling. We have shown how one model in particular has been able to qualitatively reproduce all of the features of duality with just a few basic assumptions. Today, modeling is considered to be essential for gaining insight into the physical mechanisms underlying duality.

In the future, on the experimental front, we will see more data exploring duality in various reactions, including polarized and unpolarized reactions and meson production. New duality experiments have been completed or are currently being carried out at Jefferson Lab[33,34,35], and there will be a large duality program at the 12 GeV upgrade of CEBAF[36]. On the theoretical front, models will progress to more realistic descriptions of nature. This will include using spin 1/2 quarks and explicitly modeling the decay of the excited resonances.

Acknowledgments

We thank the organizers of the 17th Annual Hampton University Graduate Studies at CEBAF (HUGS) for organizing an interesting and stimulating school.

SJ thanks Wally Van Orden for a fruitful collaboration on quark-hadron duality that led to many of the results presented here. SJ also gratefully acknowledges discussions with F. Close, R. Ent, R. J. Furnstahl, N. Isgur, C. Keppel, I. Niculescu, and W. Melnitchouk. We gratefully acknowledge the permission of the Jefferson Lab duality group, W. Melnitchouk, I. Niculescu, and S. Simula to reproduce their figures.

This work was supported in part by funds provided by the National Science Foundation under grant No. PHY-0139973.

References

1. N. Isgur, S. Jeschonnek, W. Melnitchouk, and J. W. Van Orden. *Phys. Rev.*, D64:054005, 2001.
2. E. C. Poggio, H. R. Quinn, and S. Weinberg. *Phys. Rev.*, D13:1958, 1976.
3. N. Isgur and M. B. Wise. *Phys. Rev.*, D43:819, 1991.
4. R. Lebed and N. Uraltsev. *Phys. Rev.*, D62:094011, 2000.
5. R. Rapp. *hep–ph/0201101*.
6. M. A. Shifman. *arXiv: hep–ph/0009131*.
7. Francis Halzen and Alan Martin. *Quarks and Leptons:An Introductory Course in Modern Particle Physics*. John Wiley & Sons, Inc., 1984.
8. F. E. Close. *Introduction to Quarks and Partons*. Academic Press, Inc., 1979.
9. Elliot Leader and Enrico Predazzi. *An Introduction to Gauge Theories and modern particle physics*. Cambridge University Press, 1996.
10. E. D. Bloom and F. J. Gilman. *Phys. Rev. Lett.*, 25(16):1140–1143, 1970.
11. I. Niculescu *et al. Phys. Rev. Lett.*, 85:1182–1185, 2000.
12. R. Ent, C. E. Keppel, and I. Niculescu. *Phys. Rev.*, D62:073008, (2000).
13. Otto Nachtmann. *Nucl. Phys.*, B63:237–247, 1973.

14. C. S. Armstrong, R. Ent, C. E. Keppel, S. Liuti, G. Niculescu, and I. Niculescu. *Phys. Rev.*, D63:094008, 2001.

15. A. De Rújula, Howard Georgi, and David Politzer. *Ann. Phys.*, 103:315–353, 1977.

16. I. Niculescu *et al. Phys. Rev. Lett.*, 85:1186, 2000.

17. M. Gari and W. Krümpelmann. *Phys. Lett.*, B141:295–300, 1984.

18. S. Simula. *Phys. Rev.*, D64:038301, 2001.

19. R. Ent, C. E. Keppel, and I. Niculescu. *Phys. Rev.*, D64:038302, 2001.

20. M. Jones *et al. Phys. Rev. Lett.*, 84:1398, 2000.

21. David Griffiths. *Introduction to Elementary Particles.* John Wiley & Sons, Inc., 1987.

22. C. E. Carlson. *Phys. Rev.*, D34:2704, 1986.

23. C. E. Carlson and N. C. Mukhopadhyay. *Phys. Rev.*, D58:094029, 1998.

24. E. D. Bloom and F. J. Gilman. *Phys. Rev.*, D4(9):2901–2916, 1971.

25. W. Melnitchouk. *Phys. Rev. Lett*, 86:35–38, 2001.

26. A. V. Manohar. *hep–ph/9204208.*

27. Howard Georgi and H. David Politzer. *Phys. Rev.*, D14(7):1829–1848, 1976.

28. G. Ricco, M. Anghinolfi, M. Ripani, S. Simula, and M. Taiuti. *Phys. Rev.*, C57(1):356–366, 1998.

29. S. Simula. *Phys. Lett.*, B481:14–20, (2000).

30. S. Jeschonnek and J. W. Van Orden. *Phys. Rev*, D65:094038, 2002.

31. Contributions of S. Jeschonnek and J. W. Van Orden. Proceedings of Baryons 2002.

32. R. Barbieri, J. R. Ellis, M. K. Gaillard, and G. G. Ross. *Phys. Lett.*, B64:171, 1976.

33. Spokespersons J. P. Chen, S. Choi, and N. Liyanage. Jefferson Lab experiment E01–012.

34. Spokespersons G. Dodge, S. Kuhn, and M. Taiuti. Jefferson Lab Experiment E93–009.

35. Contributions of E. Christy, C. Keppel, and I. Niculescu. Proceedings of Baryons 2002. to be published by World Scientific.

36. L. Cardman, R. Ent, N. Isgur, J.-M. Laget, C. Leemann, C. Meyer, and Z.-E. Meziani, editors. *The Science Driving the 12GeV Upgrade.* Jefferson Lab, February 2001.

WEAK INTERACTIONS IN ATOMS AND NUCLEI: THE STANDARD MODEL AND BEYOND

M.J. RAMSEY-MUSOLF

Kellogg Radiation Laboratory, Caltech, Pasadena, CA 91125

Department of Physics, University of Connecticut, Storrs, CT 06269

J. SECREST

Department of Physics, College of William and Mary, Williambsburg, VA 23187

Studies in nuclear and atomic physics have played an important role in developing our understanding of the Standard Model of electroweak interactions. We review the basic ingredients of the Standard Model, and discuss some key nuclear and atomic physics experiments used in testing these ideas. We also summarize the conceptual issues of the Standard Model that motivate the search for new physics.

1   Introduction

The quest for a unified description of all known forces of nature is something of a "holy grail" for physicists. At present, we possess a partial description, known as the Standard Model[1]. In a nutshell, the Standard Model (SM) is a unified gauge theory of the strong, weak, and electromagnetic interactions, the content of which is summarized by the group structure

$$SU(3)_C \times SU(2)_L \times U(1)_Y \quad , \tag{1}$$

where the first factor refers to the theory of strong interactions, or Quantum Chromodynamics (QCD), and the latter two factors describe the theory of electroweak interactions. Although the theory remains incomplete, its development represents a triumph for modern physics. Historically, nuclear and atomic physics have played an important role in uncovering the structure of the strong and electroweak interactions. In this series of lectures, I will attempt to give you some sense of that history, as well as some feel for the parts being played by nuclear and atomic physics in looking for physics beyond the SM. I will focus on the electroweak sector of the theory, though you should keep in mind that studies of QCD constitute one of the primary thrusts of nuclear physics today.

Before delving into the details of the SM, it is important to appreciate just what an achievement Eq. (1) represents. One way to do this is to consider some of the basic properties off the four forces of nature.

Interaction	Range(fm)	Strength	TimeScale	$\sigma(\mu b)$
Gravity	∞	$G_N M_p^2$	$\times$	$\times$
Weak	10^{-3}	$G_F M_p^2$	$\geq 10^{-8}$	10^{-8}
Strong	≤ 1	$\alpha_S(r)$	10^{-23}	10^4
EM	∞	α	10^{-20}	10

In addition to this seemingly disparate set of characteristics, each of the forces has some unique features of its own. For example, the gravitational and EM interactions get weaker the farther apart the given "charges" are, whereas the strong interaction behaves just the opposite, as the distance r increases, α_S grows. This feature is related to the fact that one never sees individual unbound quarks and that strong interaction effects at distance scales at about 1 fm are hard to compute.

Another set of unique features has to do with how each interaction obeys different "discrete" symmetries:

Parity	P	$\overrightarrow{x} \to -\overrightarrow{x}$	$t \to t$	$\overrightarrow{s} \to \overrightarrow{s}$
Time Reversal	T	$\overrightarrow{x} \to \overrightarrow{x}$	$t \to -t$	$\overrightarrow{s} \to \overrightarrow{s}$
Charge Conjugation	C	$Q \to -Q$	$e^+ \to e^-$	etc.

A simple way to visualize these symmetries is as follows. A parity transformation (P) involves an inversion of a physical system through the origin of co-ordinates. This inversion can be completed in two steps. First, reflect the system in a mirror. Second, rotate the reflected system by 180° about the normal to the mirror. However, since fundamental interactions are rotationally invariant (angular momentum is conserved), we may omit the second step. Thus, a parity transformation is essentially a mirror reflection. Note that the "handedness" of a particle – the relative orientation of its momentum and spin – reverses under mirror reflection. A time reversal transformation (T) amounts first to thinking of a physical process as being like a movie. Under T, the movie is run backwards through the projector. Finally, a charge conjugation transformation (C) turns a particle into its antiparticle.

Here's how the different interactions rate with respect to these symmetries:

Symmetry "Score Card"

Symmetry	Strong	EM	Weak
P	yes	yes	no
C	yes	yes	no
T	yes	yes	no
PCT	yes	yes	yes

Clearly, the weak interaction is a flagrant symmetry violator, whereas the strong interaction and EM interactions respect P,C, and T individually. So how can it be that all three forces fall into a unified model?

Given these differences, it is truly remarkable that physicists have figured out how to describe three out of four forces in a unified theory. It's a highly non-trivial accomplishment.

Now lets look in detail at the essential elements of the electroweak part of the Standard Model.

2 Basic Ingredients of the Standard Model

The essential building blocks of the electroweak sector of the Standard Model are the following:

1. Gauge Symmetry $\Longleftrightarrow$ gauge bosons, parity violation

2. Representations $\Longleftrightarrow$ bosons and quarks

3. Family Replication, mixing and universality $\Longleftrightarrow$ CP Violation

4. Spontaneous Symmetry Breaking $\Longleftrightarrow$ Higgs Boson, M_Z, $M_W \neq 0$

2.1 Gauge Symmetry

Lets start with the more familiar case of the electromagnetism and see how the properties can all be derived from the principle of gauge invariance.

Consider the Dirac equation for a free electron:

$$(i\not{\partial} - m)\psi = 0 \quad . \tag{2}$$

Suppose we now make the local transformation:

$$\psi(x) \to \psi(x)e^{i\alpha(x)} = \psi'(x) \quad . \tag{3}$$

This is called a U(1) transformation.

The Dirac equation is not invariant under this transformation. If ψ' satisfies (2) then one has:

$$(i\not{\partial} - m)e^{i\alpha(x)}\psi = (i\not{\partial} - m)\psi - \psi\not{\partial}\alpha = 0 \quad . \tag{4}$$

To make (2) invariant under (3), one can replace ∂_μ by the covariant derivative D_μ:

$$D_\mu = \partial_\mu - ieA_\mu \quad , \tag{5}$$

where A_μ is a gauge field identified with the photon. We require A_μ to transform in such away as to remove the unwanted term in (4).

$$A_\mu \to A'_\mu = A_\mu + \frac{1}{e}\partial_\mu\alpha \tag{6}$$

when $\psi \to \psi' = e^{i\alpha}\psi$

Thus, we obtain a new Dirac equation:

$$(i\not{D} - m)\psi = 0 \quad . \tag{7}$$

Under a gauge transformation, one has

$$(i\not{D}' - m)\psi' = 0 \quad \longrightarrow$$
$$= [i(\not\partial - ie\not{A}'_\mu) - m]\psi' = [i(\not\partial - ie\not{A}_\mu - i\not\partial\alpha) - m]\psi$$
$$= e^{i\alpha}[i(\not\partial - ie\not{A}) - m]\psi + e^{i\alpha}(\not\partial\alpha)\psi - e^{i\alpha}(\not\partial\alpha)\psi$$
$$= e^{i\alpha}(i\not{D} - m)\psi = 0 \quad . \tag{8}$$

Cancelling through the $e^{i\alpha}$ yields

$$(i\not{D} - m)\psi = 0 \quad \text{if} \quad (i\not{D}' - m)\psi' = 0 \quad . \tag{9}$$

One recognizes the replacement of $\partial_\mu \to D_\mu = \partial_\mu - ieA_\mu$ as the usual minimal substitution that gives us the interaction of the electrons with the vector potential in quantum mechanics. Apparently, requiring the Dirac equation to be invariant under U(1) gauge transformation (3) and (6) leads to a familiar result from quantum mechanics.

This idea can be generalized to other symmetry transformations. A simple generalization is a group of transformations called SU(2). Let's define the following matrices:

$$\tau_1 = \begin{pmatrix} 0 & 1 \\ 1 & 0 \end{pmatrix} \quad \tau_2 = \begin{pmatrix} 0 & -i \\ i & 0 \end{pmatrix} \quad \tau_3 = \begin{pmatrix} 1 & 0 \\ 0 & -1 \end{pmatrix} \quad .$$

$$\tag{10}$$

They satisfy:

$$\left[\frac{\tau_i}{2}, \frac{\tau_j}{2}\right] = i\epsilon_{ijk}\frac{\tau_k}{2}$$

$$\left[\frac{\tau_i}{2}, \frac{\tau_j}{2}\right]_+ = \frac{1}{2}\delta_{ij}$$

$$\text{Tr}\left(\frac{\tau_i}{2}\frac{\tau_j}{2}\right) = \frac{1}{2}\delta_{ij} \quad .$$

One defines a group of transformations – called SU(2) – whose elements are:

$$U(\vec{\alpha}) = e^{i\vec{\alpha}(x)\cdot\frac{\vec{\tau}}{2}} \tag{11}$$

where

$$\vec{\alpha}(x) = (\alpha_1(x), \alpha_2(x), \alpha_3(x))$$
$$\vec{\tau} = (\tau_1, \tau_2, \tau_3) \quad ,$$

with the α_i's being a continuously varying function of $x^\mu = (t, \vec{x})$. The transformations $U(\vec{\alpha})$ act on a two component vector. For example, let

$$\Psi_l = \begin{pmatrix} \psi_\nu(x) \\ \psi_e(x) \end{pmatrix} \tag{12}$$

denote a lepton wavefunction[a]. Then under the transformation (11)

$$\Psi_l \to \Psi_l' = U(\vec{\alpha})\Psi_l \tag{13}$$

In order for the Dirac equation to be invariant under (13), one must define a new gauge covariant derivative for SU(2):

$$D_\mu = \partial_\mu - ig\frac{\tau}{2} \cdot \vec{W}_\mu \quad, \tag{14}$$

where g is an SU(2) analog of electric charge and

$$\vec{W}_\mu = (W_\mu^1, W_\mu^2, W_\mu^3) \tag{15}$$

is a set of 3 fields analogous to A_μ. It turns out that if $\vec{W}_\mu$ transforms as

$$W_\mu^i \to W_\mu^{i\prime} = W_\mu^i - \epsilon^{ijk}\alpha_j W_\mu^k - \frac{1}{g}\partial_\mu\alpha_i \tag{16}$$

when

$$\Psi_l \to U(\vec{\alpha})\Psi_l = \Psi_l' \tag{17}$$

then the Dirac equation for ψ_ν and ψ_e will be invariant under the transformations (13) and (16):

$$(i\slashed{\partial} + g\frac{\vec{\tau}}{2} \cdot \vec{W}_\mu - \hat{M})\Psi_l = 0 \quad. \tag{18}$$

Note that the extra term in the transformation law for W_ν^i – the $\epsilon^{ijk}\alpha_j W_\nu^k$ term – is a consequence of the fact that this group of transformations is non-Abelian, that is τ_i and τ_j do not commute.

Note also that in Eq. (18) we have introduced a mass term $(\hat{M})$, whose origins I will discuss later.

As in the case of the $U(1)_{EM}$ gauge field A_μ, which we corresponds to the photon, the SU(2) gauge fields W_ν^i should also be associated with spin-one particles. To identify the character of these particles, first define:

---

[a]Here, $\psi_{\nu,e}$ are four-component Dirac spinors.

$$\tau_+ = \frac{1}{2}(\tau_1 + i\tau_2) = \begin{pmatrix} 0 & 1 \\ 0 & 0 \end{pmatrix}$$

$$\tau_- = \frac{1}{2}(\tau_1 - i\tau_2) = \begin{pmatrix} 0 & 0 \\ 1 & 0 \end{pmatrix}$$

and $W_\mu^\pm = \frac{1}{\sqrt{2}}(W_\mu^1 \pm iW_\mu^2)$. Then

$$\vec{\tau} \cdot \vec{W}_\mu = \sqrt{2}(\tau_+ W_\mu^- + \tau_- W_\mu^+) + \tau_3 W_\mu^3 \quad . \tag{19}$$

Now

$$\tau_+ \Psi_l = \begin{pmatrix} 0 & 1 \\ 0 & 0 \end{pmatrix} \begin{pmatrix} \psi_\nu \\ \psi_e \end{pmatrix} = \begin{pmatrix} \psi_e \\ 0 \end{pmatrix} \quad . \tag{20}$$

In short, acting with τ_+ on the lepton wavefunction transforms an electron wavefunction into one for a neutrino: $\psi_e \to \psi_\nu$. Similarly,

$$\tau_- \Psi_l = \begin{pmatrix} 0 & 0 \\ 1 & 0 \end{pmatrix} \begin{pmatrix} \psi_\nu \\ \psi_e \end{pmatrix} = \begin{pmatrix} 0 \\ \psi_\nu \end{pmatrix} \quad . \tag{21}$$

This turns a neutrino into an electron: $\psi_\nu \to \psi_e$. One can represent the action of $\vec{\tau} \cdot \vec{W}_\mu$ on Ψ_l in Fig. 1:

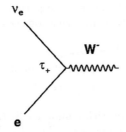

Figure 1. Charge raising weak interaction current.

From the fact that weak interactions turn $e \leftrightarrow \nu$, one is lead to identify the $W_\mu^\pm$ fields with the $W^\pm$ particles.

It would be tempting to identify the W_μ^3 with the Z^0 boson. However, it turns out to be impossible to do that and end up with the right masses for the $W^\pm$ and Z^0 bosons. Moreover, since there exists another neutral boson – the massless photon, γ – one needs two neutral fields to make the Z^0 and

γ. A nice way to produce both bosons is to *mix* the W_μ^3 with another gauge boson – called B_μ – that transforms as a singlet under SU(2) transformations. In short, $(W_\mu^3, B_\mu) \rightarrow (A_\mu, Z_\mu)$.

Before fleshing this idea out, however, we need to revisit the parity transformation discussed earlier. To that end, let's define:

$$\psi_L = \frac{1}{2}(1 - \gamma_5)\psi \tag{22}$$

$$\psi_R = \frac{1}{2}(1 + \gamma_5)\psi \quad . \tag{23}$$

One can show that if ψ is massless, then ψ_L always has negative helicity and ψ_R always has positive helicity.

In short:

$$\psi_L : \quad h = \hat{s} \cdot \hat{p} = -1$$
$$\psi_R : \quad h = \hat{s} \cdot \hat{p} = +1$$

Under a parity transformation, $\hat{s} \cdot \hat{p}$ changes sign which is equivalent to saying that $\psi_L \leftrightarrow \psi_R$.

One knows that processes like μ-decay or β-decay, the neutrinos always come with helicity h=-1, which implies parity symmetry is broken. Only left handed particles participate in these types of weak interactions.

Since the interactions of $W_\mu^\pm$ with leptons arises in Eq. (18) because of SU(2) invariance, and one needs only left handed particles to have interactions with the $W_\mu^\pm$, one must modify the transformations rule:

$$SU(2) \rightarrow SU(2)_L$$
$$\Psi_l^L \rightarrow U(\vec{\alpha})\Psi_l^L \quad \text{(doublet)}$$
$$\psi_{e_R} \rightarrow \psi_{e_R} \quad \text{(singlet)} \quad .$$

In short, only left-handed particles undergo SU(2) transformations, while right-handed particles are unaffected. Thus, the Dirac equation for $\psi_{e_R} = \frac{1}{2}(1 + \gamma_5)\psi_e$ is unchanged under SU(2)$_L$ so there is no need for the $\vec{\tau} \cdot \overrightarrow{W_\mu}$ term to maintain invariance. Hence, one has the SU(2)$_L$ in $SU(2)_L \times U(1)_Y$. Note that we have not included a transformation ψ_{ν_R} in the list of transformations. The reason is that prior to the discovery of neutrino oscillations, right-handed neutrinos were not observed to participate in low-energy weak interactions.

Now back to the Z^0 and γ. Nature gives us four bosons in the electroweak interaction: two charged particles ($W^\pm$) and two neutral particles (Z^0, γ). So far, we have identified two charged and one neutral gauge boson. We need to include one more neutral gauge boson having no accompanying particles. The way to accomplish this is to introduce one more set of U(1) transformations:

$$\psi \to e^{i\alpha(x)}\psi = \psi'$$

$$\partial_\mu \to \partial_\mu - ig'\frac{Y}{2}B_\mu$$

$$B_\mu \to B'_\mu = B_\mu + \frac{1}{g'}\partial_\mu\alpha \quad .$$

Here Y is called *hypercharge* (to distinguish it from the general EM charge). Now the Dirac equation for electrons and neutrinos is the following:

$$(i\not{\partial} + \frac{g}{2}\vec{\tau}\cdot\vec{W} + \frac{g'}{2}Y\not{B} - \hat{M})\Psi_l^L = 0 \tag{24}$$

$$(i\not{\partial} + \frac{g'}{2}Y\not{B} - m_e)\psi_e^R = 0 \quad . \tag{25}$$

Rewriting (24) slightly:

$$[i\not{\partial} + \frac{g}{\sqrt{2}}(\tau_+\not{W}_- + \tau_-\not{W}_+)$$

$$+ \frac{g}{2}\tau_3\not{W}_3 + \frac{g'}{2}Y\not{B} - \hat{M}]\Psi_l^L = 0 \quad . \tag{26}$$

Now to get EM interactions for both the e_R and e_L and to make sure the $W^\pm$ and Z^0 are different in mass, one needs the Z^0 and γ to be linear combinations of the W_3 and B. One can accomplish this by utilizing a unitary transformation:

$$\begin{pmatrix} Z_\mu^0 \\ A_\mu \end{pmatrix} = \begin{pmatrix} \cos\theta_W & -\sin\theta_W \\ \sin\theta_W & \cos\theta_W \end{pmatrix}\begin{pmatrix} W_\mu^3 \\ B_\mu \end{pmatrix} \quad . \tag{27}$$

The angle θ_W is a supremely important parameter in the electroweak standard model. It is called the "Weinberg angle", or "weak mixing angle". Inverting (27), one can substitute into (24) and (25) for W_μ^3 and B_μ to obtain:
Left-handed leptons:

$$[i\not{\partial} + \frac{g}{\sqrt{2}}(\tau_+\not{W}_- + \tau_-\not{W}_+)$$

$$+ (g\sin\theta_W\frac{\tau_3}{2} + g'\cos\theta_W\frac{Y}{2})\not{A}$$

$$+ (g\cos\theta_W\frac{\tau_3}{2} - g'\sin\theta_W\frac{Y}{2})\not{Z}^0 - \hat{M}]\Psi_l^L = 0 \tag{28}$$

Right-handed leptons:

$$[i\not{\partial} + (g'\cos\theta_W\frac{Y}{2})\not{A} - (g'\sin\theta_W\frac{Y}{2})\not{Z}^0 - m_e]\psi_{e_R} = 0 \tag{29}$$

In order to restore the original EM gauge transformation law, one needs to identify:

$$g' \cos\theta_W \frac{Y_R}{2} = eQ \tag{30}$$

$$g \sin\theta_W \frac{\tau_3}{2} + g' \cos\theta_W \frac{Y_L}{2} = eQ \quad . \tag{31}$$

This works if one takes:

$$g' \cos\theta_W = g \sin\theta_W = e$$
$$Y_R = 2Q \tag{32}$$
$$Y_L = 2(Q - T_3^L)$$

where $T_3^L \Psi_l^L = (\tau_3/2)\Psi_l^L$.
From Eqs. (32) one also has

$$\sin\theta_W = \frac{e}{g} \quad \text{and} \quad \tan\theta_W = \frac{g'}{g} \quad . \tag{33}$$

Thus, the $SU(2)_L$ and $U(1)_{EM}$ interactions depend on three parameters:
$$g, g', \text{ and } \sin\theta_W$$
or
$$g, e, \text{ and } \sin\theta_W$$
etc.

Lastly, let's rewrite our Z^0 couplings by eliminating g' in terms of g and $\sin\theta_W$:
For the right handed sector:

$$-g'\frac{Y_R}{2} = -g\frac{\sin^2\theta_W}{\cos\theta_W}Q \equiv \frac{gQ_R^W}{\cos\theta_W} \tag{34}$$

For the left handed sector:

$$g \cos\theta_W \frac{\tau_3}{2} - g' \sin\theta_W \frac{Y_L}{2} =$$
$$g \cos\theta_W \frac{\tau_3}{2} - g\frac{\sin^2\theta_W}{\cos\theta_W}(Q - \frac{\tau_3}{2}) =$$
$$\frac{g}{\cos\theta_W}(T_3^L - \sin^2\theta_W Q) \equiv$$
$$\frac{g}{\cos\theta_W}Q_L^W \quad , \tag{35}$$

where $Q_{L,R}^W$ denote the left- and right-handed "weak charges" of the leptons. To summarize, then, we have:

$$[i\partial\!\!\!/ + eQA\!\!\!/ + \frac{gQ_R^W}{\cos\theta_W}Z\!\!\!/ - m_e]\psi_{e_R} = 0 \tag{36}$$

$$[i\partial\!\!\!/ + eQA\!\!\!/ + \frac{gQ_L}{\cos\theta_W}Z\!\!\!/ \tag{37}$$

$$+ \frac{g}{\sqrt{2}}(\tau_+ W\!\!\!\!/_- + \tau_- W\!\!\!\!/_+) - \hat{M}]\Psi_l^L = 0 \quad , \tag{38}$$

where

$$g = \frac{e}{\sin\theta_W} \tag{39}$$

$$Q_R^W = -\sin^2\theta_W Q \tag{40}$$

$$Q_L^W = T_3^L - \sin^2\theta_W Q \quad . \tag{41}$$

2.2 Representations

The assignment of different $SU(2)_L$ and $U(1)_Y$ transformation properties to left- and right-handed fermions is called a choice of representations. Left-handed fermions are assigned to a doublet representations of $SU(2)_L$; right-handed fermions transform under the singlet representation.

$$\begin{pmatrix} \nu_L \\ e_L \end{pmatrix} \qquad \nu_R \qquad e_R$$

doublet   singlet   singlet

Note that although the ν_R transforms as a singlet, it has no weak interactions according to Eq. (40) since its electromagnetic charge is zero.

One can make the same assignment for quarks:

$$\begin{pmatrix} u_L \\ d_L \end{pmatrix} \qquad u_R \qquad d_R$$

doublet   singlet   singlet

Note that unlike the the right-handed neutrinos, both right-handed quarks have weak interactions since $Q_u \neq 0$ and $Q_d \neq 0$

2.3 Family Replication, Universality, and Mixing

Of course electrons, neutrinos, up quarks, and down quarks are not all the elementary leptons and quarks. They constitute the first generation. The remaining are assigned to the second and third generations:

Second:

$$\begin{pmatrix} \nu_\mu^L \\ \mu_L \end{pmatrix} \qquad \begin{pmatrix} c \\ s \end{pmatrix} \qquad \mu_R, c_R, s_R$$

Third:

$$\begin{pmatrix} \nu_\tau^L \\ \tau_L \end{pmatrix} \qquad \begin{pmatrix} t \\ b \end{pmatrix} \qquad \tau_R, t_R, b_R$$

This repeated pattern of assignment is called fermion family replication.

An important feature of the standard model is that for each family, the structure of the interactions with gauge bosons is the same. In other words, there is a *family universality*.

Moreover, the overall strength of the charged current interactions and neutral current interactions is set by the same parameter, g. This feature is known as *charged current/neutral current universality*.

Now this nice pattern of universality gets somewhat clouded for quarks because the quark eigenstates of the weak interaction can different from the quark eigenstates of the mass operator. The origin of this effect has to do with how the electroweak symmetry is broken, as discussed below. The relationship between the two sets of eigenstates can be expressed by letting each negative charged quark weak eigenstate be written as a linear combination of the three mass eigenstates:

$$\begin{pmatrix} d \\ s \\ b \end{pmatrix}_{\text{WEAK}} = \hat{V} \begin{pmatrix} d \\ s \\ b \end{pmatrix}_{\text{MASS}} \quad , \tag{42}$$

where $\hat{V}$ is a 3×3 matrix known as the Cabbibo-Kobayashi-Maskawa (CKM) matrix[2]. In general, it is unitary and parameterized by 3 angles and a phase.

It turns out that one need not to write a similar relation for the positive quarks for algebraic reasons that will not be discussed here. To illustrate, then, the weak quark doublet of the first generation is:

$$\begin{pmatrix} u \\ \tilde{d} \end{pmatrix}_L, \qquad \tilde{d} = V_{11}d + V_{12}s + V_{13}b$$

with

$$|V_{11}|^2 + |V_{12}|^2 + |V_{13}|^2 = 1 \qquad (43)$$

following from the unitarity of $\hat{V}$.

Then the $\tau_+ \not W$ term in the Dirac equation leads to the transitions shown in Fig. 2.

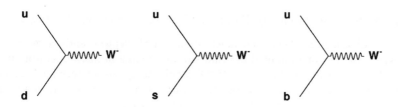

Figure 2. Transition diagrams

The parameters V_{ij} are not known *a priori*, but rather must be set by experiment. The most precise determination of any of the V_{ij} comes from a nuclear experiment – namely, nuclear β-decay[3], from which one extracts a value for V_{11}.

The study of the CKM matrix is an important component of electroweak physics. Unfortunately, we will not be able to discuss these studies in any detail in these lectures. However, a few comments are worth making here: (i) The elements of $\hat{V}$ are often labeled by the relevant charged currents transition:

$$V_{11} \rightarrow V_{ud}$$
$$V_{12} \rightarrow V_{us}$$
$$etc. \quad ,$$

that is, the element V_{11} governs the strength of the transition $d \rightarrow u + W^-$, *etc.* (ii)The phase in the matrix $e^{i\delta}$ is responsible for CP-violation. In order for such a phase to appear in $\hat{V}$, one needs at least 3 generations of massive quarks. (iii) For $m_\nu = 0$ as assumed by the Standard Model, there is no CKM matrix for leptons. The neutrinos weak and mass eigenstates are identical, and that is enough to evade the need for a CKM mixing matrix. However, now that neutrino oscillations have been observed and we know neutrinos have mass, we are forced to write down an analog of the CKM matrix for leptons. The study of the corresponding mixing angles and phase(s) are now a topic of intense study in nuclear and particle physics[b]. (iv) The neutral

current interactions are independent of the V_{ij}, and they entail no transitions among quark generations. In short, the Standard Model forbids flavor changing neutral current (FCNC) at lowest order in g.

2.4 Spontaneous Symmetry Breaking, Mass Generation, and the Higgs

So far not much has been said much about quark, lepton, and gauge boson masses. Naively, one might think that simply putting in mass operators into the Dirac equation as in Eqs. (28,29) would take care of fermion masses. However, life is more complicated than that. One needs to work with the Lagrangian from which the Dirac equation is derived:

$$L = \bar{\psi}(i\not{D} - m)\psi \quad .$$ (44)

It is straightforward to show that the mass term has the following decomposition:

$$m\bar{\psi}\psi = m\bar{\psi}_L\psi_R + m\bar{\psi}_R\psi_L \quad .$$ (45)

Now, under $SU(2)_L$ we have:

$$\psi_L \to e^{i\vec{\alpha}\cdot\frac{\vec{\tau}}{2}}\psi_L$$ (46)

$$\psi_R \to \psi_L \quad ,$$ (47)

so that the mass term breaks $SU(2)_L$ invariance in the Lagrangian. A similar problem arises for the gauge bosons. For a massless gauge boson like the γ, the Lagrangian is given by:

$$L_\gamma = -\frac{1}{4}F_{\mu\nu}F^{\mu\nu} \quad ,$$ (48)

where

$$F_{\mu\nu} = [D_\mu, D_\nu] = \partial_\mu A_\nu - \partial_\nu A_\mu \quad .$$ (49)

---

[b]If neutrinos are Majorana particles, there exist additional CP violating phases beyond the single phase associated with mixing of three generations of massive Dirac particles.

Since $F_{\mu\nu}$ is built from D_μ's, it is manifestly gauge invariant. Similarly, one could write down an $SU(2)_L$ gauge-invariant Lagrangian for the weak gauge bosons in the limit that they were massless:

$$L_{GB} = -\frac{1}{4}\sum_{\alpha=1}^{3} F^a_{\mu\nu}F^{a\mu\nu} \quad , \tag{50}$$

where

$$F^a_{\mu\nu} = \partial_\mu W^a_\nu - \partial_\nu W^a_\mu + g\epsilon^{abc}W^b_\mu W^c_\nu \tag{51}$$

The extra term in $F^a_{\mu\nu}$ is needed to maintain invariance under the non-Abelian transformation (16). To get the masses for the $W^\pm$ and Z^0, one would naively think to add the mass term:

$$L_M = \frac{1}{2}M^2 W^a_\mu W^{\mu a} \tag{52}$$

to (50). However (52) is again not invariant under (16). Since our goal is to write a Lagrangian having $SU(2)_L \times U(1)_Y$ gauge symmetry, adding mass terms as in (45) and (52) would be a disaster. How then, do we give particles their masses? The resolution is the so-called Higgs Mechanism[4]. The idea is introduce a new particle described by a two component vector which transforms as a doublet under $SU(2)_L$:

$$\Phi = \begin{pmatrix} \phi^+ \\ \phi^0 \end{pmatrix} \tag{53}$$

$$Y(\Phi) = 2(Q - T_3^L) = 1 \quad . \tag{54}$$

Using Φ, we add new terms to the gauge boson (GB)-fermion(F) $SU(2)_L \times U(1)_Y$ invariant Lagrangian:

$$L = L_{GB} + L_f + L_H + L_Y \quad , \tag{55}$$

where L_{GB} is given by (50) for both the W^a_μ and B_μ and L_f is given by (44) with the mass term removed. The new terms are:

$$L_H = (D_\mu\Phi)^\dagger(D_\mu\Phi) - V(\Phi) \tag{56}$$

$$V(\Phi) = -\mu^2\Phi^\dagger\Phi + \lambda(\Phi^\dagger\Phi)^2 \tag{57}$$

$$D_\mu\Phi = (\partial_\mu - ig\frac{\vec{\tau}}{2}\cdot\vec{W}_\mu - ig'\frac{Y}{2}B_\mu)\Phi \quad , \tag{58}$$

and

$$L_Y = f^{(e)}\bar{\Psi}_l^L\Phi\psi_e^R + f^{(u)}\bar{\Psi}_q^L\tilde{\Phi}\psi_u^R + f^{(d)}\bar{\Psi}_q^L\tilde{\Phi}\psi_d^R \quad , \tag{59}$$

where

$$\widetilde{\Phi} = i\tau_2 \Phi^* \quad \text{and} \quad Y(\widetilde{\Phi}) = -1 \quad . \tag{60}$$

There are a few observations to make: (1) The terms in L_H and L_f are all invariant under $SU(2)_L \times U(1)_Y$ transformations.
(2) The first term in (56) involves interactions of the type:

$$g^2 \Phi^\dagger \frac{\vec{\tau}}{2} \cdot \vec{W}_\mu \Phi + g'^2 \Phi^\dagger \frac{Y}{2} B_\mu \frac{Y}{2} B^\mu \Phi \tag{61}$$

plus cross terms involving $\vec{\tau} \cdot \vec{W}_\mu$ and B_μ.
(3) The potential $V(\Phi)$ has a minimum for $\Phi \neq 0$ if μ^2, $\lambda > 0$. Specifically, the minimum occurs for $\Phi^\dagger \Phi = \mu^2(2\lambda)$. The field Φ likes to sit at this point. It is energetically favorable to do so. That means the expectation value of Φ in the ground state of the universe, that is, the vacuum, should be non-zero:

$$< 0|\Phi|0 > \neq 0 \quad . \tag{62}$$

In short, Φ has a non-zero expectation value (VEV). One can arrange things to put $< 0|\Phi|0 >$ at the minimum of the potential by taking

$$< 0|\Phi|0 >= \Phi_0 = \begin{pmatrix} 0 \\ v/\sqrt{2} \end{pmatrix} \neq 0 \tag{63}$$

and letting

$$\Phi = \Phi_0 + \delta\Phi \quad , \tag{64}$$

where $\delta\Phi$ denotes fluctuations of this field, called the Higgs field, about Φ_0. Now observe what happens if one substitutes (64) into (61) one obtains

$$\frac{v^2}{8} \left[g^2(W_\mu^1 W^{\mu 1} W_\mu^2 W^{\mu 2}) \right.$$
$$\left. + (gW_\mu^3 - g'B_\mu)(gW^{\mu 3} - g'B^\mu) + O(\delta\Phi) \right.$$
$$= \frac{v^2}{8}(2g^2 W_\mu^\dagger W^{\mu -} + (g^2 + g'^2)Z_\mu Z^\mu) + O(\delta\Phi) \quad . \tag{65}$$

Note that these terms look suspiciously like mass terms if makes the identifications:

$$\frac{v^2 g^2}{4} = M_W^2 \quad \text{and} \quad \frac{v^2}{4}(g^2 + g'^2) = M_Z^2 \tag{66}$$

$$\frac{M_W}{M_Z} = \frac{g}{\sqrt{g^2 + g'^2}} = \cos\theta_W \quad . \tag{67}$$

The beauty of this idea is that (a) one gets the masses for the $W^{\pm}$ and Z^0 without spoiling the $SU(2)_L \times U(1)_Y$ invariance of the Lagrangian, and (b) the photon stays massless. One says that the $SU(2)_L \times U(1)_Y$ symmetry is *spontaneously broken* by the vacuum down to $U(1)_{EM}$ by giving the Higgs a vacuum expectation value:

$$SU(2)_L \times U(1)_Y \to U(1)_{EM} \quad .$$

Similarly one will notice from the Lagrangian L_Y that leptons and quarks get masses from Φ_0 without spoiling the $SU(2)_L \times U(1)_Y$ invariance of the fermion part of the theory.

Now, there is one last important observation. The gauge boson masses, θ_W, e, g, g' and v are all inter-related. In fact, only *three* of these parameters are independent. Once three parameters have been specified the others are now determined. For example, we may choose the gauge boson masses and EM charge as the independent parameters. Then the other parameters are determined as follows:

$$(M_W, M_Z, e) \to \cos\theta_W = \frac{M_W}{M_Z}$$
$$g = \frac{e}{\sin\theta_W} \qquad (68)$$
$$v^2 = \frac{M_W^2(M_Z^2 - M_W^2)}{\pi\alpha M_Z^2}$$
$$\approx (242 \text{ GeV})^2$$

$$(69)$$

The quantity $v \approx 242$ GeV is known as the *weak scale*, that is the scale or dimensionful parameter associated with the symmetry breakdown $SU(2)_L \times U(1)_Y \to U(1)_{EM}$ and the quantity which sets the scale of the $W^{\pm}$ and Z^0 masses. In practice, one takes the independent inputs to be α, M_Z, and G_F (the Fermi constant) measured in μ-decay which can be related to g in terms of α and M_Z.

2.5 *Additional Observations*

Let's close this section with two observations. First, the electroweak sector of the Standard Model contains a sizeable number of *a priori* unknown parameters. They are: (i) Gauge sector: g, M_Z, e (3 parameters) (ii) Higgs sector: M_H (1) (iii) Fermion sector: lepton and quark masses (9) and CKM angles and phase (4) Hence, the electroweak sector presents 17 independent parameters which must be taken from experiment. This is a fairly unsatisfying situation. In fact, one motivation for seeking a larger theory in which to embed the SM is to try and understand the origin of these parameters.

Second, you may wonder why we refer to the weak interaction as "weak", since its coupling constant g is not too different from the coupling for electromagnetic interactions. The reason for this terminology has to do with the low-energy properties of probability amplitudes for various processes. To illustrate, let's compare electron-muon scattering with muon decay $(\mu^- \rightarrow \nu_\mu + e^- + \bar{\nu}_e)$. The amplitude for the former, which is purely electromagnetic, is governed by the photon propagator, which goes as $1/q^2$, where q_μ is the momentum transfer in the scattering. The μ-decay amplitude, in contrast, is governed by the W-boson propagator, which goes as $1/(q^2 - M_W^2)$. Since the energy released in μ-decay is tiny compared to M_W, this amplitude goes as $1/M_W^2$. Thus, the ratio of the two amplitudes is

$$\frac{\text{WEAK}}{\text{EM}} \sim \frac{q^2}{M_W^2} << 1 \tag{70}$$

at low-energies. In short, low-energy weak interactions are "weak" in comparison to EM interactions because the $W^\pm$ is quite massive while the γ is massless. Note, however, that at higher energies, the strengths of the two interactions may become comparable.

3   Low-Energy Tests of the Standard Model

So far, considerable attention has been spent on the elegant structure of the Standard Model. It has been demonstrated how it provides a unified framework for weak and EM interactions based on gauge symmetry, allows masses to be generated by spontaneous symmetry breaking, accounts for parity and CP-violation, and explains the disparate low-energy strengths and ranges of the weak and EM interactions. The Standard Model also makes a number of predictions:

1. Charged current interactions are purely left-handed

2. Electroweak coupling strengths are universal

3. The charged vector currents have a simple relation to isovector electromagnetic currents

4. The weak neutral current is a mixture of $SU(2)_L$ and $U(1)_{EM}$ currents, with the degree of mixing characterized by $\sin^2 \theta_W$

5. Neutral current interactions conserve flavor

and so on. Low-energy experiments in nuclei and atoms have played important role in establishing these properties to a high degree of accuracy. The following is a review of some of the classic ways in which this has been done.

3.1 Muon Decay

The decay $\mu^- \to \nu_\mu e^- \bar{\nu}_e$ is governed at lowest order by the amplitude in Fig. 3.

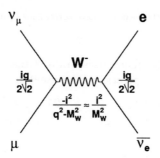

Figure 3. Muon decay

The amplitude is proportional to g^2/M_W^2, a combination of constants that is related to the Fermi constant:

$$\frac{G_F}{\sqrt{2}} = \frac{g^2}{8M_W^2} \tag{71}$$

Higher order corrections to this amplitude arise from γ exchanges as illustrated in Fig. 4.

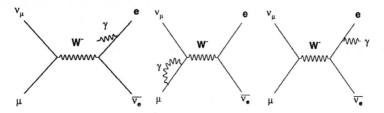

Figure 4. QED radiative corrections to muon decay

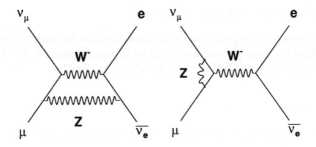

Figure 5. Electroweak radiative corrections to muon decay.

The effect on the total muon decay rate of these purely QED radiative correction can be computed precisely. The result is

$$\tau_\mu^{-1} = \Gamma = \frac{G_F^2 m_\mu^5}{192\pi^3} \left[1 - \frac{\alpha}{2\pi}(\pi^2 - \frac{25}{4}) \right] f(\frac{m_e^2}{m_\mu^2})(1 + \frac{3}{5}\frac{m_\mu^2}{m_W^2}) \qquad (72)$$

$$f(x) = 1 - 8x + 8x^3 - x^4 + 12x^2 \ln(\frac{1}{x})$$

The point of writing this expression is that if τ can be measured very accurately, G_F can be extracted extremely precisely since all the QED effects can be computed. Moreover, since G_F is related to g and M_W, it can be used as one of the three inputs into the gauge sector of the Standard Model. The muon life time is now known to the precision:

$$\tau_\mu = 2.197035(40) \times 10^{-6}$$

From which one obtains[5]:

$$G_F = 1.16637(1) \times 10^{-5} \text{ GeV}^{-2}$$

For reasons which will become apparent later, let's denote this value of $G_F \to G_\mu$. It is an experimental parameter. It can be related it to the gauge sector parameters of the Standard Model using equation (71). However, at this level of precision, one must take into account electroweak radiative corrections shown in Fig. 5: The presence of these effects, which can be computed in the Standard Model, is to modify (71):

$$\frac{g^2}{8M_W^2}(1 + \Delta r_\mu) = \frac{G_\mu}{\sqrt{2}} \quad , \qquad (73)$$

where Δr_μ denotes the corrections induced by the processes in Fig. 5. One can then use relation (73) to compute other electroweak observables where the value of g is needed very precisely, or to test the self-consistency of the electroweak measurements. Letting $g = e/\sin\theta_W$, $e^2 = 4\pi\alpha$, and treating Δr_μ as small leads to

$$G_\mu = \frac{\pi\alpha}{\sqrt{2}M_W^2 \sin^2\theta_W (1 - \Delta r_\mu)} \tag{74}$$

One can then take M_W, α, and $\sin^2\theta_W$ from experiment, compute Δr_μ, and "predict" G_μ. In this case, one obtains[6]:

$$G_\mu^{SM} = 1.1661(\mp 0.0018) \binom{+0.0005}{-0.0004} \times 10^{-5} \text{GeV}^{-2} \quad.$$

The level at agreement between this prediction and the result of τ_μ is impressive, and suggests that the way one puts together the gauge sector of the Standard Model is right.

Other features of the Standard Model can be tested with μ-decay. For example, the outgoing e^- in μ^--decay should be left-handed. In the limit where one neglects m_e, this implies $h(e^-) = -1$. Similarly, the Standard Model predicts $h(e^+) = +1$ in μ^+-decay. The experimental limits are[7]:

$$h(e^-) = -0.89 \pm 0.28$$
$$h(e^+) = +.94 \pm 0.08$$

consistent with the Standard Model.

Another handle on the Standard Model is to consider the decay of polarized muons. The rate can be written in terms of the so-called Michel parameters[8]. The rate is a function of:

$$X = \frac{|\vec{p_e}|}{|\vec{p_e}, max|}$$

and θ, the angle between $\vec{p_e}$ and the $\vec{\mu}$ spin. The parameters ρ and δ characterize the spectral shape for large X, and η characterizes the low energy region. The parameter ξ is an asymmetry parameter that characterizes the strength of a term in the rate proportional to $\cos\theta$. The Standard Model predictions (for purely left-handed weak interactions) and experimental results are[7]:

Parameter	Experiment	Standard Model
ξ	-1.0045 ± 0.0086	-1.0
ρ	0.7578 ± 0.0026	$\frac{3}{4}$
η	-0.007 ±0.013	0
δ	0.7486 ±0.0038	$\frac{3}{4}$

Again, the level of agreement with the Standard Model is quite impressive. As in the case of G_μ, one can use experimental results for the Michel parameters to put limits on possible physics "beyond" the Standard Model. Plans are currently being made to measure G_μ even more precisely, and a new experiment to determine the Michel parameters is being performed at TRIUMF by the TWIST collaboration.

3.2 Pion Decay and Lepton Universality

The reactions:

$$\pi^+ \rightarrow e^+ \nu_e$$
$$\pi^+ \rightarrow \mu^+ \nu_\mu$$

occur at lowest order when a $u\bar{d}$ quark pair annihilates into a W^+. According to the Standard Model, the coupling of the W^+ to a $u\bar{d}$ is proportional to:

$$\frac{g}{2\sqrt{2}} V_{ud}\bar{d}\gamma_\mu(1 - \gamma_5)u = J_\mu^{CC} \tag{75}$$

where "CC" stands for "charged current." The matrix element of J_μ^{CC} in the decay of the π^+ is:

$$< 0|J_\mu^{CC}|\pi^+(q) >= i\frac{F_\pi}{\sqrt{\omega_\pi}}q_\mu \quad , \tag{76}$$

where $F_\pi \approx 93$ MeV is the π decay constant. It parameterizes all the strong interaction(QCD) physics responsible for binding the $u\bar{d}$ into a π^+ – physics one can't calculate reliably. Hence, F_π is taken from experiment. The total rate for $\pi^+ \rightarrow l^+ \nu_l$ in the Standard Model is:

$$\Gamma = G_\mu^2 \frac{|V_{ud}|^2}{4\pi} F_\pi^2 m_l^2 m_\pi \left[1 - \left(\frac{m_l^2}{m_\pi^2}\right)\right]^2 \tag{77}$$

note that in the ratio of partial rates

$$R_{e/\mu} = \frac{\Gamma(\pi \rightarrow e\nu)}{\Gamma(\pi \rightarrow \mu\nu)} \tag{78}$$

the dependences on $G_\mu, |V_{ud}|$ and F_π cancel out. The only difference in the $\pi^+ \rightarrow e\nu$ and $\pi^+ \rightarrow l\nu$ rates came from the lepton masses. Thus, the Standard Model makes a very precise prediction for $R_{e/\mu}$, even after including electroweak radiative corrections of the type that enter Δr_μ.

Importantly, the ratio is insensitive to the $\pi^\pm \rightarrow W^+$ component of the decay matrix element, which is the same for both types of lepton final state. Hence, determining $R_{e/\mu}$ is a way to test the Standard Model prediction that the $W^+ \rightarrow e^+\nu_e$ and $W^+ \rightarrow \mu^+\nu_\mu$ couplings are the same, that is, universal.

Recently, $R_{e/\mu}$ has been determined very precisely at PSI[9] and TRIUMF[10]. The comparison of experiment (exp) with the Standard Model prediction (SM) yields[11]:

$$\frac{R_{e/\mu}^{\exp}}{R_{e/\mu}^{\mathrm{SM}}} = 0.9958 \pm 0.0033 (\mathrm{exp}) \pm 0.0004 (\mathrm{theory}) \quad . \tag{79}$$

Note the impressive precision both of the experiment($\sim 0.3\%$) and theory (0.04%). The result is a clear indication of the e-μ universality of the Standard Model.

3.3 Nuclear β-Decay: Lepton-hadron CC Universality and CKM Mixing

One of the classic and critically important nuclear physics processes used to test the Standard Model is nuclear β-dacay. Nuclear β-decay can occur via one of the following processes:

$$n \to p + e^- + \bar{\nu}_e \tag{80}$$

$$p \to n + e^+ + \nu_e \tag{81}$$

where the second reaction occurs when a proton is bound in a nucleus. As in the case of the π-decay, the currents on the left hand side of these diagrams are given by:

$$\mathrm{Charge\ \ lowering}: \frac{gV_{ud}}{2\sqrt{2}} \bar{d}\gamma_\mu(1 - \gamma_5)u \tag{82}$$

$$\mathrm{Charge\ \ raising}: \frac{gV_{ud}}{2\sqrt{2}} \bar{u}\gamma_\mu(1 - \gamma_5)d \quad . \tag{83}$$

For the moment, let's focus on the vector part of the quark currents, omitting the constants g,V_{ud}, etc.. One has:

$$J_\mu^+ = \bar{u}\gamma_\mu d \tag{84}$$

$$J_\mu^- = \bar{d}\gamma_\mu u \quad . \tag{85}$$

Compare these currents with the isovector EM current:

$$J_\mu^{T=1}(EM) = \frac{1}{2}(\bar{u}\gamma_\mu u - \bar{d}\gamma_\mu d) \tag{86}$$

A compact way of writing Eqs. (84-86) is to use:

$$Q = \begin{pmatrix} u \\ d \end{pmatrix} \quad \bar{Q} = (\bar{u}\bar{d})$$

So that

$$J_\mu^+ = \bar{Q}\gamma_\mu\tau_+ Q$$
$$J_\mu^- = \bar{Q}\gamma_\mu\tau_- Q \qquad (87)$$
$$J_\mu^{T=1} = \bar{Q}\gamma_\mu\frac{\tau_3}{2} Q$$

Now an astute observer will realize that the set of currents (87) form an isospin triplet, satisfying the commutation relations:

$$[I_\pm, J_\mu^{T=1,EM}] = \mp J_\mu^\pm \qquad (88)$$

$$[I_3, J_\mu^\pm] = \pm 2J_\mu^\pm \qquad (89)$$

where

$$I_k = \int d^3x \, \bar{Q}\gamma_0 \frac{\tau_k}{2} Q \qquad (90)$$

In short, $\{J_\mu^\pm, J_\mu^{T=1,EM}\}$ satisfy the same commutation relations as $\{I^\pm, I_3\}$. Consequently, matrix elements of the currents (84-86) must be related in the same way as matrix elements of the isospin operators (up to angular momentum properties). This property of the Standard Model is known as the "conserved vector current" property, or CVC. It implies that:
(i) Matrix elements of $J_\mu^\pm$ between nuclear states of different total isospin (I) must vanish.
(ii) Matrix elements of $J_0^\pm$ satisfy:

$$< I, I_z \pm 1|J_0^\pm|I, I_z > = [(I \mp I_z)(I \pm I_z + 1)]^{\frac{1}{2}} \qquad (91)$$

at $q^2 = 0$. A special set of nuclear decays sensitive to the matrix element (91) are the "superallowed" Fermi decays involving transitions:

$$(J^\pi = 0^+, I, I_z) \to (J^\pi = 0^+, I, I_z \pm 1)$$

Since the initial and final states have the same parity and zero total angular momentum, the axial current cannot connect them. Moreover, since the initial and final nuclear spins $J_f = J_i = 0$, only the vector charge operator can connect the two states. Letting H_{fi} denote the total transition amplitude, one has

$$H_{fi} = \frac{g^2 V_{ud}}{8M_W^2}[(I \mp I_z)(I \pm I_z + 1)]^{\frac{1}{2}} \qquad (92)$$

so that

$$|H_{fi}|^2 = \frac{G_\mu^2 |V_{ud}|^2}{2} [(I \mp I_z)(I \pm I_z + 1)] \tag{93}$$

at $q^2 = 0$. Amazingly, the value of $|H_{fi}|^2$ for *any* superallowed decay depends only on the muon decay Fermi constant, $|V_{ud}|^2$ from the CKM matrix, and the isospin factor. In the expression for (93), there appears no dependence on the nuclear wavefunction–no matter how complex it is! Thus, if one takes the rate for any superallowed decay and divides out the kinematical and the isospin factors, one should get the same answer as for any other superallowed decay. If this works, then the CVC prediction of the Standard Model is right. Moreover, by comparing this common rate with G_μ, one can test charged current universality for the leptons and quarks (the overall strength is $\approx G_\mu$) and extract the quark mixing parameter $|V_{ud}|$. Let's see how this works in detail:

The differential decay rate is, from Fermi's Golden Rule:

$$d\Gamma = \frac{2\pi}{\hbar} |H_{fi}|^2 \, \rho_f \, \delta(E_0 - E_e - E_\nu) \tag{94}$$

where $E_0 = $ is the energy released to the leptons and

$$\rho_f \frac{d^3 p_e}{(2\pi\hbar)^3} \frac{d^3 p_\nu}{(2\pi\hbar)^3} \tag{95}$$

is the density of final states. As an aside, consider for the moment the possibility[c] that $m_{\nu_e} \neq 0$. Putting in the factors of c, we have

$$E_\nu = (c^2 p_\nu^2 + m_\nu^2 c^4)^{\frac{1}{2}} \tag{96}$$

$$cp_\nu = (E_\nu^2 - m_\nu^2 c^4)^{\frac{1}{2}} \quad . \tag{97}$$

Integrating over $d^3 p_\nu$ gives

$$d\Gamma = \frac{2\pi}{\hbar} |H_{fi}|^2 \frac{4\pi}{c^3} (E - E_0)^2 \left[1 - \frac{(m_\nu c^2)^2}{(E - E_0)^2} \right]^{\frac{1}{2}} \frac{p_e^2 dp_e d\Omega}{(2\pi\hbar)^6} \quad . \tag{98}$$

Note that for a given detector setting which accepts all counts in a solid angle $\Delta\Omega = \sin\theta\Delta\theta\Delta\pi$, the number of electrons counted in a momentum slice Δp_e is

$$\Delta N \equiv N(p_e)\Delta p_e \propto (E_0 - E_e)^2 \left[1 - \frac{(m_\nu c^2)^2}{(E_0 - E_e)^2} \right]^{\frac{1}{2}} p_e^2 \quad . \tag{99}$$

---

[c]In the SM, one has $m_\nu = 0$. However, the results from a variety of neutrino oscillation experiments have taught us that neutrinos have mass.

Now let's think about a plot of

$$\frac{\sqrt{N(p_e)}}{p_e} \propto (E_o - E_e) \left[1 - \frac{(m_\nu c^2)^2}{(E_o - E_e)^2} \right]^{\frac{1}{4}} \quad . \tag{100}$$

This graph is known as a *Kurie plot*. The curve intercepts the x-axis at $E_e = E_0 - m_\nu c^2$. Thus, deviations from linearity at the "endpoint" would indicate $m_\nu \neq 0$. One of the most precise upper limits on neutrino mass comes from analyzing the endpoint of the decay[12]:

$$^3\text{H} \rightarrow\ ^3\text{He} + e^- + \bar{\nu}_e \quad , \tag{101}$$

which yields $m_{\bar{\nu}_e} \leq 15 eV$. So until recent observations of neutrino oscillations, nuclear β-decay confirmed at a very high level that $m_\nu = 0$ as implied by the Standard Model:

$$\frac{m_{\bar{\nu}_e}}{m_e} \leq 3 \times 10^{-5} \quad . \tag{102}$$

Now, back to super allowed decays. Converting dp_e to dE_e and putting in all the $\hbar$'s and c's, we have

$$\frac{d\Gamma}{dE_e} = \left(\frac{1}{2\pi^3 \hbar^4 c^6} \right) |H_{fi}|^2$$

$$\times E_e [E_e^2 - (m_e c^2)^2]^{\frac{1}{2}} (E_0 - E_e)^2 \left[1 - \frac{(m_\nu c^2)^2}{(E_0 - E_e)^2} \right]^{\frac{1}{2}} \tag{103}$$

after integrating over $d\Omega_e$. Letting $\epsilon = E_e/m_e c^2$:

$$\frac{d\Gamma}{d\epsilon} = \left(\frac{m_e^5 c^4}{2\pi^3 \hbar^7} \right) |H_{fi}|^2 \epsilon (\epsilon^2 - 1)^{\frac{1}{2}} (\epsilon_0 - \epsilon)^2 \left[1 - \frac{\lambda^2}{(\epsilon_0 - \epsilon)^2} \right]^{\frac{1}{2}} \quad , \tag{104}$$

where $\lambda = m_{\nu_e}/m_e$. Now, we should correct for the fact that the outgoing electron wavefunction is distorted by the nucleus and other atomic electrons. One can solve for the appropriate correction factor very precisely. Let's denote this factor by $F(Z, \epsilon)$. Multiplying (104) by this factor and integrating over ϵ gives the total rate:

$$\Gamma = \left(\frac{m_e^5 c^4}{2\pi^3 \hbar^7} \right) |H_{fi}|^2 f(Z) \tag{105}$$

where where $f(Z)$ results from including $f(Z, \epsilon)$ in the integral. For $0^+ \rightarrow 0^+$ transitions, the Standard Model value for $|H_{fi}|^2$ is ϵ-independent and is given by:

$$|H_{fi}|^2 = \frac{G_\mu^2}{2} |V_{ud}|^2 [(I \mp I_z)(I \pm I_z + 1)] \times 2 \tag{106}$$

as in Eq. (93), so that

$$\Gamma = (\frac{m_e^5 c^4}{2\pi^3 \hbar^7}) G_\mu^2 |V_{ud}|^2 [(I \mp I_z)(I \pm I_z + 1)] f(Z) \quad . \tag{107}$$

The additional factor of 2 in Eq. (106) results from the purely leptonic part of the matrix element. Now an aside on units. The quantity $G_\mu (m_e c^2)^2/(\hbar c)^3$ is dimensionless. Thus, the dimensions of (105) are

$$[\Gamma] \quad : \quad \left[G_\mu \frac{(m_e c^2)^2}{(\hbar c)^3} \right]^2 \frac{m_e c^2}{\hbar} \tag{108}$$

while

$$\frac{m_e c^2}{\hbar} = \frac{m_e c^2}{\hbar c} \times c \quad . \tag{109}$$

Thus, $[m_e c^2/\hbar] = (\text{MeV}/\text{MeV}f) \times (f/s) = 1/s$, so the dimensions work out just right, with Γ having the dimensions of a rate. The time dependence of decay is described by:

$$N = N_0 e^{-\Gamma t} \quad . \tag{110}$$

For $t = t_{\frac{1}{2}}$ denoting the half-life, at which time $N = N_0/2$ we have

$$\Gamma t_{\frac{1}{2}} = \ln 2 \tag{111}$$

or

$$f t_{\frac{1}{2}} = \left(\frac{2\pi^3 \ln 2 \hbar^7}{m_e^5 c^4} \right) \frac{1}{G_\mu^2 |V_{ud}|^2} \frac{1}{[(I \mp I_z)(I \pm I_z + 1)]} \quad . \tag{112}$$

The quantity in Eq. (112) is called the "ft" value for the decay. It turns out that for all the experimentally studied superallowed transitions, the isospin factor in the denominator of Eq. (112) is the same. Thus, the ft values should be the same for all $0^+ \to 0^+$ transition if CVC is right. In fact, nine superallowed decays have been studied. The "ft" values for the decays agree at an impressive level of precision. Thus, to an extremely high precision, the SU(2) character of the weak charged currents are confirmed by nuclear β-decay. One last important feature: the superallowed decays can test lepton-quark universality of the charged current weak interaction *and* the unitarity of the CKM matrix. To see how, let:

$$G_F^\beta = G_\mu |V_{ud}|(1 + \Delta r_\beta - \Delta r_\mu) \quad , \tag{113}$$

where Δr_β and Δr_μ denote *radiative corrections* to β-decay and μ-decay, respectively. The correction Δr_β appears because of higher-order effects in the semileptonic $(d \to ue^-\bar{\nu}_e)$ amplitude. On the other hand, $-\Delta r_\mu$ appears because the Fermi constant has been taken from the muon lifetime, and we need to subtract out radiative corrections to the muon decay amplitude because they are contained in G_μ but don't affect β-decay. These corrections must be computed from the Standard Model. Using the result of these calculations and the experimental muon lifetime and β-decay ft values, we have:

$$G_F^\beta = 1.16637(1) \times 10^{-5} \text{ GeV}^{-2} \tag{114}$$

and

$$|V_{ud}|^2 = \frac{G_F^\beta}{G_\mu(1 + \Delta r_\beta - \Delta r_\mu)} = 0.9740 \pm 0.0005 \quad . \tag{115}$$

From K_{ℓ_3} and hyperon decays we have $|V_{us}| = 0.2196 \pm 0.0023$ while B-meson decays give $|V_{ub}| = 0.0032 \pm 0.009$. Now unitarity of the CKM matrix requires:

$$|V_{ud}|^2 + |V_{us}|^2 + |V_{ub}|^2 = 1 \quad . \tag{116}$$

The experimental results give:

$$|V_{ud}|_{\text{exp}}^2 + |V_{us}|_{\text{exp}}^2 + |V_{ub}|_{\text{exp}}^2 = 0.9968 \pm 0.0015 \quad , \tag{117}$$

corresponding to a 2.2 σ deviation from the SM requirement of CKM unitarity.

Note that at the one percent level, charged current lepton-quark universality and CKM unitarity are confirmed by β-decay, μ-decay, K_{l_3}-decays, and B-decays. However, at the level of precision now achieved by experiment, these features of the Standard Model almost hang together-though not quite. There is a hint that maybe there is more to the electroweak interactions than the Standard Model, and that this "new physics" may be responsible for the very tiny derivation from lepton-quark universality and CKM unitarity. As an aside, one might wonder why one chooses to focus on nuclear decays rather than the decay of the free neutron. In fact, τ_n has been measured very precisely:

$$\tau_n = 886.7 \pm 1.9s \quad . \tag{118}$$

However, it is a more complicated matter to extract G_F^β from these decays. Because the neutron has spin 1/2, both the vector and axial vector weak quark currents contribute to the decay rate. The axial vector current is not protected by a CVC type symmetry, and it gets important renormalizations due to the strong interaction. At present, we cannot compute these strong interaction effects with the kind of precision we'd need in order to extract $|V_{ud}|^2$ from τ_n alone. In order to circumvent this problem, one can perform measurements of parity-violating asymmetries associated with, *e.g.*, the direction of the outgoing e^- relative to the direction of neutron spin. Knowing both τ_n and one of these asymmetries allows one to determine separately the vector and axial vector contributions. From the former, we can obtain a value for $|V_{ud}|^2$ without having to worry about incalculable strong interaction effects. It is only recently that experiments have begun to determine these asymmetries with the kind of precision needed to determine $|V_{ud}|^2$ with the same precision as obtained from $0^+ \to 0^+$ decays. At present, there is an active research program underway at Los Alamos that will use polarized, ultracold neutrons to measure the β-decay asymmetries. The goal of this program is to match or even exceed the precision on $|V_{ud}|^2$ obtained from the superallowed decays.

3.4 Parity violating DIS and Weak Neutral currents

So far, we've seen how low energy experiments have provided important confirmation of several features of the charged current weak interaction. What about the weak neutral current?

There exists a basic difficulty in this case. Nature has given us two neutral currents, the electromagnetic (EM) and weak neutral current (NC):

$$J_\mu^{EM} = \sum_f Q_f \bar{f} \gamma_\mu f \tag{119}$$

$$J_\mu^{NC} = \sum_f \bar{f} \gamma_\mu (g_V^f + g_A^f \gamma_5) f \quad, \tag{120}$$

where the sum runs over all species of fermions, Q_f denotes the EM charge of fermion f, and g_V^f (g_A^f) is the vector (axial vector) coupling of f to the Z^0 boson. In a low energy charge neutral process, amplitudes associated with both enter. So the problem will be how to separate them.

To be concrete, consider the scattering of electrons from quarks inside a hadronic target. The total amplitude for the process of eq scattering is:

$$M = M_{EM} + M_{NC} \quad, \tag{121}$$

while the cross section is $\propto |M^2|$:

$$|M|^2 = |M_{EM}|^2 + 2Re(M_{EM}^* M_{NC}) + |M_{NC}|^2 \quad. \tag{122}$$

Consequently, the neutral current cross section can be separated into three terms:

$$\sigma^{tot} = \sigma^{EM} + \sigma^{int} + \sigma^{NC} \quad . \tag{123}$$

Here, σ^{EM}, σ^{int}, and σ^{NC} denote the purely electromagnetic contribution, the part arising from the interference of EM and weak NC amplitudes, and a purely weak NC contribution (corresponding to the three terms in (122). The coupling strengths entering the amplitudes are e for the EM amplitude and $gM_Z/4M_W$ for the weak NC amplitude. With this normalization one has :

$$g_V^f = 2T_3^f - 4Q_f \sin^2 \theta_W \tag{124}$$

$$g_A^f = -2T_3^f \quad . \tag{125}$$

This allows us to write:

$$|M_{EM}|^2 \propto \left(\frac{e^2}{q^2}\right)^2 = \left(\frac{4\pi\alpha}{q^2}\right)^2 \quad , \tag{126}$$

$$2Re(M_{EM}^* M_{NC}) \propto 2\left(\frac{e^2}{q^2}\right)^2 \left(\frac{gM_Z}{4M_W}\right)^2 \frac{1}{M_Z^2 - q^2} \tag{127}$$

$$\rightarrow 2 \times \frac{4\pi\alpha}{q^2} \frac{G_\mu}{2\sqrt{2}}$$

at $q^2 = 0$, and

$$|M_{NC}|^2 \propto \left[(\frac{gM_Z}{4M_W})^2 \frac{1}{M_W^2 - q^2}\right]^2 \propto \frac{G_\mu^2}{8} \tag{128}$$

at $q^2 = 0$. Now consider the relative strengths of the corresponding cross sections at low-energies. The ratio of the interference cross section σ^{int} to the EM cross section σ^{EM} goes as

$$\frac{\sigma^{int}}{\sigma^{EM}} = \frac{2Re(M_{EM}^* M_{NC})}{|M_{EM}|^2} \propto \frac{G_\mu}{\sqrt{2}} \frac{q^2}{4\pi\alpha} \tag{129}$$

Letting $|q^2| \propto (1\text{GeV})^2 \propto m_p^2$ (units where $\hbar = c = 1$) be a typical momentum transfer for a low-energy scattering reaction, we have

$$\alpha \approx \frac{1}{137} \tag{130}$$

$$G_\mu \sim \frac{10^{-5}}{m_p^2} \quad , \tag{131}$$

leads to:

$$\frac{\sigma^{\text{int}}}{\sigma^{EM}} \approx 10^{-4} \quad . \tag{132}$$

The magnitude of σ^{NC}/σ^{EM} is even smaller. So how is one ever going to see weak neutral current effects at low energies? The basic idea is to use the fact that M_{NC} contains pieces that are odd under parity, where as M_{EM} is parity even. Letting

$$V_\mu^f = \bar{f}\gamma_\mu f \tag{133}$$

$$A_\mu^f = \bar{f}\gamma_\mu\gamma_5 f \tag{134}$$

one has

$$J_\mu^{EM} = \sum_f Q_f V_\mu^f \tag{135}$$

$$J_\mu^{NC} = \sum_f g_V^f V_\mu^f + \sum_f g_A^f A_\mu^f \tag{136}$$

Now, let's look at the e-q amplitudes again, as illustrated in Fig. 6. In

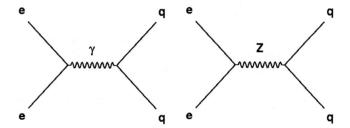

Figure 6. Electron-quark scattering diagrams

schematic terms, one has

$$M_{EM} \sim Q_e Q_q V_e \cdot V_q \tag{137}$$

$$M_{NC} \sim \frac{G}{2\sqrt{2}}[g_V^e g_V^q V_e \cdot V_q + g_A^e g_A^q A_e \cdot A_q$$
$$+ g_V^e g_A^q V_e \cdot A_q + g_A^e g_V^q A_e \cdot V_q] \tag{138}$$

The last two terms in M_{NC} transform as pseudoscalars; they are odd (change sign) under parity. All the other terms in Eqs. (137,138) are parity-even. Thus, to the extent that one can experimentally isolate these terms by measuring an observable which is parity-odd, one has a way to "filter out" the much larger EM interaction and get one's hands on the effects of the weak NC.

Now consider the following experiment. A beam of electrons with with spin and momentum vectors parallel (helicity $h = +1$), elastically scatter from a nucleon. Subsequently, the incident electron's spin is flipped, so that its spin and momentum vectors are anti-parallel ($h = -1$). One may then form the helicity-difference or "left-right" asymmetry:

$$A_{LR} = \frac{N(h = +1) - N(h = -1)}{N(h = +1) + N(h = -1)} , \tag{139}$$

where $N(h)$ denotes the number of events for electrons having incident helicity h. Since $h = \hat{s} \cdot \hat{k} \rightarrow -\hat{s} \cdot \hat{k}$ under parity, the asymmetry A_{LR} must be proportional to the parts of $|M|^2$ which contain pseudoscalars. In fact, one has

$$A_{LR} = \frac{2Re(M_{EM}^* M_{NC}^{PV})}{|M_{EM}|^2 + \cdots} \approx \frac{2Re(M_{EM}^* M_{NC}^{PV})}{|M_{EM}|^2}$$
$$= \frac{G_\mu q^2}{4\sqrt{2}\pi\alpha} \frac{g_V^e g_A^q V_e \cdot A_q + g_A^e g_V^q A_e \cdot V_q}{Q_e Q_q V_e \cdot V_q} \tag{140}$$

Here I have not performed a sum over quark spins nor integrated over energies, angles of the outgoing quarks, *etc.* The expression is essentially schematic, illustrating the basic physics ingredients in the asymmetry. Now, it takes a little algebra to perform this calculation, but it is doable. In addition, one can analyze e^-N or e^-A scattering *as if* one were scattering off individual quarks and then summing over different types of quarks in the nucleon. To do this one needs to work in a kinematic regime where

$$W^2 = (P_{target} + q)^2 = q^2 + 2M_t\nu + M^2 >> M_T^2 \tag{141}$$

where $\nu = q_0$ is the energy transfer to the target (having mass M_T), which is just the energy loss of the electron.

This kind of experiment is called deep inelastic scattering (DIS). One of the first experiments to test the neutral current structure of the Standard Model was carried out at SLAC in the 1970's using parity-violating (PV) DIS from a deuterium target[13]. The L-R asymmetry for this scattering is:

$$A_{LR}^{DIS}(\vec{e}D) = \frac{G_\mu q^2}{4\sqrt{2}\pi\alpha}(\frac{9}{10})\left[\tilde{a}_1 + \tilde{a}_2\frac{1 - (1-y)^2}{1 + (1-y)^2}\right] \tag{142}$$

where:

$$\tilde{a}_1 = \frac{1}{3}g_A^e(2g_V^u - g_V^d) = 1 - \frac{20}{9}\sin^2\theta_W \tag{143}$$

$$\tilde{a}_2 = \frac{1}{3}g_V^e(2g_A^u - g_A^d) = 1 - 4\sin^2\theta_W \tag{144}$$

and

$$y = \frac{\nu}{E_e} \tag{145}$$

in the lab frame. The SLAC experiment was carried out at four energies in the range:

$$16.2 \geq E_e \geq 22.2 \ \text{GeV}$$
$$< Q^2 > = 1.6 \ (\text{GeV}/c)^2 \ .$$

The A_{LR} was measured as a function of y, which allowed a separation of $\tilde{a}_1$ and $\tilde{a}_2$. The best fit to the results can be used to extract a value of $\sin^2\theta_W$:

$$\sin^2\theta_W = 0.224 \pm 0.020 \ . \tag{146}$$

Note that $\sin^2\theta_W \approx \frac{1}{4}$ which implies that $\tilde{a}_2 \approx 0$. This result corresponds to A_{LR} having almost no dependence on the kinematic variable y. Thus, the SLAC data is consistent with the Standard Model picture of SU(2)$_L$ and U(1)$_Y$ mixing in the neutral current sector with the mixing parameter, $\sin^2\theta_W$, being close to $\frac{1}{4}$. At the time the experiment was performed, there existed competing electroweak models which predicted a more dramatic y-dependence of A_{LR}. These models were ruled out by the results of this experiment.

The SLAC value for $\sin^2\theta_W$ was later confirmed in a series of purely leptonic experiments[14]:

$$\nu_\mu + e \rightarrow \nu_\mu + e$$
$$\bar{\nu}_\mu + e \rightarrow \bar{\nu}_\mu + e$$
$$\bar{\nu}_e + e \rightarrow \bar{\nu}_e + e$$
$$e^+e^- \rightarrow \mu^+\mu^- \ ,$$

where the last measurement involved studying the the forward-backward asymmetry. Taken together, these leptonic experiments implied that $\sin^2\theta_W \approx 0.22$, in agreement with the SLAC result. Thus, lepton-quark universality also holds for the neutral current sector of the $SU(2)_L \times U(1)_Y$ theory, with a common set of couplings (g,g') and mixing parameter $(\sin^2\theta_W)$ governing both leptonic and semileptonic interactions.

3.5 Atomic PV and Weak Neutral Currents

Additional confirmation of the Standard Model structure of the weak neutral currents comes from atomic physics, where one looks for PV asymmetries associated with the weak interaction between atomic electrons and the nucleus. In fact, one of the most precise determinations of the weak neutral current eq interaction has been performed by the Boulder group using atomic parity-violation (APV) in cesium[15]. The idea behind these experiments was developed by the Bouchiats[16] in Paris in the mid 1970's-about the same time the SLAC experiment was underway.

As in the case of PV DIS, the use of APV to get at the weak neutral current relies on an interference effect between the parity violating weak neutral current atomic matrix elements and the electromagnetic matrix elements. However, unlike PV DIS, APV relies on a coherent sum over the individual electron-quark amplitudes. The basic physics is that an atomic electron interacts with the nucleus by exchanging both a γ and a Z^0 boson. The probability amplitude for the PV part of the latter is

$$M_{PV} \sim \frac{G_\mu}{2\sqrt{2}} [g_A^e A_e \cdot < N| \sum_q g_V^q V_q |N> + g_V^e V_e \cdot < N|g_A^q A_q|N>] \quad . \quad (147)$$

This amplitude, which contains two distinct terms, causes the atomic states of the opposite parity to mix. In the atomic Hamiltonian derived from this amplitude, one finds that the first term is dominated by the time components of the currents, whereas the second part is dominated by the space components. For the first term, we have

$$< N| \sum_q g_V^q V_q^{\mu=0} |N> = < N| \sum_q g_V^q q^\dagger q |N> \quad . \quad (148)$$

Now $q^\dagger q$ just counts the number of quarks of flavor q. Thus the matrix element (148) gives:

$$< N| \sum_q g_V^q V_q^{\mu=0} |N> = (2N+Z)g_V^d + (2Z+N)g_V^u \quad (149)$$

$$= N(2g_V^d + g_V^u) + Z(2g_V^u + g_V^d) \equiv Q_W \quad ,$$

where Q_W is the "weak charge" of the nucleus. Now

$$g_V^u = -1 + \frac{4}{3}\sin^2\theta_W \quad (150)$$

$$g_V^d = 1 - \frac{8}{3}\sin^2\theta_W \quad (151)$$

so that the weak charges of the proton and neutron, respectively, are

$$Q_W^p = 2g_V^u + g_V^d = 1 - 4\sin^2\theta_W \tag{152}$$
$$Q_W^n = 2g_V^d + g_V^u = -1 \quad . \tag{153}$$

In terms of these quantities, the weak charge of an atomic nucleus is

$$Q_W = NQ_W^n + ZQ_W^p \tag{154}$$
$$= -N + Z(1 - 4\sin^2\theta_W) \quad . \tag{155}$$

Note that $\sin^2\theta_W \approx 0.231$ so that $Q_W^p = 1 - 4\sin^2\theta_W \sim 0.1$ whereas $Q_W^n = -1$. Hence, in contrast to the γ charge couplings $Q_{EM}^p = 1$ and $Q_{EM}^n = 0$, the Z^0 has a vector coupling to the neutron of strength unity and a tiny coupling to protons. In short, the γ see mostly protons, whereas the Z^0 sees mostly neutrons. Now using $g_A^e = 1$ the first term in (147) becomes:

$$\frac{G_\mu}{2\sqrt{2}} Q_W e^+ \gamma_5 e \tag{156}$$

The second term is more complicated, since $< N|\sum_q A_q^\mu|N >$ is not coherent and depends on the nuclear spin(the first term does not). Moreover, $g_V^e = -1 + 4\sin^2\theta_W$ further suppresses this term.

To get some intuition into the structure of the resulting Hamiltonian, one can take the limit of a non-relativistic electron in the field at a point like nucleus:

$$\hat{H}_W^{PV} = \hat{H}_W^{PV}(NSID) + \hat{H}_W^{PV}(NSD) \tag{157}$$

where

$$\hat{H}_W^{PV}(NSID) = \frac{G_\mu}{4\sqrt{2}}\frac{1}{m_e c} Q_W\{\vec{\sigma}\cdot\vec{p}, \delta^3(\vec{r})\} \tag{158}$$

$$\hat{H}_W^{PV}(NSD) = \frac{G_\mu}{4\sqrt{2}}\frac{1}{m_e c}(1 - 4\sin^2\theta_W)g_A^N\{\vec{\sigma}\cdot\vec{p}, \delta^3(\vec{r})\} \quad . \tag{159}$$

So how does an experiment which probes these interactions actually work? The most precise result is from APV in cesium, which has $Z = 55$ and $N = 78$. The ground state of the atom is a 6S state. The PV interaction (157) mixes a bit of the 6P state into this S state. Similar mixing occurs for the excited states. Now, what the Colorado experiment does is apply an external electric field, which causes mixing of the S and P states due to the Stark effect. In this case, one can obtain an E1 transition which is proportional to $\vec{E}$. The atoms are excited into the 7S state with circularly polarized light and the transition rate is measured. This rate can be expressed as:

$$\Gamma(6S \to 7S) = \beta^2 E^2 \epsilon_z^2 \left[1 + K \frac{E1_{PV}}{\beta E} \frac{\epsilon_x}{\epsilon_z}\right] \quad , \tag{160}$$

where β is the Stark-induced amplitude, K is a geometric factor (dependent on the m-quantum numbers), $\epsilon_{x,z}$ are laser polarization components, and $(E1)_{PV}$ is the PV-induced amplitude. By reversing $\vec{E}$ or the laser polarization, one can experimentally isolate the interference term containing $E1_{PV}$.

It also turns out that one can experimentally isolate the effects of H(NSID), containing Q_W, and H(NSD) by looking at sums and difference of the rates for different E1 hyperfine transitions. When all is said and done, one then can extract the following quantity:

$$\xi Q_W \quad , \tag{161}$$

where ξ is a quantity which depends on an atomic calculation of the PV atomic mixing matrix element

$$< P_{1/2} | \{\vec{\sigma} \cdot \vec{p}, \delta^3(\vec{r})\} | s_{1/2} > \quad . \tag{162}$$

Taking into account the latest atomic theory computations of ξ, the result of the Boulder measurement gives[17]:

$$Q_W^{\text{exp}} = -72.81 \pm 0.28(\text{exp}) \pm 0.36(\text{atomic theory}) \quad , \tag{163}$$

whereas the Standard Model prediction for Q_W is[5,18]:

$$Q_W^{SM} = -73.17 \pm 0.03 \quad . \tag{164}$$

Until recently, the values of Q_W^{exp} and Q_W^{SM} differed by up to more than two standard deviations. However, in the past year, atomic theorists have included have included some rather subtle, nucleus-dependent effects in the radiative corrections that have changed the value of ξ and moved the weak charge into agreement with the SM prediction. Thus, taken together with the SLAC experiment and the low-energy charged current measurements discussed earlier, cesium APV provides a substantial vote of confidence in the essential ingredients of the Standard Model.

4 Physics beyond the Standard Model

As the previous sections have tried to illustrate, low-energy experiments in nuclear and atomic physics have played an important role in verifying some of the basic ingredients of the Standard Model. Of course, tests have been performed over a wide range of energy scales, with some of the most decisive having been carried out at high energy colliders. The existence of the $W^{\pm}$ and Z^0 bosons was discovered in collider experiments at CERN, while measurements in e^+e^- collisions at center of mass energies $\sqrt{s} \approx 90$ GeV – both at CERN and at SLAC – have tested the properties of the neutral current sector of the Standard Model with sub-one percent precision. Similarly, the discovery of the top quark in $p\bar{p}$ collisions at the Tevatron represented an important triumph for the Standard Model, since the mass of the top quark is consistent with what one expects based on the m_t-dependence of electroweak radiative corrections to a variety of other measured electroweak observables. From these standpoints, the Standard Model has been an enormously successful theory.

Nevertheless, there exist many reasons for believing that the Standard Model is not the end of the story. Perhaps the most obvious is the number of independent parameters that must be put in by hand. As we saw earlier, the electroweak sector of the theory alone contains 17 *a priori* unknown parameters. The $SU(3)_C$ sector (QCD) introduces two more, the strong coupling, g_s, and θ. The latter parameterizes a term in the Lagrangian:

$$L_\theta^{QCD} = \theta \frac{\alpha_s}{2\pi} G_a^{\mu\nu} \tilde{G}_{\mu\nu}^a \tag{165}$$

$$\tilde{G}_{\mu\nu} = \epsilon_{\mu\nu\alpha\beta} G^{\alpha\beta} \quad , \tag{166}$$

with $G_{\mu\nu}^a$ being the gluon field strength tensor and $a = 1, \ldots, 8$. Note that L_θ is both a pseudoscalar and odd under time reversal. It is also even under charge conjugation, so the interaction (165) is CP-violating. Measurements of the electric dipole moment of the neutron and neutral atoms imply that $\theta \leq 10^{-9} - 10^{-10}$. This seems "un-natural", given the size of the other parameters in the Standard Model. Hence the questions:

What is the origin of the various parameters in the Standard Model?

Why do they have the values one observes them to have?

Why is θ_{QCD} so tiny? (This is the "strong CP Problem")

There exists already some need to go beyond the Standard Model simply to answer these questions. But there is even more motivation:

1.Coupling Unification

There exists a strongly held belief among particle physicists and cosmologists that in the first moments of the life of the universe, all the forces of nature were "unified", that is, they all fit into a single gauge group structure whose interaction strengths were described by a single coupling parameter, g_U. It is a remarkable idea, and an intellectually appealing one.

The scenario goes in the following manner: as the universe cooled down, spontaneous symmetry breaking occurred, giving gauge bosons masses and changing the way the interaction strength for various forces evolved or "ran" down to lower energies/temperatures. To see how this idea works, one must work out how the couplings(which are not constants!) run with the energy scale, μ. The origin of this running is renormalization. Fermion and gauge boson wave functions, as well as interaction vertices, get the following contributions as seen in Fig. 7 for the vertex, Fig. 8 for the fermion wave function and Fig. 9 for the gauge boson wavefunction.

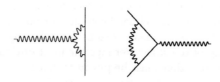

Figure 7. Lowest order diagrams for the renormalizing the gauge boson-fermion vertex.

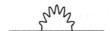

Figure 8. Lowest order diagrams renormalizing the fermion wavefunction.

Each of these diagrams is infinite. This means that the theory was not constructed correctly to begin with, and some redefinitions must be made:

$$\psi \to \psi_0 = \sqrt{Z_2}\psi$$
$$W_\mu^a \to W_{\mu 0}^a = \sqrt{Z_3}W_\mu^a \tag{167}$$
$$g \to g_0 = g\frac{Z_1}{Z_2\sqrt{Z_3}}$$

Figure 9. Lowest order diagrams renormalizing the gauge boson wavefunction.

The constants Z_i are defined in such a way that when one computes the loops with the redefined quantities and add them to the tree level quantities, one gets a finite answer. The only hitch is that in making these definitions one has to specify the energy scale μ at which one is working. Hence:

$$Z_i = Z_i(\mu) \tag{168}$$

This is because the graphs depend on μ. Now, the Lagrangian can't depend on μ; it must hold for all scales. Since the Lagrangian depends on the "bare parameters", ψ_0, $W^a_{\mu 0}$, and g_0, these parameters can't depend on μ either. Thus, one has:

$$\mu \frac{\partial}{\partial_\mu} g_0 = 0 \tag{169}$$

Comparing (167) and (169) one sees that g must vary with μ in order to compensate for the μ-dependence of the Z_i. Thus, one obtains

$$\mu \frac{\partial g}{\partial \mu} = \beta(g) \tag{170}$$

where $\beta(g)$ depends on the group structure and fermion content of the theory. It dictates how $g(\mu)$ runs with μ. This relationship between $g(\mu)$ and $\beta(g)$ is known as a renormalization group equation. One can just as well convert Eq. (170) to an equation for the running of:

$$\alpha_k(\mu) = \frac{g_k(\mu^2)}{4\pi} \quad ,$$

where the subscript k indicates the gauge group to which g_k pertains. The renormalization group equations for the running of α_k are ,

$$\frac{d}{dt}\alpha_k^{-1} = -\frac{b_k}{2\pi}$$

where $t = \ln(\mu/\mu_0)$, μ_0 is a reference scale, and

$$b_1 = \frac{41}{20}, \qquad g_1 = \sqrt{\frac{5}{3}}g'$$
$$b_2 = -\frac{19}{6}, \qquad g_2 = g$$
$$b_3 = -7, \qquad g_3 = g_s$$

We can see that $\alpha_2(\mu)$ and $\alpha_3(\mu)$ – the couplings for the non-Abelian groups – decrease with μ whereas the $U(1)_Y$ coupling increases with μ. In the case of QCD, this feature is known as asymptotic freedom.

If one were to plot the running couplings as a function of μ, one would see that the three Standard Model couplings almost meet at a common point around $\mu \sim 10^{16}$ GeV, but not quite. This result is tantalizing from the standpoint of unification. It is one of the motivations for believing something else is out there, as this something else could modify the running of the couplings and produce unification[d].

An aside on some terminology: One often hears reference made to the high-energy "desert." This desert is the region in μ between the weak scale,

$$M_{\text{weak}} \sim 250 \ \text{GeV}/c^2 \tag{171}$$

and the scale where one believes gravity becomes strong. The obvious parameter which defines this scale is Newton's gravitational constant,

$$G_N = 6.67259(85) \times 10^{-11} \ m^3 \ kg^{-1} \ s^2 \tag{172}$$

From this one may define the Planck mass:

$$M_{\text{pl}} = \sqrt{\frac{\hbar c}{G_N}} \cong 1.22 \times 10^{19} \ \text{GeV}/c^2 \quad . \tag{173}$$

The "desert" then refers to the region

$$M_{\text{weak}} \leq \mu \leq M_{\text{pl}} \quad . \tag{174}$$

The interest in looking for physics beyond the Standard Model, or "new" physics, is really about learning what else lies in the desert. If, in fact, the

---

[d]I have not discussed how the gravitational interaction gets incorporated into unification. That is a separate, very difficult problem, for which string theory may ultimately provide a solution.

desert is just that – a particle physics wasteland devoid of anything new – then one seemingly cannot get unification.

2. The hierarchy problem

Suppose there does exist some new particle or particles in the desert having mass $m \gg M_{\text{weak}}$. To be concrete, suppose the particle is a fermion. Presumably, this particle interacts with the Higgs boson, which is responsible for particle masses. This interaction will affect the mass of the Higgs through higher order diagrams. After renormalization, this diagram yields a finite contribution to the mass of the Higgs:

$$\delta m_H^2 = \frac{3|\lambda_f|^2}{8\pi^2} m_f^2 \ln\frac{\mu}{m_f} + \dots \qquad (175)$$

Now, we know that m_H itself must be on the order of M_{weak} or below. The reason is that one can rewrite the Higgs potential as:

$$V(\Phi) = -\mu^2 \Phi^\dagger \Phi + \lambda(\Phi^\dagger \Phi)^2$$
$$= -\frac{1}{2} m_H^2 \Phi^\dagger \Phi + \lambda(\Phi^\dagger \Phi)^2 \qquad (176)$$

with

$$M_{\text{weak}} = v = \left(\frac{m_H^2}{2\lambda}\right)^{\frac{1}{2}} \sim 250 \ \text{GeV}/c^2 \qquad (177)$$

or $m_H \sim \sqrt{2\lambda} \times 250 \ \text{GeV}/c^2$. It would be unnatural for λ (a dimensionless quantity) to be significantly different from unity, so one expects that $m_H \sim M_{\text{weak}}$. On the other hand, if $m_f \gg M_{\text{weak}}$, one has $\delta m_H^2 \gg m_H^2 \sim M_{\text{weak}}^2$! In short, any particles which exist deep in the desert give huge corrections to m_H, making it unbelievable that m_H comes out close to M_{weak}.

To put it another way, the electroweak scale is destabilized by radiative corrections involving heavy particles. It starts out at $\sim 250 \ \text{GeV}/c^2$ at tree level but grows as heavy particles come into the theory. This is not a desirable situation for any good theory. The problem is known as the "hierarchy problem": How does the weak scale remain stable if the desert becomes populated?

Another aspect of the hierarchy problem is the spectrum of the Standard Model masses themselves:

$$M_{\text{weak}} \sim M_{W,Z} \sim m_{top} \gg m_b \gg m_\tau \gg m_e \gg m_\nu \qquad (178)$$

How does one explain this hierarchy of masses? The Standard Model gives us no clue as to how to handle the hierarchy problem. Evidently, something new is needed.

3. Quantization of Electric Charge

Recall from Eq. (32) that

$$Q = T_3^L + \frac{Y}{2} \tag{179}$$

For any particle, T_3^L is quantized in integer or half-integer units. This follows from the algebra of SU(2):

$$[T_i, T_j] = i\epsilon_{ijk}T_k \quad . \tag{180}$$

One the other hand, a U(1) group has no such algebra, and therefore the eigenvalues of the group generator Y are not restricted. Equation (179), however implies that Y must take on integer values for leptons and fractional values for quarks, in order to reproduce the known fermion charges. This seems rather arbitrary from the standpoint of symmetry. Why, then, is Y – and therefore Q – also quantized? The Standard Model does not motivate electromagnetic charge quantization, but simply takes it as an input. The deeper origin of Q quantization is not apparent from the Standard Model.

4. Discrete Symmetry Violation

By construction, the Standard Model is maximally parity-violating; it was built to account for observations that weak c.c. processes only involve left handed particles (or right handed anti-particles). But why this mismatch between right-handedness and left-handedness? Again, no deeper reason for the violation of parity is apparent from the Standard Model.

Similarly, the Standard Model allows CP-violation to creep in two places:
(i) a phase in the CKM matrix
(ii) the QCD θ term
Already the θ mystery has been discussed; it is incredibly tiny for no obvious reason in the SM. What about δ, the CKM phase factor? Where did it come from? In some sense, its appearance is an artifact of the mathematics for three generations of massive quarks. But the reason for the existence of three generations, and again the magnitude of the phase factor is not explained by the Standard Model. It would be desirable to have answers to these questions, but it will take some new framework to provide them.

5. Baryon Asymmetry of the Universe (BAU)

Why do we observe more matter than anti-matter? This is a problem for both cosmology and the Standard Model. To quantify this problem, let $n_B = n_b - n_{\bar{b}}$, difference in the number of baryons and anti baryons per unit volume, and let n_γ be the photon number density at temperature T. Standard cosmology makes very accurate predictions for the cosmological abundance of

H, ^{3}He, ^{4}He, ^{2}H, B, and ^{7}Li given that $\eta \equiv n_B/n_\gamma$ has been constant since nucleosynthesis. The primordial abundances of ^{2}H and ^{3}He imply:

$$3 \times 10^{-10} \leq \eta \leq 10 \times 10^{-10} \tag{181}$$

If $\eta = 0$ at the Big Bang, then standard cosmology implies that $\eta \leq 10^{-18}$ – much smaller than the range in (181). Hence, the early universe must have $\eta \neq 0$ to explain primordial element abundances.

What is the connection to the Standard Model? It was provided by Sakharov, who pointed out that a non-vanishing η may exist in the early universe if: (i) Baryon number is violated (ii) Both C and CP are violated (iii) At some point, there has bee a departure from thermal equilibrium. These are known as the Sakharov criteria[19]. The reason for (i) is clear. The violation of both C and CP is needed so that:

$$\Gamma(\text{baryon production}) \neq \Gamma(\text{anti} - \text{baryon production}) \quad .$$

The third Sakharov criterion is needed to get a non-zero thermal average of B. Now, it is known that in the Standard Model, B+L is broken by instanton effects. Similarly, the Standard Model has maximal C-violation. Consider, for example, the charged current interaction

$$\bar{u}\gamma^\mu(1 - \gamma_5)d\, W_\mu^- \quad .$$

The axial vector part of this interaction is C-odd because

$$\bar{u}\gamma^\mu\gamma_5 d \overset{\rightarrow}{\underset{C}{}} +\bar{u}\gamma^\mu\gamma_5 d$$
$$W_\mu \overset{\rightarrow}{\underset{C}{}} -W_\mu \quad .$$

CP violation enters via the CKM phase factor δ as well as the θ parameter. As noted earlier, θ is incredibly tiny – far too small to provide the necessary amount of CP-violation for the baryon asymmetry. Similarly, the magnitude of δ's contribution to the baryon/antibaryon asymmetry is significantly smaller than needed to produce the required value of η.[e] Thus, if one is going to live within standard cosmology, one needs additional sources of large CP-violation beyond the Standard Model to explain BAU.

To summarize, despite the triumphant successes of the Standard Model, there exist conceptual motivations for believing that there is something more, that the high energy desert is not so barren after all.

One of the goals in experimental high energy physics – as well as in precision low energy electroweak experiments – is to go looking for new physics.

---

[e]The magnitude of this contribution depends not only on δ but also on the other angles appearing in the CKM matrix.

In fact, the results of these searches can constrain the new physics scenarios people have invented, or at least dictate what some of the parameters in these scenarios must be. Some of the most popular such scenarios include supersymmetry (SUSY), extended gauge symmetry, and extra dimensions. The appeal of SUSY is that it provides a natural solution to the hierarchy problem, produces unification of gauge couplings, and contains new CP-violating effects that could help produce a sufficiently large BAU. Extended gauge theories, on the other hand, can also produce gauge unification, provide a natural mechanism for electric charge quantization, and and can account for the violation of parity invariance in low-energy weak interactions. Finally, the idea that we live in more than four spacetime dimensions – which has been motivated by string theory – gives an alternative solution to the hierarchy problem than contained in SUSY. The implications of this paradigm for the phenomenology of electroweak interactions is now a lively area of research in particle physics.

Given the scope of these lectures, I do not have the time and space to discuss these scenarios in any depth. At the very least, however, I hope to have provoked your curiosity and motivation for learning more about them. We should always keep in mind, that however one seeks to extend the Standard Model, one should take care to respect the basic ingredients of the Standard Model and its phenomenological successes:

Gauge symmetry

Universality

V-A dominance

Quark Mixing

neutral currents and $\sin^2 \theta_W$

conserved vector currents

etc.

Any deviations from the Standard Model predictions based on these ideas must be small, and any new physics scenario must explain why it only produces small deviations in a natural way.

Acknowledgements

This work was supported in part under U.S. Department of Energy contracts DE-FG02-00ER41146, DE-FG03-02ER41215, DE-FG03-88ER40397, DE-FG03-00ER41132 and NSF Award PHY-0071856.

References

1. S. Weinberg, Phys. Rev. Lett. **19**, 1264 (1967); A. Salam, In *Elementary particle physics (Nobel Symposium No. 8)*, Ed. N. Svartholm; Almqvist and Wilsell, Stockholm (1968); S.L. Glashow, Nucl. Phys. **22**, 579 (1961); D. Gross and F. Wilczek, Phys. Rev. Lett. **30**, 1343 (1973); S. Weinberg, Phys. Rev. Lett. **31**, 494 (1973); H. Fritzsch, M. Gell-Mann, H. Leutwyler, Phys. Lett. B **47**, 365 (1973);
2. N. Cabibbo, Phys. Rev. Lett. **10**, 531 (1963); M. Kobayashi and M. Maskawa, Prog. Theor. Phys. **49**, 652 (1973).
3. See, *e.g.*, J.C. Hardy and I.S. Towner, Eur. Phys. J. A **15**, 223 (2002).
4. P.W. Higgs, Phys. Rev. Lett. **12**, 132 (1964); Phys. Rev. **145**, 1156 (1966).
5. The Particle Data Group, Review of Particle Physics, Phys. Rev. D **66**: 010001 (2002).
6. W. Marciano, Phys. Rev. D **60**: 093006 (1999).
7. For a discussion and references, see *e.g.*, E.D. Commins and P.H. Bucksbaum, *Weak Interactions of Leptons and Quarks*, Cambridge University Press, Cambridge (1983).
8. L. Michel, Proc. Phys. Soc. London, Sect. A63, 153 (1950); C. Bouchiat and L. Michel, Phys. Rev. **106**, 170 (1957).
9. G. Czapek, *et al.*, Phys. Rev. Lett. **70**, 17 (1993).
10. D.I. Britton, *et al.*, Phys. Rev. Lett. **68**, 3000 (1992).
11. W. Marciano and A. Sirlin, Phys. Rev. Lett. **71**, 3629 (1993).
12. V.M. Lobashev, *et al.*, Phys. Lett. B **460**, 227 (1999); Ch. Weinheimer, *et al.*, Phys. Lett. B **460**, 219 (1999).
13. C.Y. Prescott, *et al.*, Phys. Lett. B **77**, 347 (1978); Phys. Lett. B **84**, 524 (1979).
14. For a discussion and references, see *e.g.*, T.-P. Cheng and L.-F. Li, *Gauge Theory of Elementary Particle Physics*, Oxford University Press, Oxford (1984).
15. C.S. Wood *et al.*, Science **275**, 1759 (1997); S.C. Bennett and C.E. Wieman, Phys. Rev. Lett. **82**, 2484 (1999).
16. M.A. Bouchiat and C. Bouchiat, Phys. Lett. B **48**, 111 (1974); J. Phys. (Paris) **35**, 899 (1974); J. Phys. (Paris), **36**, 493 (1975).
17. A.I. Milstein, O.P. Sushkov, and I.S. Terekhov, [hep-ph/0212072]; V.A. Dzuba, V.V. Flambau, and J.S.M. Ginges, Phys. Rev. D **66**: 076013 (2002).
18. J. Erler, A. Kurylov, and M.J. Ramsey-Musolf, [hep-ph/0302149].
19. A.D Sakharov, JETP Lett. **5**, 24 (1967).

THE IMPORTANCE OF FLAVOR PHYSICS*

P. RANKIN

Physics Department,
University of Colorado,
Boulder, CO 80309-0390, USA
E-mail: Patricia.Rankin@Colorado.edu

The recent measurements at B-factories confirming a non-zero value of $\sin^2 \beta$ show that CP violation is not restricted to the kaon system. This discovery is the latest of many that support the current Standard Model of particle physics. However, particle physicists believe that the Standard Model is not a complete explanation of nature. I discuss the questions that particle physicists are seeking to answer and why precision measurements in flavor physics are one way to search for the physics beyond the Standard Model. I begin with a brief overview of the CKM matrix and its parameters. I will explain why improving the precision of measurements of the CKM matrix elements is important, and why better measurements and more measurements of CP violation are needed to help probe beyond the Standard Model. Some possible extensions of the Standard Model are described. I will discuss some of the experimental issues and then make the case for the need for a broad program of research in heavy flavor physics, comparing and contrasting different experimental techniques. The measurement of $\sin 2\beta$ is discussed in some detail to give the reader a feel for some of the experimental complications that need to be considered.

1. Introduction

What do particle physicists want to do? The Standard Model [1] of particle physics provides a nice framework within which to understand a wide range of phenomena and it unifies the electromagnetic and weak forces. However, it is not a complete theory. The Higgs particle has yet to be found. If it exists it should be discovered within the next ten years, either at the Tevatron at Fermilab near Chicago or at the Large Hadron Collider (LHC) at CERN near Geneva. If it does not exist experiments at these machines will be exploring alternative mechanisms of electroweak symmetry breaking. Whether the Higgs exists or not particle physicists will still have lots to do

---

*This work is supported by DoE grant DE-FG03-95ER-40894

in the future. The Standard Model raises as many questions as it answers. Why are there three generations of particles? Why do quarks and leptons have the masses they do? Why do the electron and the proton have equal but opposite charges? Why do we live in a matter-dominated universe when the Big Bang produced equal amounts of matter and anti-matter?

The large number of free parameters in the Standard Model and the large number of things it leaves unexplained are driving the search for what lies beyond it. Particle physicists have used two basic techniques to look for new physics and evidence of more unification. They have continually sought to reach the highest energies possible to exploit $E = mc^2$ and make heavier and heavier fundamental particles. Alternatively, they have worked to increase the precision of measurements to better probe the internal consistency of the Standard Model. It is this latter approach which is the subject of this paper.

Since the field of flavor physics is so vast, and since there are frequent reviews and frequent updates of the field's status,[2] I have decided to focus my efforts in this contribution towards filling in some of the background rather than surveying the data. The choice of topics and the amount of detail given grows out of the revealed interests of the 2002 HUGS school participants during the presentations which formed the basis of this paper. The opinions presented here are the authors own.

2. The Standard Model

The Standard Model unifies the weak and electromagnetic interactions. It has an SU(2)xU(1) gauge structure. The angle $\sin^2 \theta_w$ measures the mixing between the electromagnetic and weak sectors of the theory. Measurements of the width of the Z^0 resonance imply that there are three generations of quarks and leptons. Particle masses in the theory are generated via the Higgs mechanism. The Higgs particle - a remnant of the Higgs mechanism is the one particle that is part of the Standard Model that has yet to be discovered experimentally.

2.1. *The CKM Matrix.*

The quark eigenstates of the strong interactions and the weak interactions are different. The Cabibbo-Kobayashi-Maskawa (CKM) matrix [3,4] elements tell us how strongly a 2/3 charged quark couples to each of the −1/3 charged quarks - for example, V_{tb} gives the strength of the top quarks coupling to the bottom quark.

$$V_{CKM} = \begin{bmatrix} V_{ud} & V_{us} & V_{ub} \\ V_{cd} & V_{cs} & V_{cb} \\ V_{td} & V_{ts} & V_{tb} \end{bmatrix} \quad (1)$$

Alternatively, the CKM matrix (V_{CKM}) connects the physical quark mass eigenstates of definite flavor (unprimed) to the weak eigenstates (primed).

$$\begin{bmatrix} d' \\ s' \\ b' \end{bmatrix} = V_{CKM} \begin{bmatrix} d \\ s \\ b \end{bmatrix} \quad (2)$$

In the framework of the Standard Model, the elements can be complex and the CKM matrix is unitary. This means that any pair of rows or columns are orthogonal. As a result, there are six basic relationships between the elements which can be represented as triangles in the complex plane - these are known as the *unitarity* triangles. It is customary to label the triangles by indices specifying the rows and columns used. Fig. 1 shows these six unitarity triangles.[5] All of the triangles have the same area and that area is only non-zero because elements of the CKM matrix are complex. The diagrams indicate the expected lengths of the sides of the triangles - experimentalists have focused on the **bd** triangle since the sides are expected to be of similar lengths and therefore easier to make measurements of.

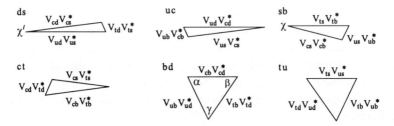

Figure 1. The six CKM "unitarity triangles". The bold labels, e.g. **ds**, refer to the rows or columns used in the unitarity relationship. This plot is taken from the Snowmass 2000 proceedings, Report of the P2 working Group.

A very popular parameterization of the CKM matrix is the Wolfenstein Approximation.[6]

$$V_{CKM} = \begin{bmatrix} 1 - \lambda^2/2 & \lambda & A\lambda^3(\rho - i\eta(1 - \lambda^2/2)) \\ -\lambda & 1 - \lambda^2/2 - i\eta A^2\lambda^4 & A\lambda^2(1 + i\eta\lambda^2) \\ A\lambda^3(1 - \rho - i\eta) & -A\lambda^2 & 1 \end{bmatrix}$$

(3)

The Wolfenstein A,λ parameters are relatively well known. The **bd** unitarity triangle is often redrawn to produce a triangle in the ρ,η plane of unit base and with an apex position dependent on the less well constrained Wolfenstein ρ, η parameters. This rescaling is based on the fact that V_{tb} and V_{ud} are very close to one and $V_{cd} \approx \lambda$. References to measuring $\sin^2 \beta$ in $B \to J/\psi K$ decays refer to measurements of the angle β of this triangle, as shown in Fig. 2. In the Wolfenstein parameterization framework, instead of thinking of measurements as relating directly to CKM matrix elements, they are thought of as constraining the position of the ρ,η vertex.

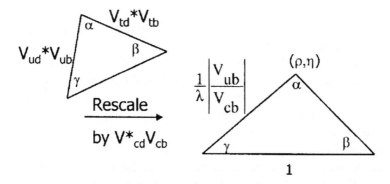

Figure 2. This figure shows the relationship between the **bd** unitarity triangle and a rescaled triangle defined in terms of the Wolfenstein parameters. β indicates the angle that has recently been measured at the B factories.

3. Searching for New Physics

The flavor physics program has as one of its goals a precise measurement of all nine of the CKM matrix elements. A longer-term goal is to measure the parameters in many different ways and to over constrain the CKM matrix - the hope is that this procedure will reveal inconsistencies in the Standard Model approach and will point to an improved theory. For example, if the matrix is not unitary (triangles do not close) then that is evidence for the existence of additional generations of particles. However, if the angles of the triangle do add up to 180 deg; this does not exclude the possibility that new physics is present, new physics may decrease one angle and increase another by an equal amount.

Measuring CKM matrix elements is conceptually straightforward, different matrix elements can be measured by studying decays which depend on them - for example, neutron decay allows a probe of V_{ub}. However, in real life things are much more complex. Quarks are never found in the free state and any extraction of CKM parameters requires an understanding of strong interaction effects. Also, while tree level diagrams provide a way to make basic estimates of decay rates, other diagrams contribute to the overall rate, often depending on other CKM matrix parameters, and their effects need to be taken into account. This means that the success of any experimental program depends also on related theoretical advances - in areas such as lattice QCD. Some measurements, which may not appear to relate directly to measurements of CKM matrix elements, are needed precisely because they help to reduce the theoretical uncertainty associated with other measurements.

The existence of higher order diagrams is not always considered to be a nuisance. In fact, since the internal lines in these higher order diagrams represent virtual particles, there are no limits on the masses of particles that can contribute to determining a reaction rate. This means that in principle when a measurement is made, the result depends on all particles that exist. In practice some effects decouple - such as if a standard doublet of quarks (one up-like, one down-like) of similar masses is added to the theory. However, other effects do not - such as those related to the Higgs particle. The result is that the effective strengths of the coupling constants become energy dependent, and that energy dependence depends on the types of particles which the theory contains and their masses.

The "running" of coupling constants (their change in value as a function of energy) has been exploited in recent years to enable particle physicists to put limits on the mass of the Higgs. These limits assume that the contributions to the running of the coupling constants by all particles other than the Higgs are known. These limits also assume that all the measurements made so far of $\sin^2 \theta_w$ should be consistent with each other and give an allowed range of masses for the Higgs. These predictions imply that the Higgs mass is comparatively light and put the discovery of the Higgs within the reach of current experiments at the Tevatron.

Another way of looking at this, is that all of the measurements we have made so far combine the effects of known physics (the Standard Model) and unknown physics/new physics. The possible ways that new physics can manifest itself are constrained by the current measurements. Since measurements which depend on the same Standard Model parameters at tree level, may depend on different new physics entering through higher order (loop) diagrams, it is important to measure as many different processes as possible. The Standard Model tree processes can be considered as a background to any search for new physics.

If the Standard Model tree contribution is large, the effects of small enhancements due to new physics will be difficult to isolate. It should be easier to spot the effects of new physics where the new physics significantly enhances the rate of a process that is suppressed in the Standard Model. Many experiments are trying to measure the branching rates of rare decays or set limits on them for this reason. Mixing measurements and CP violation studies are interesting because while the Standard Model predicts the existence of CP violation in the B system, there have to be sources of CP violation beyond those in the Standard Model.

3.1. CP Violation

The charge conjugation operator (C) reverses the additive quantum numbers of particles such as their electric charge and their lepton number, that is, it changes particles into their anti-particles. It does not alter a particles spin or its direction of travel. The parity operator (P) inverts spatial coordinates - that is, it takes a mirror image. It reverses linear momentum (a true vector) but not spin (like angular momentum spin is a pseudo or axial vector). Systems are CP violating if the combined effect of the C and P operators on a system produces an unphysical system or if the CP of a system is not conserved when it decays. The time reversal operator (T) converts particles moving forward in time to particles moving backwards. Gauge invariance requires systems to be CPT invariant and field theories must be gauge invariant to make physical sense. This means that CP violation requires a compensating T violation.

The Big Bang produced equal numbers of particles and anti-particles but today, matter dominates our solar system. It also appears from cosmic ray data that our galaxy is matter dominated. Finally, the baryon/photon ratio is high enough that there is no evidence for annihilation boundaries between matter and anti-matter universes. Explaining the lack of anti-matter requires that baryons and anti-baryons do not decay identically.

While baryons and anti-baryons are required by the CPT theorem to have the same lifetimes, CP violation allows for the required differences in partial decay rates. CP violation is especially interesting because if it is to explain the observed matter-antimatter asymmetry of the universe then new physics must be present. First, the mass scale associated with the Standard Model is to low to explain the observed cosmological effects. Second, there are no baryon number violating decays in the Standard Model and these must exist to explain the change in the net baryon number of the universe. Since new physics must be present and connected to the phenomena of CP violation a natural place to look for new physics is where CP violation has already been observed.

3.1.1. CP Violation in the B system

One reason for the interest in rare decays and in mixing is that they provide ways to probe CP violation. A requirement of observing Standard Model CP violation is that there are (at least) two routes for a particle to decay to a given final state. Interference between these routes can lead to CP violation. In principle, loop diagrams involving for example t, c, u quarks can interfere but usually the size of these contributions varies enough that it is difficult to observe this interference. It is better to find two diagrams of approximately equal strength. Rare decays that can proceed via penguin and/or highly suppressed tree diagrams are one place to look (see Fig.10). Another way is to look for decays to CP eigenstates which can be reached through the decay of either a B or a $\bar{B}$. B mixing (see Fig. 11) then supplies the two routes needed.

Lets look at the basic formalism for neutral B's, the CP eigenstates $(\mid B_1^0\rangle, \mid B_2^0\rangle)$ are given by

$$\begin{aligned}
\mid B_1^0\rangle &= \tfrac{1}{\sqrt{2}}(\mid B^0\rangle + \mid \bar{B}^0\rangle), CP \mid B_1^0\rangle = \mid B_1^0\rangle \\
\mid B_2^0\rangle &= \tfrac{1}{\sqrt{2}}(\mid B^0\rangle - \mid \bar{B}^0\rangle), CP \mid B_2^0 >= - \mid B_2^0\rangle
\end{aligned} \tag{4}$$

However, since neutral B's mix, the mass eigenstates $(\mid B_L\rangle, \mid B_H\rangle)$ that are superpositions of the flavor eigenstates $(\mid B^0\rangle, \mid \bar{B}^0\rangle)$ may not be CP eigenstates.

$$\begin{aligned}
\mid B_L\rangle &= p \mid B^0\rangle + q \mid \bar{B}^0\rangle \\
\mid B_H\rangle &= p \mid B^0\rangle - q \mid \bar{B}^0\rangle
\end{aligned} \tag{5}$$

where

$$p = \frac{1}{\sqrt{2}} \frac{1+\epsilon_B}{\sqrt{1+|\epsilon_B|^2}} = \frac{e^{-i\phi}}{\sqrt{2}}$$
$$q = \frac{1}{\sqrt{2}} \frac{1-\epsilon_B}{\sqrt{1+|\epsilon_B|^2}} = \frac{e^{i\phi_m}}{\sqrt{2}} \tag{6}$$

this leads to CP violation if

$$\epsilon_b \neq 0, ie, |\frac{q}{p}| \neq 1 \tag{7}$$

$$\frac{q}{p} = e^{2i\phi_m} = \frac{V_{tb}^{\star} V_{td}}{V_{tb} V_{td}^{\star}} \tag{8}$$

The states evolve according the standard Schrodinger equation

$$i\frac{d}{dt}\begin{pmatrix}p\\q\end{pmatrix} = H \begin{pmatrix}p\\q\end{pmatrix} = (m - \frac{i}{2}\Gamma)\begin{pmatrix}p\\q\end{pmatrix} \tag{9}$$

The mass eigenstates evolve straightforwardly (as mass eigenstates should),

$$|B_L(t)\rangle = e^{\frac{-\Gamma_L t}{2}} e^{\frac{im_L t}{2}} |B_L(0)\rangle$$
$$|B_H(t)\rangle = e^{\frac{-\Gamma_H t}{2}} e^{\frac{im_H t}{2}} |B_H(0)\rangle \tag{10}$$

The flavor eigenstates have a more complex time evolution (which is not purely exponential)

$$|B^0(t)\rangle = e^{(im+\frac{\Gamma}{i})}(cos\frac{\Delta m t}{2} |B^0(0)\rangle + i\frac{q}{p} sin\frac{\Delta m t}{2} |\bar{B}^0(0)\rangle)$$
$$|\bar{B}^0(t)\rangle = e^{(im+\frac{\Gamma}{i})}(cos\frac{\Delta m t}{2} |\bar{B}^0(0)\rangle + i\frac{q}{p} sin\frac{\Delta m t}{2} |B^0(0)\rangle) \tag{11}$$
$$\text{where} \quad m = \frac{m_L+m_H}{2}, \Delta m = m_H - m_L, \text{ and }, \Gamma = \Gamma_L \approx \Gamma_H$$

CP violation due to mixing - ie where one of the two routes to a final state is provided by the mixing is referred to as "indirect". "Direct" CP violation occurs through differences in the actual decay amplitudes to a final state

$$\text{ie} \quad \text{if} \quad A = \langle f_{cp} | H | B^0 \rangle, \bar{A} = \langle f_{cp} | H | B^0 \rangle, \text{ and } |\frac{A}{\bar{A}}| \neq 1 \tag{12}$$

Each source of CP violation has an associated phase, which means that what is observed depends on a mixture of phases (especially if many diagrams contribute). Different final states correspond to different phases, different combinations of CKM matrix elements, and different angles of the unitarity triangles. The different phases can be untangled by making many measurements and doing a global fit to all the possible parameters.

3.2. *Possible New Physics Scenarios*

What sort of new physics might we expect to find? One possible extension of the Standard Model is an expansion of the SU(2) doublet structure into an SU(5) quintet that connects quarks and leptons.

$$\begin{pmatrix} e^- \\ \nu_e \end{pmatrix} \rightarrow \begin{pmatrix} e^- \\ \nu_e \\ \bar{d}_r \\ \bar{d}_b \\ \bar{d}_g \end{pmatrix} \tag{13}$$

One reason to do this is that linking quarks and leptons together in the same multiplet provides an explanation for the fact that the electron and the proton have equal but opposite charges. At the same time, this theory predicts the existence of gauge bosons that would mediate baryon number changing decays. Attractive as SU(5) is as a theory however, it is ruled out in its simplest form by the limits on proton decay.

Variants of SU(5) survive however, usually as part of a supersymmetric theory.[7] Supersymmetric theories go one step further than SU(5) and connect fermions and bosons. This breaks the dichotomy of having fermion constituents and boson gauge particles and gives all of the known particles super-partners. For example, the photon is partnered by the spin 1/2 photino, and the electron by the spinless selectron. At the time of writing (late 2002), no superpartners have been seen experimentally.

One of the reasons that supersymmetry survives as a theory is that without it the fermionic loop corrections to the Higgs mass make the Higgs mass unphysical. The Standard Model fermionic contributions are cancelled in supersymmetric theories by the addition of the contributions of their superpartners which add to the amplitude with an opposite sign due to their different spins. Indirect evidence of the existence of supersymmetry (or something like it) can be found by looking at the running of the coupling constants. Adding in superpartners makes these converge. Supersymmetry or an alternate is also required to explain why the masses of the known particles are much lower than the Planck mass M_{PL} derived from the speed of light(c), Planck's constant(h) and Newton's gravitational constant(G),

$$M_{PL} = \sqrt{\frac{hc}{G}} \approx 10^{19} GeV \tag{14}$$

Recently, the fact that some string theory variants predict the existence of supersymmetry, has also bolstered support for the concept.

It is clear that supersymmetry cannot be an exact symmetry and must be a broken symmetry because the super-partners have different masses than the particles they partner - for example, a selectron with the mass of an electron has been excluded experimentally. However, if the masses of the superpartners become too big then they decouple and the cancellation of the undesirable terms in the Higgs mass calculation fails.

Any supersymmetric motivated extension of the Standard Model adds many new parameters, including contributions to CP violation. As a result, supersymmetric theories are already tightly constrained by existing measurements. Two especially important constraints are due to the limit on the electric dipole moment of the neutron, and the need to suppress flavor-changing transitions.

The additional CP violating phases (ϕ_A, ϕ_B) that supersymmetry adds contribute to electric dipole moments. Elementary fermions are not expected to have a static electric dipole moment in the Standard Model because of the requirement that the interactions of particles with the electromagnetic field be invariant under P,T operations. The only situations which yield electric dipole moments are those where degenerate energy states of opposite parity exist, as in the case of Hydrogen, for example. The experimental limit on the electric dipole moment of the neutron $(d_N < 1.1 \times 10^{-25} e\ cm)$ is one of the strongest constraints. This limits the possible values of the phases and supersymmetry mass scale $(\tilde{m})$ according to the relationship

$$\left(\frac{100 GeV}{\tilde{m}} \right) \sin \phi_{A,B} \leq 10^{-2} \frac{d_N}{10^{-25} e\ cm} \tag{15}$$

The Minimal Supersymmetric Model (MSSM), as its name suggests, is the simplest supersymmetric extension of the Standard Model. This model adds a superpartner for every Standard Model particle but keeps coupling strengths the same. However, the natural values to assume in the MSSM for $\sin \phi$ is of order one. The natural value for the mass scale is the electroweak symmetry breaking scale $(\tilde{m} \approx m_z \approx 100 GeV/c^2)$. These values predict a neutron electric dipole moment about a hundred times bigger than the current limit. This means that the MSSM is effectively ruled out and the surviving supersymmetric theories have to avoid this limit by either effectively increasing the mass scale significantly or by making the phases associated with supersymmetry very small. There are lots of models in the literature which do this.

The possible models differ by how much they suppress the electric dipole moment of the neutron, their effects on the rate of rare decays, their effects on mixing rates, and their effects on CP violation. So, not only does a broad program of flavor research improve the chances of seeing evidence of new physics, a broad program will be needed to determine exactly what type of new physics is being seen.

4. Experimental Techniques and Issues

This section will explain some of the common terms used in the field of flavor physics. I will review some basic techniques and some key experimental issues. The goal is to enable the interested reader to go further and to understand a detailed flavor physics paper or proposal.

4.1. *Cross Sections and Integrated Luminosity*

Knowing the cross section for a reaction at a given machine and energy is not enough to allow one to predict the number of events that will be produced, the integrated luminosity is also needed. If a machine produces an integrated luminosity of $1fb^{-1}$, then one event will be produced at the machine if the cross section for the reaction is $1fb$ (and a 1000 events if the cross-section is $1pb$).

4.2. *Tagging*

B mixing measurements and CP violation studies both require that one can determine the b quark content, or flavor, of the decaying B meson. There are several ways to do this, that is, to "tag" the B's flavor. One of the cleanest methods is to use the fact that a negatively charged lepton is produced when a b quark decays semi-leptonically. However, as shown in Fig. 3, b quark decays are not the only source of leptons in $B\overline{B}$ events. It is also possible to produce leptons by the semileptonic decays of secondary particles. These cascade decays can produce leptons of either charge. In order to cleanly separate leptons from the primary b decay and those from the secondary charm decays, a momentum cut must be applied. The leptons from the primary decay are more energetic and have a harder spectrum than those from secondary decays. This is shown in Fig. 4 which gives the spectra of primary and secondary electrons obtained by the CLEO [8] experiment. Events were selected for this analysis if they contained two leptons. If both leptons had the same sign it was assumed that one lepton came from a primary decay, the second from a secondary decay. Data from both neutral and charged B meson decays were used. The neutral meson data was corrected for the effects of B_d mixing which produces wrong sign correlations.

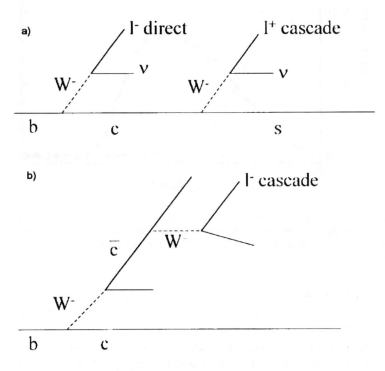

Figure 3.   Lepton production by semileptonic decays.

One limitation of using lepton tagging is that only a small fraction of B's (about 10%) decay semi-leptonically. The fraction of usable decays is further decreased by the momentum cut applied to get a clean signal. This has led to the development of a variety of other tagging techniques which vary in efficiency and cleanliness.

Another way to infer the flavor of the B is to use the flavor of the charmed meson produced in the decay. In some cases the decay chain is fully reconstructed - for example $B \to D^{*-}l^+, D^{*-} \to D^0\pi^-$. On the $\Upsilon(4S)$ it is possible to use a partial reconstruction technique - inferring the momentum of the D^* from the momentum and direction of the slow pion that is produced in the decay.

224

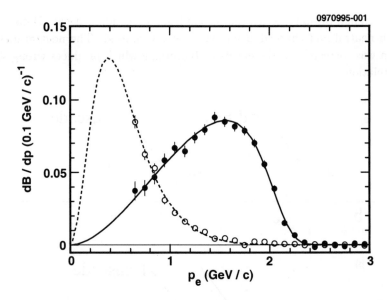

Figure 4.   Cleo data showing the spectra of electrons from $B \rightarrow Xe\nu$(filled circles) and $b \rightarrow c \rightarrow Ye\nu$ (open circles after continuum and fake subtraction and mixing corrections. The curve is an example of a fit used to estimate the uncertainty in the extrapolation from 600 MeV to the origin.

It is also possible to use the sign of the kaon produced in the secondary decay of the charmed meson as a flavor tag.   In this case, the position of the vertex of the B decay may be harder to estimate accurately (since the tagging track/tracks do not come from the primary vertex).   Also, an allowance must be made for doubly Cabibbo suppressed decays which lead to the production of the "wrong sign" of kaon and care has to be taken in dealing with events were the decay chain leads to the production of multiple strange particles.   As a general rule, the cleaner the tag is, the fewer the events that will be taggable.

4.2.1.   *Combining Information*

Since there are many tagging variables which can be used to determine the quark content of the decaying B meson it is worth thinking a bit about how to maximize the usefulness of the data.   Tagging decisions can be made based on a single tag parameter or by using some combination of variables. There are many ways to combine the information from several variables. One option is to choose a set of cuts to apply to the variables of interest and to use only events that pass all the cuts. However, such sets of fixed cuts do not necessarily give the optimal separation of events into two classes (b flavor or $\bar{b}$ flavor). Over time, several methods have been developed to use the information in an event as efficiently as possible. These approaches include the use of genetic algorithms, parameterizations, Fisher variables, and the use of neural network techniques.

Genetic algorithms work by encoding the cuts and treating them as genes. Each set of cuts is defined to be a "chromosome". The genetic algorithm searches through a "population" of chromosomes (sets of cuts). The concepts of Darwinian survival of the fittest are then employed to find the wanted solution. The user controls the measure to be used as a fitness value - e.g. the signal/noise ratio associated with an individual set of cuts. The genetic algorithm calculates the fitness of all the individuals in a group and removes the least fit individuals from the population. A new generation is then spawned from the survivors of the culling, using three genetic operators; reproduction, crossover, and mutation. The fittest individuals may just be copied ("elitism"). Reproduction or selection gives higher priority to fitter individuals in selecting parent chromosomes. Crossover or recombination then takes genes from each of the parents to form offspring. The genes in these offspring may then be randomly mutated. Studies show that on average, the overall fitness of the population improves with each successive generation. There are several nice web sites which are set up to teach people how genetic algorithms operate. [9]

Parameterized approaches are based on relative likelihood calculations. Suppose that events are to be separated into two classes and that the density distributions of the N variables which are to be used to distinguish between them are known (g^A, g^B).

The ratio of likelihoods (X_{pa}) is then used to characterize the event.

$$X_{pa} = \frac{g^A(x_1, x_2....., x_N)}{g^A(x_1, x_2, ..., x_N) + g^B(x_1, x_2, ..., x_N)} \tag{16}$$

Instead of using an N-dimensional density distribution, it can be assumed that the variables are uncorrelated and that the N-dimensional density distribution can be approximated by taking the product of the N one-dimensional density distributions. Variables can be included in this approach which have little discriminating power without decreasing the overall effectiveness of other variables. This approach is optimal if there are no correlations between the variables. If the variables are correlated then information is lost. A variation on this technique is to use two-dimensional density distributions for highly correlated variables.[10]

The Fisher or Mahalanobis approach is more complex.[11] The technique involves combining the N variables chosen to describe the events linearly (this approach is also known as a Linear Discriminant Analysis or LDA). Discrimination between events is performed by searching for the axis in the R^N space of the discriminating variables which maximally separates the two classes. This can be done if the mean value of each variable is known for the combined sample and separately for each of the two classes. In addition, the covariance matrix is also required. The Fisher technique uses a covariance matrix $W_{\mu\nu}$ which gives "within class" variances - that is it reflects the dispersion of events relative to the center of gravity of their own class. The full covariance matrix, used by the Mahalonbis method, adds a "between class" matrix $(B_{\mu\nu})$ to the within class one to form the covariance matrix. The between class components represent the distance of a class from the total center of gravity.

Event discrimination can be performed by comparing the value of a discriminating function

$$X_{FI} = \frac{\sqrt{n_A n_B}}{n} (\overline{x_A} - \overline{x_B})^T W^{-1} x \qquad (17)$$

with some threshold value

$$\theta_0 = \frac{\sqrt{n_A n_B}}{n} (\overline{x_A} - \overline{x_B})^T W^{-1} \frac{(\overline{x_A} + \overline{x_B})}{2} \qquad (18)$$

where n_A, n_B, n are the number of events in each sample and the total number of events respectively. The distance between projected points is a maximum along the direction defined by the line between x_A and x_B. The line segment $\overline{x_A}, \overline{x_B}$ is a projection axis. The probability for an event to be in each class is usually calculated based on the value of X_{FI}.

Finally, instead of a linear approach there are non linear approaches such as neural networks. These, like genetic algorithms take their inspiration from biology. Human beings are known to be extremely adept at pattern recognition - just think about peoples ability to read other peoples handwriting! Neural networks are designed to mimic the behaviors of the human brain and to produce flexible algorithms for problem solving. Many networks are based around the "perceptron"[12]. The perceptron is an analog of a neuron. It essentially sums several weighted inputs and bases a decision to "fire" on some threshold function. Perceptrons can be grouped into layers with the outputs of one layer being the inputs to the next. This architecture is known as the multi-layer perceptron (MPL) and is one of the common types of networks used by particle physicists. The main advantage of neural networks is that they can adapt to deal with unknown probability distributions, even if these are strongly correlated. Neural networks divide broadly into two classes, those that are supervised and those that are unsupervised. Supervised training requires the use of a set of known signal and known background events which the network is tuned to discriminate between. One of the complications of using neural networks is in deciding just how to select these samples. If Monte Carlo data is used then how can one be sure that the samples are realistic? If real data sets are used - how are these selected? How can errors be assigned? Unsupervised networks do not need such samples to be provided and can in principle be used if reliable training samples cannot be obtained - the complication here is that one does not have control over the features of the sample used for separation. A possibly apocryphal account is given of such a network being used to select between tanks and armored cars. The unsupervised network performed excellently in the laboratory but was a complete failure in field testing. Further study revealed that all the photographs of tanks used for training had been taken on a sunny day while those of the armored cars had been taken in cloudy conditions.

4.3. Background Rejection at the $\Upsilon(4S)$

Since so many measurements have been done at the $\Upsilon(4S)$ lets look a bit at why. The hadronic cross section in the region of the $\Upsilon(4S)$ is shown in Fig. 5; details of the cross section are given in Table 1. The $b\bar{b}$ cross section is about 1.15nb, the exact value depends on the energy spread of the colliding beams. The resonance sits on a background of continuum light quark production with a cross section of about three times the signal.

The $\Upsilon(4S)$ has a mass only slightly above the threshold for the production of $B_d\bar{B}_d$ pairs so no additional fragmentation hadrons are formed in the $\Upsilon(4S)$ decay. The result is that the B mesons that are produced at the $\Upsilon(4S)$ have known energies and are produced almost at rest in the $\Upsilon(4S)$ rest frame, with values of $\beta \approx 0.06$. The distance they travel before decay is given by $\beta\gamma c\tau$ which corresponds to about 26μ in the $\Upsilon(4S)$ center of mass.

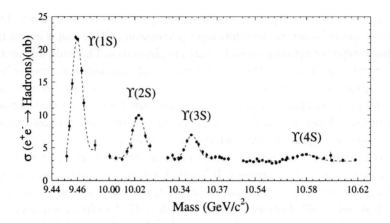

Figure 5. Hadronic cross section in the region of the $\Upsilon(4S)$ resonance.

Table 1. Cross Sections at the $\Upsilon(4S)$

Channel	Cross Section (nb)
$b\bar{b}$	1.15
$c\bar{c}$	1.30
$s\bar{s}$	0.35
$u\bar{u}$	1.39
$d\bar{d}$	0.35
$\tau^+\tau^-$	0.94
$\mu^+\mu^-$	1.16
e^+e^-	≈ 40

4.3.1. Kinematic Constraints at the $\Upsilon(4S)$

The tightly constrained kinematics of B meson production at the $\Upsilon(4S)$ make it possible to use a variety of techniques to reconstruct B mesons which cannot be used by experiments working at other energies.

The energy and momentum of the B's in the $\Upsilon(4S)$ center of mass can be determined exactly from the energies of the colliding beams. Figure 6 shows the power of using the mass and momentum of the B to separate events from the continuum background. Several variables can be constructed from the B's four-momentum to use in background suppression. Only two of these are independent of each other.

The energy of the B depends on the beam energy which can shift with time, either intentionally due to energy scans of the peak or unintentionally due to hardware fluctuations or changes. It is easier to understand plots

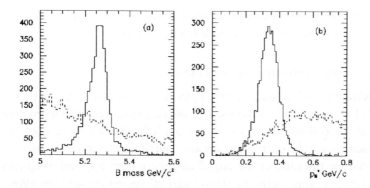

Figure 6. a)B invariant mass, b)B momentum in $\Upsilon(4s)$ restframe (signal : solid, light quark background: dashed).Taken from the BaBar Physics Book.

done in terms of the variable ΔE,

$$\Delta E = E_B - E_{beam}. \tag{19}$$

The distribution of this variable is centered around zero for real B's.

A common choice for the second independent variable to use is the beam-constrained mass or more properly the substitute energy mass (m_{se}), since the use of this variable does not involve a constrained fit.

$$m_{se} = \sqrt{E_{beam}^2 - p_B^2}. \tag{20}$$

This variable is useful because the smallness of the momentum of the B's (a few hundred MeV/c. in the $\Upsilon(4S)$ center of mass). This means that the resolution in this quantity is dominated by the beam energy spread (about 2.7 MeV at CESR). Since the beam energy is being used, the beam constrained mass does not depend on the mass hypotheses assigned to the tracks in the candidate B.

The momentum of a track in the lab frame is measured directly. However, the energy of the B candidate is calculated and depends on the mass hypotheses assigned to the tracks combined to form it. For $B^0 \rightarrow \pi^+\pi^-$ taking one of the pions to be a kaon shifts the calculated energy of the B by an average of about 40MeV.

Experiments at symmetric colliders such as CESR and DORIS work in a situation in which the laboratory and center of mass frames match. Experiments at KEK-B and PEP-II have to boost between these frames. Since the boost depends on the energy of the state, there is an additional source of error in measuring the candidate energy in the center of mass if the wrong mass assignments have been made to tracks.

4.3.2. Continuum Subtraction at the $\Upsilon(4S)$

Another advantage of running at the $\Upsilon(4S)$ is that data taken a little below the $\Upsilon(4S)$ resonance can be used to estimate and correct for random backgrounds to B decay signals coming from light quark decays since below the resonance only continuum decays contribute. Scaling the energy of the tracks (by the ratio of the on/off resonance energies) and weighting by the integrated luminosities of the on and off resonance samples allows for "continuum subtraction". One issue worth discussing is how to decide how much on/off resonance data should be taken. If there is no background to signals of interest then no off resonance data is needed. As the backgrounds become more significant then the amount of off resonance data needed increases - but if too much off resonance data is taken the statistical significance of the signal will drop too much. The balance between on:off resonance running depends on the background to the physics of interest. CLEO, an experiment at a symmetric $\Upsilon(4S)$ collider ran mostly with a ratio of 2:1 (on:off). BaBar and Belle - the B-factory experiments now running at asymmetric $\Upsilon(4S)$ colliders have an added advantage of a boost to separate out the B decay vertexes. As a result, both experiments take more data on resonance, (closer to 5:1), partly because of the value of using vertex separation as a continuum suppression tool, and partly because of their focus on measurements of CP violation which primarily involve decay modes which can easily be separated from backgrounds.

4.3.3. Shape Variables

Variables based on the shape of an event are often used to separate out B decays from the continuum background. One such variable is based on the Thrust of an event. The Thrust (T) of an event is defined in the center of mass frame by

$$T = \frac{\sum_i |\hat{T} \cdot p_i|}{\sum_i |p_i|} \tag{21}$$

where $\hat{T}$ is the Thrust axis of the event, that is, the direction which maximizes the sum of the longitudinal momenta of the particles. The Thrust tends to one for events which are comprised of two jets (continuum decays), and to a half for isotropic events (B decays). The angle between the decay axis of the B and the Thrust axis of the rest of the event can also be used to separate signal and background decays. These axes are roughly collinear for candidate two body B decays coming from the background (background candidates are usually comprised of tracks from both jets) and is uncorrelated for signal events.

Two other shape variables relate to the sphericity tensor

$$S^{\alpha\beta} = \frac{\sum_i p_i^\alpha \cdot p_i^\beta}{\sum_i p_i^2} \tag{22}$$

where $\alpha, \beta = 1, 2, 3$ correspond to x,y,z components respectively.

The Sphericity(S) of an event is defined by

$$S = \frac{3}{2}(\lambda_2 + \lambda_3) \tag{23}$$

where $\lambda_{2,3}$ are the two larger eigenvalues of the diagonalized sphericity tensor Isotropic events have a sphericity close to one and jetty events have a sphericity close to zero. The sphericity axis of an event is defined to be the direction of the eigenvector with the largest eigenvalue (c.f thrust axis).

A closely related variable is aplanarity (A) which depends on the smallest eigenvalue of the sphericity tensor and is defined as

$$A = \frac{3}{2}\lambda_1 \tag{24}$$

This measures the component of transverse momentum directed out of the event plane. The aplanarity of a planar event is zero ($\lambda_1 = 0$) and has a maximum value of 0.5 for an isotropic event (when all three eigenvalues are equal to 1/3).

Figure 7 shows the power of Thrust and Sphericity in suppressing the continuum background.

4.3.4. Fox-Wolfram Moments

This variable can be used to assess if the tracks used to reconstruct a candidate B in the center of mass frame at e^+e^- colliders really come from a B decay. The vector sum of the transverse components of the momenta for the tracks not used to reconstruct the B candidate w.r.t. the B candidate direction is small for jetty (background) events, but there is no such correlation for true B decay events.

Fox-Wolfram moments (H_l) are defined by

$$H_l = \sum_{i,j} \frac{|p_i| \cdot |p_j|}{E_{vis}^2} P_l(cos\theta_{ij}) \tag{25}$$

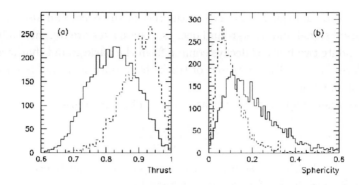

Figure 7. Examples of event shape variables for the whole event, (a) thrust, (b) sphericity (signal: solid line, light quark continuum: dotted line).Taken from the BaBar Physics Book

where the P_l are the Legendre polynomials, $p_{i,j}$ are particle momenta, θ_{ij} is the opening angle between the particles and E_{vis} is the total visible energy in the event. If particle masses are small enough to neglect, energy-momentum conservation gives $H_0 = 1$. Two jet events have $H_{odd} = 0$, and $H_{even} = 1$. The ratio H_2/H_0 of Fox-Wolfram moments is often used as a discriminator between B events and the background (it tends to zero for B events which are isotropic/spherical in the center of mass).

4.4. Rapidity

Rapidity (y) or more commonly pseodorapidity (η) is commonly used as a variable when working at hadron machines at high energies. It is a measure of longitudinal momentum.

$$y \approx \eta = \frac{1}{2}ln\left(\frac{E + p_L}{\sqrt{p_t^2 + m^2}}\right) \qquad (26)$$

In the center of mass (CM) system $\eta = 0$ corresponds to $p_L = 0$ and

$$\eta_{max} = \tfrac{1}{2}ln\left(\frac{S}{\sqrt{p_t^2 + m^2}}\right) \qquad (27)$$

$$\eta_{max} = -\eta_{min}$$

where S is the square of the CM energy and m is the mass of the particle of interest. One of the advantages of rapidity as a variable is that a change of reference shifts the overall rapidity distribution of all the particles produced by adding a fixed constant to every rapidity - that is, the shape of the distribution is unchanged. Other variables used in production physics studies include x_F and p_t^2.

4.5. Decay Constants

Mesonic decays to leptons (eg $D \to l\nu$) are the easiest to understand theoretically, as all of the strong interaction uncertainties can be parameterized in terms of a single "decay constant"(f_M). Since the weak interactions are essentially point-like at low energies, the decay constant is a measure of the probability that the two quarks in the meson have zero separation.

4.6. Spectators and Form Factors

One way to treat the decays of mesons containing a heavy quark is to assume that it is the heavy quark which decays and that the presence of the light quark can be ignored - that is that the light quark "spectates". For example, the decay, $D \to Ke\nu$, is termed "semi-leptonic" as the W emitted in the heavy quark decay itself decays to leptons.

Semi-leptonic decays depend only on a single diagram at tree level and this makes these decays theoretically tractable. The strong interaction effects at the quark-W vertex are described by variables termed "form factors". The number of form factors needed to fully describe the decay depends on the number of independent Lorentz variables needed to describe the decay. In the limit where lepton masses are neglected, only one form factor is needed to describe decays where a pseudoscalar meson decays to a pseudoscalar. This is termed the vector form factor. Three form factors $(A_1, V,$ and $A_2)$ are required to describe the decays of pseudoscalars to vector particles $(e.g. D \to K^* l\nu)$. [14]

4.7. Systematic errors

There is a lot more to a successful analysis than the size and quality of the real data set. As data sets grow in size and as the experimental error decreases it is becoming increasingly important to reduce systematic errors. There are many uncertainties associated with the use of Monte Carlo data to estimate backgrounds and efficiencies. These include concerns about how well the B decays are modeled. Are input parameters to the Monte Carlo such as B lifetimes and the amount of B mixing known well enough? Are

Δm_d, and the fraction of neutral to charged B pair production measured to sufficient accuracy? How well are the light quark continuum backgrounds modeled and understood? Are specific CP violating decays modeled well?

In addition, the larger and larger data sets allow for more and more precise measurements to be made, which introduces issues that could be neglected previously. It is now important to understand how well the machine backgrounds are modeled and how well they need to be modeled. One approach to this issue is to mix real background events with Monte Carloed "signal" events. One issue that experiments always need to face is the required level of sophistication of the simulation versus the time taken to generate events.

There are also uncertainties associated with the reconstruction of the events. What are the true vertex and drift chamber resolutions? How well understood are particle identification efficiencies and mis-identifications? What uncertainties in tagging performance are there? Are there any charge dependent tracking asymmetries?

It is becoming common to separate out different types of systematic errors and to explain how the results depend on the assumptions made. This is especially useful if the result depends on something which can change as a function of time as better measurements are made. Branching fraction measurements, for example, depend on the ratio of the number of neutral B meson pairs to charged B meson pairs produced at the $\Upsilon(4S)$.

5. Experiments and Machines

Experiments in flavor physics roughly divide into those at hadron machines and those at e^+e^- machines. Those at e^+e^- machines subdivide into those at symmetric colliders (both beams have the same energy) and those at asymmetric colliders (beams are of different energies). Experiments at hadron machines sub-divide into collider experiments and fixed target experiments.

The program of work for the last few years has concentrated on studies of rare kaon decays, ϵ'/ϵ and the B system. The large data sets available mean that the B system now represents a precision frontier for studies of weak decays and of CP violation. However, there is a need for studies to be done at lower energies, in particular of the charm system, to help understand and reduce theoretical uncertainties and allow full value to be extracted from the data.

5.1. e^+e^- Collider Experiments

Two "B Factories" have been built, at SLAC [15] and at KEK. [16] These machines are both two ring machines. The asymmetric colliders began operation in 1999. KEK-B has an 8.0 GeV electron beam which collides with a 3.5 GeV positron beam giving a $\beta\gamma = 0.42$. PEP-II has a 9.0 GeV electron beam and a positron beam with an energy of 3.1 GeV resulting in $\beta\gamma = 0.56$. The asymmetric machines running at the $\Upsilon(4S)$ have been very successful, so successful that the CESR collider, a symmetric machine that used to operate at the $\Upsilon(4S)$ is now running at lower energies as a charm factory.

5.1.1. CLEO

The CLEO experiment has been around for many years in its various incarnations. Until the advent of the B factories in 1999, it was the workhorse of B physics studies at the $\Upsilon(4S)$. Over the years it has been upgraded several times and is now being converted to CLEO-c, an experiment that will study charm decays. One of the most important motivations for CLEO-c is that the measurements made of decay constants and form factors will help constrain the theoretical inputs used in B physics. There are many theoretical calculations of decay constants based on Lattice QCD, potential models, and QCD sum rules. More details of the CLEO-c program can be found on the web, see Ref.15.

5.1.2. B Factory Detectors

BaBar and Belle, the currently operating B factory detectors have very similar designs. The BaBar detector is shown as an example in Fig. 8. Since much of the interesting flavor physics is being produced right now by B factory detectors lets look at the key components of these detectors.

The innermost detector (closest to the interaction point) has to be capable of tracking charged tracks with high resolution. This inner tracker, usually made of silicon, is used to separate B^0 and $\overline{B}^0$ vertices and to find tracks with low transverse momenta (less than $\sim p_t \approx 100 MeV/c$). Some dE/dx information may be provided by these detectors to help with particle identification.

The outer tracker covers a much larger volume and is usually a drift chamber. A low density, helium based gas mixture is often used to reduce multiple scattering and preserve the p_t of tracks passing through. This detector is primarily used to reconstruct the momentum of charged tracks but it also helps with particle identification by providing information on the energy lost by particles traversing it.

Next comes a system devoted to particle identification. The performance of this detector component is often key in determining how broad and how successful an experiment will be. The separation of pions and kaons at asymmetric machines needs to be done over a wide range of momenta - up to around $4.0 GeV/c$ for tracks from candidate $B \to \pi\pi$ decays. BaBar and Belle differ slightly in the technology choices they made for the particle identification system.

A Caesium Iodide Electromagnetic Calorimeter with high granularity is standard. This is used for π^0 reconstruction and must provide a high efficiency for detecting low energy photons. The system needs to have good energy and angular resolutions to help with π^0 and B^0 mass resolutions. This detector also helps tagging by providing lepton identification.

Leptons can be identified using the ratio of the measured energy (E) in a shower in a calorimeter to the measured momentum (p). This ratio, designated as E/p, is close to one for electrons which tend to deposit all of their energy and very small for muons which deposit little energy since they are minimum ionizing particles. The values of E/p for hadrons lie in between these two extremes, hadrons typically interact but don't lose all their energy. Leakage reduces the value of E/p for electrons but bremsstrahlung can increase it if the electron momentum is measured after the photon emission but the emitted photon(s) merges with the electron shower.

After the energy deposition clusters associated with charged tracks are eliminated, the remaining clusters can be studied to see if they are associated with neutral particles. Shower shape variables can be used to determine if single clusters are associated with a single photon or not. These variables give a qualitative description of the distribution of energy within a cluster that provides information on the type of incident particle. Electromagnetic showers due to an incident photon, tend to be regular. They are usually cylindrically symmetric, with an exponential fall off in the amount of energy deposited from the center of the shower. The showers associated with hadronic particles tend to be irregular, and often there is more than one cluster associated with the particle. The $\pi^0 \to \gamma\gamma$ decay mode can be reconstructed by taking pairs of clusters to form candidates and requiring the shape of showers used to be consistent with an electromagnetic origin. A complication is that the opening angle between the photons produced in the decay decreases as the π^0 energy increases. Thus, above $\sim 1.5 GeV$, the two photons from the decay may enter adjacent crystals, or only one crystal, forming "merged" candidates.

It is also usual to instrument the magnetic flux return. This provides π/μ separation and helps with K_L detection.

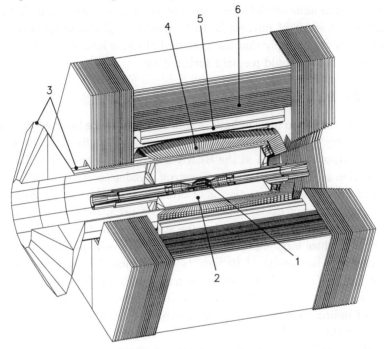

Figure 8. The BaBar Detector. 1.Silicon Vertex Tracker, 2.Drift Chamber, 3.Particle Identification Subsystem (DIRC - Detector of Internally Reflected Cerenkov Light, 4.Electromagnetic Calorimeter, 5.Magnet, 6.Instrumented Flux Return.

The physics program of both of these detectors centers around studies of CP violation in the B system (see section 6.3) and searches for rare decays. Ideally, one would like to measure all three angles of the unitarity triangle but measurements of the angle γ in particular, will require very large data sets and are more easily made at experiments using hadron beams.

5.2. Hadron Machine Experiments

5.2.1. RSVP

The planned RSVP program at the Brookhaven AGS is comprised of two experiments which will each search for a rare decay.[18] KOPIO is designed to discover and study the decay $K \to \pi^0 \nu \bar{\nu}$. A sample of these decays will allow them to probe CP violation in the kaon system with a new level of precision and explore the direct component of CP violation in the kaon system. MECO (Muon to Electron COnversion) is searching for evidence of muon lepton number violation with a planned sensitivity to enable them to detect one event of interest in 10^{17}. Evidence of lepton number non-conservation would imply that the Standard Model is incomplete since none of its gauge bosons could mediate such a decay.

5.2.2. CDF, D0

The B cross section in hadron-hadron collisions is a strong function of center of mass energy. In collisions of $\sim$900 GeV protons with a fixed target, which can be produced at HERA in the HERA-B experiment or at Fermilab, the cross section is of order 10 nb. At Tevatron collider energies, 2 TeV in the center of mass, the cross section is of order 100 μb. At LHC energies, 14 TeV in the center of mass, the cross section is predicted to be about 500 μb. Given the available luminosities, the strength of hadron colliders is the ability to produce large numbers of b hadrons. For example, the Tevatron, operating at a luminosity of $10^{32} cm^{-1} s^{-1}$, produces 10^{11} b pairs per year (about $1 fb^{-1}$ of data). Moreover, hadron machines simultaneously produce all species of B's: B_d, B_u, B_s, b baryons of all sorts, and B_c states. A good source of information about the current and proposed Tevatron B physics program is the web.[19]

CDF[20] and D0[21] are running collider experiments at the Tevatron with broad physics programs which include searching for the Higgs particle, searching for evidence of supersymmetry, and heavy flavor physics studies of the top and bottom quarks. Both experiments have made significant contributions to flavor physics during earlier runs at the Tevatron and are now running with upgraded detectors. The high rates of interactions mean that trigger design is an important part of achieving their physics goals. While these experiments will measure $\sin 2\beta$ they can also search for rare decays such as $B \to K \mu^+ \mu^-$ and study B_s mixing.

5.2.3. BTeV, LHC-B

BTeV and LHC-B will be dedicated B physics experiments that will make use of the Tevatron and LHC respectively. Both experiments are shaped like fixed-target experiments rather than central, high P_t experiments such as CDF,D0,ATLAS, and CMS. This geometry allows the reconstruction of high momentum b particles that are therefore well separated from the production vertex and suffer very little multiple coulomb scattering. The BTeV[22] experiment is shown in Fig. 9 and is currently awaiting P5 approval[23]. Dedicated experiments usually have a rate advantage over general purpose experiments when looking at specific decays and that is the case here. The number of $B \rightarrow K\mu^+\mu^-$ decays expected in a year of BTeV running is around 2500 compared to 50 at CDF. BTeV is capable of measuring values of x_s, the B_s mixing parameter of up to 70 in one year or running. While the layout of components is different to that at an e^+e^- collider detector, the basic functions that must be covered is the same. However, the need for an efficient trigger and the high rates impact the technology choices. The BTeV vertex detector uses planar pixel arrays for example and the particle ID system makes use of two different Cerenkov radiators to cover the full momentum range.

BTeV Detector Layout

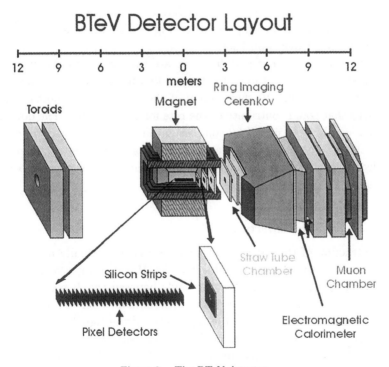

Figure 9.   The BTeV detector

6. Some Interesting Measurements

This section describes some of the key measurements to be made and their motivation.

6.1. *Rare decays*

The term "Rare Decays" usually refers to a decay in which the rate is suppressed relative to the dominant tree diagram. These decays can involve off diagonal CKM elements (Cabibbo suppressed), penguin diagrams (one loop diagrams where a W is emitted and absorbed) or box diagrams. Penguin diagrams can be gluonic (see Fig. 10(b)) or photonic (see Fig. 10(d)). Rare decays are interesting ways to probe for the existence of new physics that could significantly enhance their rates. They also provide a way to probe the values of CKM matrix elements. Recent attention has been focused on several rare charmless B decays such as $b \to s\gamma$ (an electromagnetic penguin), $B \to (\pi, \rho, \omega)l\nu$, $B \to J/\psi K_s$, $B \to K\pi$ (a gluonic penguin), and $B \to \pi\pi$ (Cabibbo suppressed, eg Fig. 10(a) which shows a $b \to u$ transition that depends on V_{ub}).

The decays $B \to K\pi$ and $B \to \pi\pi$ are two body and therefore have a clear kinematic signature consisting of two back to back tracks in the B rest frame. Figure 10 shows some of the potentially important decay diagrams that could contribute to the rate for these decays. The dominant diagram can be determined by comparing the rates of various rare decays. In particular, results on the branching fractions for $B \to K\pi$ and $B \to \pi\pi$ decays now strongly support the importance of the gluonic penguin diagram.

6.1.1. *Signal Yield Extraction*

The extraction of the signal yield for rare decays is usually done by performing an unbinned maximum-likelihood (ML) fit using several variables such as ΔE, m_{se}, track dE/dx, the angle between the B meson momentum and the beam axis, and a Fisher discriminant. The Fisher discriminant is designed to maximize the separation of the signal and continuum backgrounds. A separate fit is performed for each charged topology. The likelihood of the event is parameterized by the sum of the probabilities for all relevant signal and background hypotheses. The relative weights are determined by maximizing the likelihood function (L) and the probability of a particular hypothesis the given by the product of the probability density

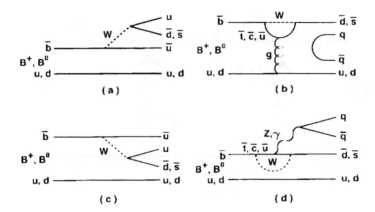

Figure 10. Charmless Rare Decay Diagrams. (a) External W emission, (b) Gluonic Pengiun, (c) Internal W emission, and (d) External Electroweak penguin.

functions (PDF) for the input variables. These are estimated using MC and independent data samples (limited statistics dominate the uncertainties).

6.2. B Meson Oscillation rates

Flavor mixing is mediated by box diagrams of the type shown in Fig. 11. The probability of a B mixing and decaying as a B of the other flavor is given by

$$r = \frac{\Gamma(B^0 \to \overline{B}^0)}{\Gamma(B^0 \to B^0) + \Gamma(B^0 \to \overline{B}^0)} = \frac{x_d^2 + y_d^2}{2 + 2x_d^2} \qquad (28)$$

where $x_d = \Delta m_d / \Gamma_d$, and $y_d = \Delta \Gamma_d / 2\Gamma_d \approx 0$ for B decays since the mixing is expected to be dominated by the mass difference (unlike the kaon case). x is given by

$$x = \frac{\Delta M}{\Gamma} \propto V_{tb}^2 V_{td}^2 f_B^2 B_B m_t^2 F(m_t x / m_w)^2 \qquad (29)$$

f_B is the related decay constant and B_B is the "bag" parameter which describes the amount that the box diagrams contribute to the mixing.

Time dependent measurements of B mixing require that the presence of a B meson be identified and its flavor tagged at production and decay. The decay time of the state must also be reconstructed. Experiments differ in the details of the measurements but not in these basics. Tagging has been discussed in some detail in section 4.2. The decaying B meson can

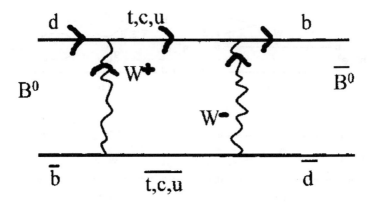

Figure 11. A box diagram showing how a B^0 becomes a $\bar{B}^0$

be identified by a high p_t lepton tag, by partial reconstruction, or by full reconstruction. The tagging of the meson flavor at production is harder - tagging the flavor of the other B may give the wrong answer if this B has already oscillated. Mistag probabilities due to physics and to experimental errors (misidentification of particles, mis-assignment of tracks to vertexes, etc.) have to be estimated and accounted for in the analysis. The B boost calculation uses information on the momentum of the tagging lepton and/or any other identified decay products. The resolution obtained on the decay time varies from around 0.3ps for inclusive lepton analyses at LEP to about 0.06ps for $D_s l$ events studies using the SLD detector.

B_d meson oscillations are well established. Figure 12 shows the results obtained by various experiments for measurements of the B_d mixing rate. However, the oscillation rate between neutral B_s states is predicted to be about 20 times faster than the oscillation rate between B_d states. Since B_s mesons are not produced at the $\Upsilon(4S)$ this measurement will likely be made at a hadron collider.

6.2.1. Possible Running at the $\Upsilon(5S)$

Occasionally there is discussion of the possibility of running B factories at the $\Upsilon(5S)$. The $\Upsilon(5S)$ mass is $10.568 \; MeV/c^2$ which is above the energy threshold for production of $B_s \bar{B}_s$ pairs. However, the cross section at the peak is quite small ($< 0.3nb$) and there are several possible final states for the $\Upsilon(5s)$ to decay into including $B_d \bar{B}_d$, $B^* \bar{B}$, $B^* \bar{B}^*$, $B_s^* \bar{B}_s$ and $B_s^* \bar{B}_s^*$ pairs. There is also a background from B pair decays accompanied by a

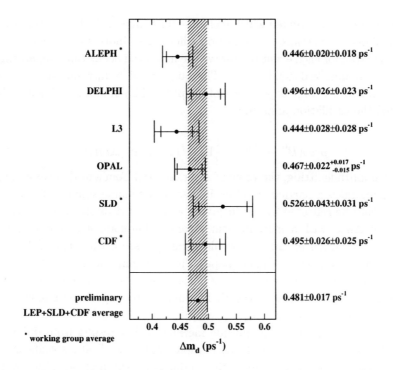

Figure 12. Experiment averages for Δm_d. LEP B Oscillations working group.

pion. Adding in the fact that phase space prefers decays to lighter modes the predicted $B_s \overline{B}_s$ cross section is likely to be $< 0.1nb$. The B_s mixing rate is also fast and for the boost values typical of the current asymmetric machines it will be difficult to improve upon current limits. The boost between the lab and the cm frames could be increased to compensate for the increased oscillation rate. However, changing the boost will complicate the switching of operations between the $\Upsilon(4S)$ and $\Upsilon(5S)$ and could require significant alterations to a detector. All these reasons make it unlikely that B_s mixing rates will be measured at an e^+e^- machine.

6.2.2. Amplitude Fitting

Experiments looking for evidence of B_s mixing can use amplitude fitting[24] to place limits on the oscillation rate. Time dependent mixing generates a periodic signal which is well suited to Fourier analysis. A log likelihood distribution is constructed which includes effects due to the finite resolution of the detector, mistags, and selection efficiencies. Individual probabilities are evaluated for each event to observe the measured decay time, separating mixed and unmixed candidates. The quantity A is introduced into the expression for the time dependence of the probability of mixing. A is usually termed the oscillation amplitude.

$$prob(B^0 \to \overline{B^0}) = \frac{1}{2}\Gamma e^{-\Gamma t}(1 - A\cos(\Delta mt)) \tag{30}$$

In amplitude fitting, the value of A is extracted from a fit done assuming that Δm_s takes a particular value. If there is a genuine signal in the data, this should be reflected by the fit giving a value of A of one at the correct value of Δm_s, within measurement errors. Therefore, if a plot is made of A as a function of Δm one would expect to see a variation of A with Δm which has a Breit-Wigner shape. The maximum of the Breit-Wigner will occur at the exact value of the mass difference. The full width will correspond to the inverse of the B_s lifetime. So far experiments have only set limits on the value of Δm_s.

One advantage of the amplitude method is that results from different experiments for the value of the oscillation amplitude, A, at a specific Δm_s can be easily combined. Figure 13, produced by the LEP working group on B oscillations, shows an example of such a fit, setting a limit on Δm_s of around $14ps^{-1}$.

6.2.3. Canceling uncertainties by taking ratios

An important reason to measure both B_d and B_s mixing is that by taking ratios of these measurements a lot of the theoretical uncertainties can be eliminated or at least minimized, for example,

$$\frac{\Delta m_s}{\Delta m_d} = \left[\frac{B_s}{B_d}\right]\left[\frac{f_{B_s}}{f_{B_d}}\right]\left[\frac{m_{B_s}}{m_{B_d}}\right]\left|\frac{V_{ts}}{V_{td}}\right|^2 \tag{31}$$

6.3. Measuring a CP asymmetry

In 2001, evidence was presented for the existence of CP violation in B decays by each of the two e^+e^- b-factory experiments, BaBar [26] at PEP-II and Belle [27] at KEK-B. The values measured were

$$\begin{aligned}
\sin(2\beta) &= 0.59 \pm 0.14(\text{stat}) \pm 0.05(\text{sys}) \text{ [BaBar]} \\
\sin(2\beta) &= 0.99 \pm 0.14(\text{stat}) \pm 0.06(\text{sys}) \text{ [Belle]}
\end{aligned} \tag{32}$$

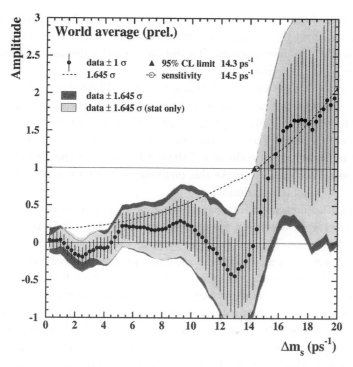

Figure 13. Amplitude fit - World Data, Produced by LEP working group on B Oscillations.

This is an important enough measurement to consider in some detail. At the $\Upsilon(4S)$ the $B_d \overline{B}_d$ pair is produced in a p-state and will remain in this coherent state (even though the neutral mesons can mix) until one of the mesons decays. The time evolution of a CP asymmetry at the $\Upsilon(4S)$ due to mixing is governed by the time difference $t_2 - t_1$, t_2 is proper time at which one B decays into a CP eigenstate and t_1 is proper time at which other B decays into a tagging mode. The double differential decay rate is given by

$$\frac{dN}{dt_1 dt_2} \propto e^{-(t_1 - t_2)/\tau}[1 \pm 2\sin(2\theta)\sin(\Delta m(t_2 - t_1))] \qquad (33)$$

where Δm corresponds to $m_{B_H} - m_{B_L}$ and θ is the relevant unitarity triangle angle. The sign in front of the oscillatory term depends on whether the flavor tag corresponds to a B or $\overline{B}$. As equation (33) makes clear there is no asymmetry if the measurement is integrated over time. This means that one must know the time ordering of the decays to be able to study CP violation through mixing. This is not possible if one is running at a symmetric e^+e^- collider since the B's will only travel a distance of around 26μ in the $\Upsilon(4S)$ rest-frame.

In principle, a time integrated measurement could be made by shifting to slightly higher machine energies and running at the threshold for BB^* production, where the initial system starts out in the opposite CP state ($C = +1$ in this case). However, measurements of the cross section at this energy [25] indicate that it is about 1/7 that of $B\overline{B}$ pairs at the $\Upsilon(4S)$. This cross section is too small to make this measurement practical at present or planned e^+e^- machine luminosities.

Instead, the measurements have been made at asymmetric colliders. The asymmetry in the beam energies means that the $\Upsilon(4S)$ rest frame (center of mass frame) is no longer the same as the laboratory frame. The relativistic boost between the frames produces an average decay separation between the B's of about 250 microns in the laboratory frame at SLAC which makes it possible to study the time ordering of the decays.

To determine an asymmetry we need to fit the time difference, $\Delta t = t_{cp} - t_{tag}$, in the decay $\Upsilon(4s) \to B\overline{B}$, where one B decays into a particular CP eigenstate, such as $B \to J/\psi K_s$, and the decay of the other allows a determination of the flavor of the b-quark it contains, that is, it decays into a tagging mode.

B factory experiments actually measure the difference in the positions in the lab frame of the two decays

$$\Delta z = z_{cp} - z_{tag} = \gamma\beta c\Delta t \approx 0.56 \times 468\mu m \tag{34}$$

This expression is exact for decays along the z-axis. However, there is a small correction for other decay axes. This can usually be ignored due to the boost which makes this correction small even for decays which are perpendicular to this axis.

If measurement errors are neglected, the Δz distribution has the form

$$\Delta z = Ne^{-\Gamma_z|\Delta z|}(1 + a_f(\Delta z)) \tag{35}$$

where

$$a_f(\Delta z) = C_f cos(\Delta m\Delta z) + S_f \sin(\Delta m\Delta z) \tag{36}$$

The decay $B^0 \to J/\psi K_s$ is often called the "gold-plated mode" for studies of CP violation in the B system. This decay has $S_f = \sin 2\beta$, and $C_f = 0$. It has a relatively large branching fraction ($\sim 5 \times 10^{-4}$), can be reconstructed with high efficiency and with only a small background, and the results can be interpreted with negligible theoretical uncertainty. Reconstruction of the $J/\psi \to e^+e^-, \mu^+\mu^-$ decays is achieved by looking for two tracks consistent with coming from a common vertex and with at least one track identified as an electron or muon. The invariant mass is calculated (the error on this is dominated by the detector resolution). The invariant mass calculation is complicated for electron channel by bremsstrahlung, especially in the beam-pipe and support tubes, which gives a long tail towards lower invariant masses. In principle, information from the calorimeter may be used to correct for this.

6.3.1. Some Experimental Complications

How accurately can an angle such as $\sin 2\beta$ be extracted from the data? In practice, these measurements will not be made with perfect detectors. Clearly, the resolution on the spatial separation of the decays will help to determine how precise an answer is obtained. In addition to the statistical error associated with finite event samples there will be errors associated with the ability to separate the CP eigenstate decay from its background (needed to a level of $\approx 10^{-5}$ at least) and the ability to correctly flavor tag the other decay.

The vertex of the B which decays to the CP eigenstate is usually well defined since there is a fully reconstructed B decay to work with. This means that tracks which are known to come from the decay vertex can be used to establish its position. As an example, the J/ψ decay vertex can be used for $B \to J/\psi K_s$ decays.

It's harder to get an accurate location for the position of the tagging decay's vertex. First, one may not have fully reconstructed the decay of this B and so it may be unclear which tracks belong to which part of its complete decay chain. One option is to use all the remaining tracks in estimating the B decay vertex. If this approach is taken then one will almost certainly include daughter tracks from long lived intermediate particles, even after tracks clearly coming from secondary decays (for example from K_s decays) are removed. In practice, a variety of methods are used - depending on the type of flavor tag. If a lepton tag is used, then the impact parameter of this lepton, that is the point on the leptons tag closest to the interaction point can be taken as an estimator of the decay vertex. If other flavor tags are used then the method depends on how many tracks remain after track quality cuts are made. If there is a single track, an impact parameter approach can again be taken. If there are more tracks then vertex fits can be made - tracks can be added and removed from these fits and comparisons made of the goodness of the fit.

The uncertainty in the measurement of the separation of the decays, σ_z, will contribute to the error in the determination of the asymmetry. How big the effect is depends on the size of the asymmetry - the error is smaller for larger asymmetry values. Table 2, which is taken from the BaBar Physics Book [13] shows the dependence of the single event asymmetry error σ_0, for A=0.7, $\Delta m/\Gamma = 0.75$.

Table 2.   Dependence of errors on spatial resolution

$\sigma_z/\gamma\beta c\tau$	0.	.25	.50	.75	1.0	2.0
$\sigma_0(\sigma_z)$	1.36	1.42	1.59	1.85	2.17	3.66

The best tagging strategies optimally balance the need for a high tagging efficiency against the need for high tag purity so that the number of "wrong-sign" tags is small. If one gets the flavor of the tagging B incorrect then the asymmetry will be reduced.

The measured asymmetry is related to the true asymmetry by a "dilution" factor (D)

$$a_{obs}(t) = D \cdot a_{true}(t) \tag{37}$$

where

$$D = (1 - 2w) \tag{38}$$

where w is the probability of a decay being mis-tagged. The dilution of the asymmetry effect due to the finite resolution of the vertexing is not included in this correction since it is considered separately. The tagging separation $\langle s^2 \rangle$, which measures the tagging purity is given by

$$\langle s^2 \rangle = (1 - 2w)^2 \tag{39}$$

if the cuts being used to decide if an event has been tagged are fixed, and by

$$\langle s^2 \rangle = \left\langle \left(\frac{p_R - p_D}{p_R + p_D} \right)^2 \right\rangle = \langle (1 - 2p_D)^2 \rangle = \langle (1 - 2p_r)^2 \rangle \tag{40}$$

if instead events are assigned a probability that they have been tagged correctly.

The statistical uncertainty in the measured asymmetry for events tagged in a given category (c) is given by

$$\sigma \propto \frac{1}{\sqrt{\epsilon_{tag,c} \langle s_c^2 \rangle}} \tag{41}$$

Every CP asymmetry measurement is dependent on the tagging algorithms used. This means that it is better not to rely on Monte Carlo data to estimate the right and wrong sign tag probabilities - or at least that it is important to be able to estimate the error associated with using the Monte Carlo to do this. Luckily, real data can be used to check that the gross features of Monte Carlo data distributions, such as the track multiplicity, and momenta spectra are correct. It is also possible to cross check the tagging algorithm using a sample of independently tagged decays (events where one B^0 is fully reconstructed). The tagging algorithm can then be applied to the rest of the event and its result compared to the expected result. This tagged sample can also be used to extract distributions for discriminating variables, and to develop a training set for neural networks.

7. Advantages/Disadvantages

There are many reasons for experimenters to choose to run at e^+e^- machines if they are want to make precision measurements. These machines provide a clean environment with good signal to noise ratios for the decays of interest. This is especially true of machines operating at the $\Upsilon(4S)$ where the kinematics of B production allows the use of many kinematical constraints. It is also possible to study decay modes involving neutral particles - which will be very difficult (if not impossible) to do at hadron colliders. The triggers are straightforward compared to those at hadron machines (rates are typically in the range of 100 Hz). The decays are easy to identify with high reconstruction efficiencies.

Studies of CP violation through B_d mixing are best made at asymmetric machines running at the $\Upsilon(4S)$. However, not all measurements are best made at such machines. It is easier to suppress continuum backgrounds at asymmetric machines since a separated vertex cut can be used to enhance the fraction of $B_d\overline{B}_d$ events. Conversely, some measurements are complicated by the fact tracks are concentrated into a forward cone and the overall detector acceptances are slightly lower.

The main disadvantage to working at an e^+e^- machine is that the cross sections are much lower (nb compared to μb at a hadron machine). Another complication is that studies at the $\Upsilon(4S)$ are restricted to studies of B_d mesons. Experiments at hadron machines are needed to add studies of B_s and baryons containing b-quarks. Cross sections for the rare processes of interest are much higher. The triggers are more complicated but rate can be traded for cleanliness.

Table 3, which is taken from the report of the NSF special emphasis panel on B physics [28] compares the ability of electron-positron colliders and hadron machines to make some key measurements. This table makes clear that a complete set of measurements on the B system will require a variety of techniques to be used.

Table 3. Strengths and weaknesses of machines for important physics topics. In the table, a $+$ indicates belief that significant measurements can be made; $-$ indicates that they cannot; and ? indicates that the capabilities are uncertain. This table is adapted from the 1998 Report of the NSF Special Emphasis Panel on B Physics.

topic	Asymmetric e^+e^- at the $\Upsilon(4S)$	hadron collider
$\sin 2\beta$	+	+
α	+	+
Direct CP violation	+	+
γ	+	+
x_s	−	+
Absolute branching fractions B_d	+	−
Absolute branching fractions B_s	?	?
General properties of B_s decays	?	+
B_c physics	−	+
b-baryon physics	−	+
Rare exclusive $B_{u,d}$ decays with γ's	+	?
Rare exclusive $B_{u,d}$ decays with π^0's	+	?
Rare exclusive $B_{u,d}$ decays with l^+l^-	+	+
Rare inclusive $B_{u,d}$ decays with γ's	+	?
Rare inclusive $B_{u,d}$ decays with π^0's	+	−
Rare inclusive $B_{u,d}$ decays with l^+l^-	+	?
Very rare exclusive $B_{u,d}$ decays	−	?
Rare exclusive B_s decays with l^+l^-	−	+
Semileptonic decays ($B_{u,d} \to c$)	+	+
Semileptonic decays ($B_{u,d} \to u$)	+	?
Semileptonic decays ($B_s \to c$)	−	+
Semileptonic decays ($B_s \to u$)	−	?
Leptonic decays of $B_{u,d}$	+	−
Leptonic decays of D and D_s	+	−

8. Personal Comments

Particle Physics is expensive. Priorities have to be set and it is unlikely that every experiment that people want to do can be done. So how do we balance conflicting needs? In a frontier science like particle physics it is impossible to predict where the real breakthroughs will come from so pursuing a variety of activities is a sound strategy. Not all measurements need or can be done with infinite precision. Some experiments allow us to benefit from large amounts of infrastructure already provided, while some take us into new areas. Some experiments will provide answers earlier than we might get them using other methods. Some experiments duplicate others. We need to be careful to balance commitments to ensure future opportunities with keeping the current program that will train the next generation healthy. Flavor physics is important and funding for upcoming opportunities must be part of a sensible planning process.

9. Conclusions

Much of what we now know about CKM matrix elements, B mixing, and rare decays has come from experiments at e^+e^- machines, but there is still much to learn and it may not all be learned at e^+e^- machines. In fact, it is likely that experiments at hadron machines, with their ability to look at a wide range of B species, will become increasingly important in the future. If all goes well, we can expect that experiments will provide us with much more information over the next few years. Will CP violation measurements agree perfectly with Standard Model predictions? Will rare decay rates always be consistent with expectations? We can hope not!

Acknowledgments

I benefited from many sources of information in preparing these lectures. I appreciate the helpful comments of my colleagues at Colorado, Kevin Stenson and Eric Zimmerman. I have had many interesting and pleasant discussions with Marina Artuso, an experimentalist co-convenor of the Snowmass 2001 Flavor Physics group and with Boris Kayser who has shed much light on many theoretical issues. I'd like to thank the HUGS students for their willingness to allow me to experiment with a more interactive style of lecturing and for the enthusiasm they showed. This work was supported by funding from the Department of Energy.

References

1. There are many books which give a detailed overview of the Standard Model. A standard text is "Introduction to High Energy Physics" by Donald H. Perkins, published by Addison Wesley.
2. Most conference proceedings are now at least partially available on the web. Good starting points are the proceedings of the Snowmass 2001 meeting

(http://www.slac.stanford.edu/econf/C010630/) and the most recent Division of Particles and Fields (DPF)meeting (http://www.dpf2002.org).

3. M. Kobayashi and T. Maskawa, Prog. Theor. Phys. 49 (1973) 652.

4. N. Cabibbo, Phys. Rev. Lett. 10 (1963) 531.

5. The report of the P2 working group can be found at the Snowmass 2001 proceedings web site (http://www.slac.stanford.edu/econf/C010630/).

6. L. Wolfenstein, Phys. Rev. Lett. 51 (1983) 1945.

7. A good starting point for learning about supersymmetry in more detail is http://www.slac.stanford.edu/gen/meeting/ssi/2001/

8. CLEO Collaboration, B. Barish et al, Phys.Rev.Lett.76(1570)1996.

9. The web site http://cs.fet.cvut/cz/ xobitko/ga/ is a useful starting point for genetic algorithm studies.

10. F. LE Diberder, et.al "Treatment of Weighted Events in a Likelihood Analysis of B_s Oscillations or CP Violation." BaBar Note 132.

11. R. A. Fisher, Annals of Eugenics, **7**,179(1936). Also see "An Introduction to Applied Multivariate Statistics",M.S.Srivastava and E. M. Carter. North Holland, Amsterdam(1983).

12. If you want more information Scientific American is a good resource. A basic description of the field can be found in "An introduction to neural computing" by Igor Aleksander and Helen Morton. Chapman and Hall 1990.

13. The BaBar Physics Book, Physics at an Asymmetric B Factory, P.F.Harrison, H.R.Quinn, editors, SLAC-R-504

14. See, for example,"Semileptonic B decays" by Sheldon Stone in B Decays, World Scientific, Revised 2nd edition (1994).

15. BaBar Collaboration, D. Boutigny et al, SLAC-R-0497. 1995.

16. BELLE Collaboration, M.T. Cheng et al, BELLE-TDR-3-95.

17. www.lns.cornell.edu/public/CLEO/spoke/CLEOc

18. For information on RSVP see www.bnl.gov/rsvp

19. For the report of a recent workshop on B-physics at the Tevatron see http://www-theory.lbl.gov/Brun2/report

20. For information on CDF see www-cdf.fnal.gov

21. For information on D0 see www-d0.fnal.gov

22. For information on BTeV see www-BTeV.fnal.gov

23. The P5 process is intended to determine the relative funding priorities for experiments based at different laboratories and with different goals.

24. Mathematical Methods for $B\bar{B}$ Oscillation Analyses, H.G.Moser and A.Roussarie, OPEN-99-030

25. CLEO Collaboration, D.S. Akerib, Phys.Rev.Lett.67(1692)1991.

26. B. Aubert et al. hep-ex/0201020.

27. K. Abe et al. (BELLE), Phys. Rev. Lett. 87 (2001) 091802.

28. Report of the NSF Elementary Particle Physics Special Emphasis Panel on B Physics, Robert Cahn, chair (July 1998)

ASPECTS OF QCD [a]

A. P. SZCZEPANIAK
Physics Department and the Nuclear Theory Center,
Indiana University, Bloomington IN 47405

Some aspects of QCD, the underlying theory of nuclear interactions are discussed. Emphasis is put on simple and intuitive explanation of various phenomena.

[a] Lectures presented at the 2001 Hampton University Graduate Studies (HUGS) summer school, Jefferson Lab.

1 Introduction

Nuclear physics has been developed for well over half a century and spans a large research area. It covers phenomena separated by over 10^{20} orders of magnitude in distance scales. For example, in its domain are processes occurring at sub-nuclear scales governing the structure of protons and neutrons, the building blocks of atomic nuclei, as well as processes at astronomical scales, for example related to the structure of neutron stars and stellar evolution. These seemingly unrelated systems follow the same fundamental laws of quantum mechanics with dynamics evolution determined by the same fundamental interactions, the nuclear forces.

The strong nuclear force which is responsible for binding nucleons in atomic nuclei, the weak nuclear force which is responsible for example for the neutron β decay and the electromagnetic interactions determining the chemistry of elements form the Standard Model which unifies all known matter and all know interactions except for gravity.

At the Jefferson Lab essentially all aspects of nuclear physics are being studied, even the ones with applications to nuclear astrophysics. For example precision measurements of neutron skin distribution in heavy nuclei will help better understand the dynamics of neutron stars.

In these lectures, I will focus on some aspects of nuclear physics which have to do with the underlying quark and gluon structure of "small" systems such as the proton, neutron, or the pion, *i.e.* elementary hadrons. Instead of giving a detailed description of a particular process or a particular theoretical approach I will discuss a few examples with the goal of trying to present a simple and intuitive description of the underlying mechanisms.

I hope that after these lectures some of the concepts that you often hear about in seminars or find in research papers will become less obscure and that you will be able to see connections between various phenomena and techniques.

2 The Hadronic Zoo

After over forty years we know that the fundamental degrees of freedom participating in strong nuclear interactions are quarks and gluons. The concept of quarks was introduced in the mid 1960's Gell-Mann and independently by Zweig to systematize the multitude of known "elementary" hadrons *i.e.* particles that interact strongly [1]. The nature of particle interactions is reflected, for example in the life-times of their excitations, which for strong decays are of the order of $10^{-23}s$ or in the sizes of scattering cross-sections, which for strong interactions are typically of the order of 10^{-3} barn [2]. Hadrons can be classified, just like any other "elementary" particles by they internal quantum numbers such as spin, or intrinsic parity. Spin not only describes how a particle behaves under spatial rotations but it also determines the type of quantum statistics a particle obeys; integer spin particles are bosons and in the hadronic world are referred to as mesons and half-integer spin, fermions are known as baryons [3]. By mid sixties there were approximately 100 baryons and mesons known.

In a search for an underlying theory the first thing is to identify the underlying symmetries. These can be manifested through degeneracies in the spectrum or through similarities of certain reactions, $i.e.$ scattering amplitudes or decay patterns [4].

To illustrate how symmetries work consider a nonrelativistic spinless particle bound by a three-dimensional spherically symmetric harmonic oscillator potential. The energy levels are given by (here we work with a natural system of units with $\hbar = c = 1$),

$$E_{nl} = \omega \left(2n + l + \frac{3}{2} \right), \tag{1}$$

with n and l labeling the radial and orbital excitations respectively. The energy does not depend on the magnetic quantum number, m which together with n, and l is needed to completely determine an eigenstate. Thus a given orbital l-state is $2l+1$ degenerated (the possible values of m are, $m = -l, -l+1, \cdots, l-1, l$) $i.e.$ for given l all $2l + 1$ states have the same energy. This is because starting from a state specified by a set of quantum numbers, (n, l, m) the other m-substates, $(n, l, -l), (n, l, -l + 1), \cdots, (n, l, l)$ may be obtained by rotating the system of coordinates which, for rotationally inveriant interacton does not affect the energy. Note however, that states with different angular momentum l do not have the same energy. This is because angular momentum is conserved $i.e.$ a 3-dimensional rotation cannot change this eigenvalue. Therefore states with different $l's$ are not related to each other by a rotation, $e.g.$ the p-orbitals have different shapes then the s-orbital and there is no a $priori$ reason for them to have the same energy.

All this can be simply formalized if instead of a finite rotation one considers infinitesimal rotations. This is not a limitation since any finite rotation can be expressed as a series of infinitesimal rotations. For example a rotation around $\hat{n} = \mathbf{n}/|\mathbf{n}|$, by an angle $\alpha = |\mathbf{n}|$ can be obtained by $N \to \infty$ successive rotations around $\mathbf{n}/N$,

$$R(\hat{n}) = \lim_{N \to \infty} \left[1 - i \frac{\mathbf{n}}{N} \cdot \mathbf{L} \right]^N = e^{-i\mathbf{n} \cdot \mathbf{L}}. \tag{2}$$

Here $-i\mathbf{L}$ defines an operation which generates infinitesimal rotation and the factor of $-i$ is introduced by convenience. For R to be a three-dimensional rotation, *i.e.* to satisfy, $R^{-1} = R^T$ the operators **L** have to satisfy the canonical commutation relations of the angular momentum,

$$[L_i, L_j] = i\epsilon_{ijk}L_k. \tag{3}$$

Invariance of the Hamiltonian under rotations means that, $RHR^{-1} = H$ or $[H, \mathbf{L}] = 0$. Similarly invariance of the angular momentum under rotations implies that $[|\mathbf{L}|, L_i] = 0$. The type of degeneracy, *e.g* the fact that for given l there are exactly $2l + 1$ degenerate levels is not a property of the Hamiltonian (other then the symmetry requirement) but it follows from the type of symmetry operation itself. Therefore degeneracy patterns can be abstracted from the underlying dynamics and may be used to determine the symmetry group and the underlying commutation relations of generators of symmetry transformations, (*c.f.* Eq. (3)). The number of states which can be related by symmetry (in this case $2l + 1$) specifies the *representation* of the symmetry transformation.

Strong interactions are naturally expected to be invariant under Lorentz transformations of the coordinate system which include three-dimensional rotations and Lorentz-boosts. Obviously, proton in a free space has the same mass regardless of its polarization, and this s not affected by Lorentz boosts. But inspecting the spectrum of light hadrons, such as p, n, Δ, Σ *etc.* one finds that there must be another, *internal* symmetry relating these seemingly different particles. For example masses of the proton and neutron are almost the same and are not much different from masses of the Λ, Σ and the Ξ particles [2].

Gell-Mann has argued that these similarities can be explained if one assumes that the underlying degrees of freedom of the strong interactions carry a specific intrinsic charge, now refereed to as *flavor* and that the underlying dynamics invariant (symmetric) under rotations in the internal space spanned by this flavor charge. The dimension of this space is simply given by number of flavors. In order to account for symmetries in the light hadron spectrum with masses $m < 2\mathrm{GeV}$, this flavor quantum number should take one of the 3-values, $f = u, d, s$. Now we know that these are just the three lightest of the known six flavors. The observed patterns in light hadron spectroscopy can be explained by assuming that the underlying theory is invariant under unitary transformations – rotations in this 3-dimensional, internal, flavor space. A set of unitary, unimodular transformations of a complex 3-dimensional space is referred to $SU(3)$.

Gell-Mann's scheme introduces a new degree of freedom which he called *quark* which at this stage is only needed to introduce the flavor quantum number: one flavor – one quark. A simple calculation based on the properties of the $SU(3)$ group shows that by combining flavor quantum numbers of three quarks ($3 \times 3 \times 3 = 27$ possibilities) it is possible to explain the observed patter of masses in baryon spectrum and similarly by combining flavors of a quark and antiquark (3×3 possibilities) one can explain the nonet-like pattern of mesons masses[5]. These follow from the rules of combining representations of the $SU(3)$ group, analogous to the Clebsh-Gordon decomposition in the case of angular momentum (*i.e.* the $SU(2)$ group). All hadrons in a given multiplet have the same masses (*i.e* analogous the $2l+1$, m-substates of the harmonic oscillator example discussed above) but in general have different electric charge. This implies that the flavor quantum number and the electric charge have to be connected. In particular one of the predictions of the $SU(3)$-flavor model is that quarks have fractional, ($2/3e, 1/3e$) electric charge. This made searches for free quarks even more interesting, but no positive results were obtained. Nowadays we know that the underlying dynamics prevents quarks from becoming isolated, free particles.

The idea that a hadron can be described as a collection of a small number of valence quarks quickly developed into a sophisticated, phenomenological model commonly referred to as the constituent quark model (CQM)[6]. Similar to the shell model in nuclear physics, in which nuclear properties are described by quasi-particle proton and neutrons excitations in a phenomenological potential, in CQM the interactions between quarks were postulated in terms of effective (constituent) quark degrees of freedom: relative positions, spins and flavors. This lead to a variety of effective Hamiltonians, in general quite successful in describing hadron spectrum and many other static properties of hadrons, *e.g* charge distributions, magnetic moments *etc.*. The success of the quark model, which as we know now, is a rather simplified description of hadronic structure means that there may indeed be a deeper connection between this effective, quasi-particle like-picture and the underlying theory of strong interaction – Quantum Chromodynamics, (QCD).

Soon after, the $SU(3)$-flavor classification, in the early 1970's a series of SLAC experiments shed new light on hadronic structure. In those experiments the electrons were used to produce *virtual* photons which would then inelastically scatter of a proton target, *i.e* the reaction $ep \rightarrow eX$[7]. Here X stands for any particle created from the hadronic target after the inelastic collision, with its properties (mass, spin, momentum) not being measured. Thus only the energy and the scattering angle of the final electron was measured allowing to *include* all possible final states. This type of processes is therefore referred to as an inclusive reaction as opposed to an exclusive reaction where an exclusive final state is selected *e.g* $ep \rightarrow ep$, or $ep \rightarrow ep\pi$ *etc.* The underlying reaction mechanisms is shown in Fig 1, and can be thought off as a second-order (in electromagnetic interaction) contribution to a time-order perturbative expression

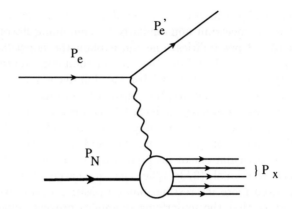

Figure 1: Leading order contribution to deep inelastic $ep \to eX$ scattering amplitude.

of the scattering amplitude.

Given initial electron energy E, the scattering amplitude (and the cross-section) for this process depends only on the final electron energy, E' and scattering angle, θ, These can also be related to Lorentz invariant quantities, $Q^2 = -q^2$ the "mass" of the virtual photon, and the so called Bjorken scaling variable, x_{BJ}, defined by,

$$Q^2 = -(p'_e - p_e)^2, \quad x_{BJ} = \frac{Q^2}{2p_N \cdot q} = \frac{Q^2}{2M\nu}. \tag{4}$$

Here, $\nu = E - E'$ is the energy transferred to the proton in the rest frame of the nucleon. Alternatively one can replace one of the two kinematical variables (Q^2 or x_{BJ}) by W – the invariant mass of the produced hadronic system, X,

$$W^2 = p_X^2 = (p_N + q)^2 = m_N^2 + 2p_N \cdot q + q^2 = m_N^2 + Q^2 \frac{(1 - x_{BJ})}{x_{BJ}}. \tag{5}$$

The allowed kinematical region for this reaction is $Q^2 > 0$ and $0 < x_{BJ} < 1$. Note that the limit $x_{BJ} \to 1$ corresponds to elastic scattering since in this case W becomes equal to the mass of the proton.

Now suppose that by transferring energy from electron to the proton, ($\nu = E - E' > 0$) it is possible to excite a proton resonance much like the hydrogen atom levels can be excited by electromagnetic radiation. This will produce a state of mass $m_X^2 = m_N^2 + \Delta m^2 > m_N^2$, corresponding to

$$x_{BJ} = \frac{Q^2}{Q^2 + \Delta m^2} \sim 1 - \frac{\Delta m^2}{Q^2}. \qquad (6)$$

Thus in the deep inelastic limit when $Q^2 >> m_N^2$, the contribution of proton's excited states move to the region $x_{BJ} \sim 1$. But since $Q^2 = -q^2 = \mathbf{q}^2 - q^{0^2}$ as Q^2 increases, photon probes proton's structure at shorter distances, $O(1/|\mathbf{q}|)$. Since the cross-section is proportional to the charge distribution integrated over a region in space "seen" by the photon, $\delta \mathbf{x} < 1/|\mathbf{q}|$, it is expected that for fixed excitation energy, the contribution to the cross-section from a single resonances will quickly decrease with increasing Q^2. Thus if no other processes were present one would expect the cross section to rapidly decrease with Q^2. The data looks quite different. As Q^2 increases, the resonance contribution indeed decreases in magnitude and get pushed to $x_{BJ} \to 1$ region but the net cross-section stays large and essentially Q^2 independent.

In principle one should be able to account for the residual cross-section by summing over background inelastic process $e.g.$ multiple pion production. However, since as explained above an individual exclusive channel is expected to give a small and strongly Q^2 dependent contribution to the cross section, a Q^2-independent result can only be obtained by adding contributions from a large number of channels. This simply means that the physical, hadronic degrees of freedom are not the $natural$ ones to describe deep inelastic scattering. And alternative description was suggested by Feynman[8] and is based on the assumption that the photon scatters off constituents of the proton. The existence of in principle two alternative descriptions, one bases on hadronic and the other on quark degrees of freedom is referred to as duality. From the Q^2-independence it follows that photons probe almost free and thus point-like constituents – otherwise one has the situation described above. As shown by Feynman, the quasi-free behavior of proton's constituents can be understood by analyzing DIS in a frame in which proton is moving with a large (infinite) momentum. The partons actually do interact with each other, however, in this frame this interaction time is much longer than time it takes for a parton to absorb the photon. Therefore the photon effectively scatters off $incoherently$ one of the partons at the time. For any finite Q^2 photon sees a quasi-free distribution of partons and as $Q^2 \to \infty$ the distribution approaches an asymptotic one. Since at finite but large Q^2 partons are quasi-free, one can predict the change in parton distribution as a function of Q^2 –photon's resolution, using perturbation theory. Since the change in parton distribution is associated with interactions which make partons radiate other partons it turns out this the distribution has the property of self-similarity $i.e$ the form of the distribution is the same regardless the resolution at which it is measured[9].

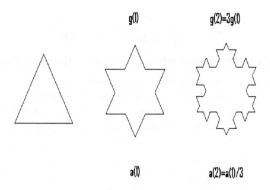

Figure 2: van Koch snowflake

A self-similarity is a common phenomena and it is at hart of any local field theory. It is related to such concepts as scale invariance and its anomalous breaking, renormalization, regularization, and in case of QCD – asymptotic freedom. It can be simply illustrated using fractal geometry as an example. Consider a fractal known as the van Koch snowflake. The "procedure" for drawing the snowflake is as follows, i) start with an equilateral triangle, ii) divide each side into three equal pieces, iii) remove the middle side and on top of it draw another equilateral triangle, iv) repeat. After as few as 3-4 steps the figure starts resembling a real snowflake, see Fig. 2!

Let $a(1)$ be the length of each side after firts iteration. After each iteration the step the number of triangles, $g(n)$ increases by three, and their size decrease by three,

$$g(n+1) = 3g(n), \quad a(n+1) = \frac{a(n)}{3}. \tag{7}$$

These can be combined into the *renormalization group equation*

$$a\frac{dg(a)}{da} = \gamma g(a). \tag{8}$$

with $\gamma = \log 3/\log(1/3)$ being the equivalent of what in field theory is referred to as the *anomalous dimension*. The snowflake is obviously self-similar – at any

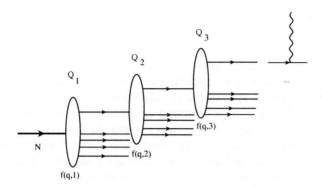

Figure 3: Q^2 dependence ("evolution") of parton distribution, $f(q, Q)$ as seen by a photon in DIS.

step, "zooming into the picture" revels the same basic triangle shapes. Because of this self-similarity one can "predict the properties" of the snowflake, *i.e.* its perimeter, area covered or number of triangles, as a function of resolution. If γ was equal to zero, the 'snowflake would not "evolve" – it would display the same structure *e.g* number of triangles regardless resolution. In the case of DIS this corresponds to $Q^2 \to \infty$ limit when proton no longer evolves with Q^2 and a photon sees an asymptotic distribution of partons (quarks) in the proton. The analogy between this simple fractal self-similarity and the structure of proton as measured in DIS is now straightforward. As illustrated in Fig. 3, as resolution of the photon increases (Q^2) the number of partons seen by the photon increases. This is because with increasing resolution photon "sees" more processes in which parent partons radiate daughter partons. The underlying interaction however is self-similar (scale invariant) *i.e.* all sub-process look alike. The number of partons is therefore analogous to the number of little triangle on the perimeter of the snowflake.

In practice there is only a finite number of steps one can take in drawing the snowflake, *e.g.* a finite size of the printer's head makes is it impossible to resolve sizes smaller then the corresponding printer's resolution. Thus there is a natural cutoff on $a(n)$, $a > a_{min}$ and $g(a)$. We can say that beyond this cutoff our "theory of the snowflake" becomes inadequate. This may be the case with fundamental interactions and the Standard Model for which, from dimensional analysis it follows that the natural cut-off is the Planck mass. Beyond this scale, (and quite possible several orders of magnitude below) we can expect to resolve what now is considered as a pointlike, elementary particle *e.g.* a quark, and discover limitations of the Standard Model. In the Standard Model, however, there is in principle no need for a "resolution" or "cutoff" scale. This is a property of a renormalizable field theory. In such case renormalization group relation like the one in Eq. (8) can in principle be tested to arbitrary high precisions.

From the DIS data one finds that only about 50% of proton's momentum is carried by quarks. In later experiments done by the EMC [10] collaboration it was also found that spin of the quarks does not add up to the total spin of the proton. This can be interpreted as evidence for a missing, electrically neutral component of the proton. Such component is needed for a consistent relativistic theory of parton interactions. Unlike in the nonrelativistic case the relativistic theory cannot have static potentials and interactions need to be introduced by exchanges of quanta carrying both momentum and energy and possibly other internal quantum numbers. In the case of strong interactions these quanta are referred to as gluons [11]. In order for gluons to play a dynamical role there should be a new charge that gluons would couple to, just like photon couples to the electric charge. The need for an additional, internal charge – degree of freedom besides flavored quarks is also apparent from the description based on the quark counting scheme of Gell-Mann [12]. Consider for example the Δ^{++} particle which in many respects behaves like the proton except it has spin $S = 3/2$. In the Gell-Mann classification it would have the following quark substructure,

$$|\Delta^{++}, S = 3/2, S_z = +3/2\rangle = \sum_{ijk} |u(+1/2, i)u(+1/2, j)u(+1/2, k)\rangle \Psi_{ijk}, \quad (9)$$

where i, j, k refer to other quantum numbers of an individual quark $e.g.$ momentum or position. If one makes the assumption that quarks in the low mass hadrons are all in an orbitally-symmetric ground state the wave function Ψ_{ijk} will be symmetric under exchange of the quark spatial quantum numbers. This however, would violate the Pauli principle and can be avoided if quarks carry another internal degree of freedom which can be used to differentiate between them, $e.g.$ for each flavor, $q(i) \rightarrow q(i, \alpha)$ with $\alpha = 1, 2, 3$. A completely anti-symmetric Δ^{++} state consistent with Pauli principle can then be given by

$$|\Delta^{++}\rangle = \sum_{\alpha\beta\gamma} \sum_{ijk} \epsilon_{\alpha\beta\gamma} |u(i, \alpha)u(j, \beta)u(k, \gamma)\rangle \Psi_{ijk}. \quad (10)$$

This extra degree of freedom is referred to as color. Once we have a new "charge" there will currents associated with this charge and therefore radiation. The quanta of this radiation can now be associated with gluons.

In summary if there are three colored quarks then there is a total of 8 colored gluons since, for example a red quark can radiate a red-anti-blue gluon and become a blue quark making all together $3 \times 3 = 9$ possibilities. However a combination of equal contribution of color and anti-color in a gluon makes them color-neutral and therefore not interacting. With gluons carrying color charge one expects the spectrum of strongly interacting particles to be much richer then that of a pure quark system. For example there could be bound states of pure gluons – glueballs, or hybrids containing both quark and gluons degrees of freedom. This is different from quantum electrodynamics, where i) photons are electrically neutral, and ii) the relative interactions determined by the magnitude of the electric charge is weak.

3 Strong interactions as a field theory: QCD

The proper way to describe a relativistic quantum mechanical system is to formulate it as a relativistic field theory [3]. The ingredients of a field theory are quantum fields (operators) which represent a collection of an ∞ number of degrees of freedom interacting via *local* interactions. If interactions were nonlocal obviously there would have to be a more fundamental mechanisms needed to explain the nature of the non-locality. In QCD quarks are represented by,

$$q \rightarrow \psi(\mathbf{x}, t, \alpha), \tag{11}$$

where α stands for all internal degrees of freedom, spin, flavor, and color. Similarly for the gluons we have,

$$g \rightarrow A(\mathbf{x}, t, \alpha). \tag{12}$$

At the classical level, ψ and A are c-number functions. In quantum theory they become operators acting on an (abstract) Hilbert space of physical states.

One then proceeds by writing a Hamiltonian with interactions between quark and gluons given by operators which are local products of ψ and A and satisfy underlying symmetry requirements. The disadvantage of the Hamiltonian formulation, however, is that it is not manifestly covariant, and it is much easier to start with a Lagrangian and the action,

$$S = \int dt L(t) = \int dt \int d\mathbf{x} \mathcal{L}(\mathbf{x}, t) = \int d^4 x \mathcal{L}(x^\mu). \tag{13}$$

From the Lagrangian one can calculate the Hamiltonian using the Legendre transformation. Insisting on Lorentz symmetry means we have to deal with a Lagrangian rather then Hamiltonian, and make sure the interactions a constructed in such a way that the *Lagrangian density*,

$$\mathcal{L}(\mathbf{x}, t) = \mathcal{L}(x^\mu), \tag{14}$$

is a Lorentz scalar. The QCD Lagrangian is given by [13]

$$\mathcal{L} = - \sum_{a=1\cdots 8} \frac{1}{4} F^{\mu\nu,\, a} F^a_{\mu\nu} + \sum_{ij=1,2,3, f=u,d,s,\cdots} \bar{\psi}_{i,f}(iD_{ij,\mu}\gamma^\mu - m_f\delta_{ij})\psi_{j,f}, \quad (15)$$

where

$$F^a_{\mu\nu} = \partial_\mu A^a_\nu - \partial_\nu A^a_\mu + g f_{abc} A^b_\mu A^c_\nu, \quad (16)$$

and

$$D_{ij,\mu} = \partial_\mu \delta_{ij} - i T^a_{ij} A^a_\mu. \quad (17)$$

Here a and ij refer to the gluon and quark colors respectively, $T^a_{ij} = \lambda^a_{ij}/2$ are the generators of the $SU(3)$-color group, and f_{abc} are the $SU(3)$ structure constants,

$$[T^a, T^b]_{ij} = i f^{abc} T^c_{ij}. \quad (18)$$

The 3×3, Gell-Mann matrices λ^a are the $SU(3)$ analog of the 2×2 Pauli matrices σ^i.

The spin content of the quark and gluon fields is determined by relativistic covariance. This implies for example that quark fields are Dirac spinors rather then Pauli spionrs and gluons are Lorentz vectors rather then 3-dimensional vectors. In notation of Eqs. (11) and (12) $\psi(\mathbf{x}, t, \alpha) \to \psi_{i,f}(x)$ and $A(\mathbf{x}, t, \alpha) \to A^a_\mu(x)$.

We have "reduced" the theory of strong interactions to the QCD Lagrangian. As it stands the theory depends on only a handful of parameters, the strong coupling constant, g and quark masses, one for each flavor, $m_f, f = u, d, s, c, b, t$, however as will discuss later this "naive" counting will have to be revisited.

It is quite amazing that such a simple looking Lagrangian is expected to describe all phenomena having to do with strongly interacting particles. But what exactly does it men to solve QCD? It is unrealistic to expect for example, that starting from Eq. (15) we would be able to describe, with arbitrary precision, structure and dynamics of complex nuclei. After all, the conventional description in terms of interactions between nonrelativistic protons and neutrons has proven to be quite successful. This is because, indeed it is protons and neutrons which make nucleons to look "natural". However, we would certainly like to be able to understand from first principles as many features of the QCD spectrum as possible. For we would like to put the ideas of the quark model and the Feynman parton model on a solid ground and find a connection between them, or find if there are novel phases of quark gluon systems and possible new forms of matter (quark stars ?).

In order to understand the physical content of QCD and what it means to "solve" a field theory, we will consider a simple theory of a scalar, noninteracting field in one space dimension, *i.e.* a quantum string. The corresponding Lagrangian is given by,

$$\mathcal{L} = \frac{1}{2}\partial^\mu\phi\partial_\nu\phi - \frac{1}{2}m^2\phi^2. \tag{19}$$

Here $\phi = \phi(x,t)$ plays the role of a continuum of dynamical variables, *i.e.* in each space point x there is a "displacement" $\phi(x,t)$. The momenta canonically conjugate to those "displacements" are defined as usual via,

$$\Pi(x,t) \equiv \frac{\partial\mathcal{L}}{\partial\dot\phi(x,t)}. \tag{20}$$

In terms of ψ and Π one can calculate the Hamiltonian from,

$$H(t) = \int dx \mathcal{H}(x,t) = \int dx \left[\Pi(x,t)\dot\phi(x,t) - \mathcal{L}\right], \tag{21}$$

and the the corresponding *Hamiltonian density*, $\mathcal{H}(x,t)$ is given by,

$$\mathcal{H}(x,t) = \frac{1}{2}\left[\Pi(x,t)^2 + (\nabla\phi(x,t))^2 + m^2\phi(x,t)^2\right]. \tag{22}$$

Since time dependence of any operator, $O(t)$ is given by $\dot O(t) = i[H, O(t)]$ it follows that H is time-independent and we can use $\phi(x) \equiv \phi(x, t = 0)$ and $\Pi(x) \equiv \Pi(x, t = 0)$ to define H. This Hamiltonian describes a collection of (an infinite number of) coupled harmonic oscillators. This becomes obvious if instead of a continuum of dynamical variables $\phi(x)$ one discretizes space by setting $x = ai, i = \cdots, -2, -1, 0, 1, 2, \cdots$ and then obtains a system, of still infinite but denumerable "displacements", $\phi_i = \phi(x_i)$ The corresponding, discretized Hamiltonian is given by,

$$H = \frac{a}{2}\sum_i \left[\Pi_i^2 + \frac{1}{a^2}(\phi_{i+1} - \phi_i)^2 + m^2\psi_i^2\right] = \sum_i \frac{p_i^2}{2\mu} + \frac{1}{2}\sum_{ij} q_i A_{ij} q_j, \tag{23}$$

where $\mu \equiv 1/a$, $\omega \equiv m$ and we have redefined the canonical variables introducing, $p_i \equiv \Pi_i$, and $q_i \equiv a\phi_i$. The matrix $A = A_{ij}$ is given by,

$$A_{ij} = (\mu\omega^2 + 2\mu^3)\delta_{ij} - \mu^3\left(\delta_{i,j+1} + \delta_{i+1,j}\right). \tag{24}$$

A set of coupled harmonic oscillators can be solved (at both classical and quantum levels) by finding the normal modes. This is equivalent to finding an orthogonal linear transformation

$$q_i \to \tilde{q}_i = S_{ij}q_j, \tag{25}$$

that diagonalizes A. The quantum nature of the Hamiltonian is reflected in the operator nature of p_i and q_i which are defined to satisfy canonical commutation relations,

$$[p_i, q_j] = -i\delta_{ij}. \tag{26}$$

In the continuum limit this becomes,

$$[\Pi_i, \phi_j] = \frac{1}{a}[p_i, q_j] = (-i)\left(\frac{1}{a}\delta_{ij}\right) \to -i\delta(x - y). \tag{27}$$

Since the string Hamiltonian is translational invariant one can immediately guess the transformation in Eq. (25) which leads to the normal modes – it is equivalent to a Fourier transform.

To make things a bit simpler we restrict the x-space to a finite interval (*i.e.* put a system in a box) $-L < x < L$, with $L = Na$ and impose periodic boundary conditions which assures that the system in the box continues being translationally invariant. Now we are dealing with a finite collections of "beads" representing vibrations of a closed string. One can easily diagonalize the Hamiltonian and find the energy spectrum to be given by,

$$E = \pi \sum_{q=-N}^{N} \omega_q(n_q + 1/2), \tag{28}$$

where

$$\omega_q = \sqrt{m^2 + \frac{2}{a^2}(1 - \cos(\frac{2\pi}{(2N+1)}q))}. \tag{29}$$

Here n_q is the number of normal modes with momentum q and each mode contributes energy ω_q to the spectrum. The zero point energy is given by

$$E_0 = \frac{\pi}{2} \sum_{q=-N}^{N} \omega_q \propto N, \tag{30}$$

and it grows linearly with the size of the system. Lets consider two limits, i) $m = 0$ and ii) $m >> 1/a$. The first case is relevant because of the, *chiral* symmetry which arises in QCD for massless quarks. The other case is relevant for constructing an *effective* approximation to QCD when dealing with heavy quarks.

In the first case the energy of a single mode is,

$$\omega_q = \frac{2}{a} |\sin \frac{\pi}{2N+1} q|, \tag{31}$$

and the zero point energy is given by,

$$E_0 = \pi \sum_{q=-N}^{N} \omega_q = \frac{2N}{a} = \frac{L}{a^2}, \tag{32}$$

and it diverges both, as the size of system grows $(L \to \infty)$ and as the lattice spacing decreases $(a \to 0)$, *i.e.* as one approaches the continuum limit. The energy density, is given by $E_0/L = 1/a^2$. In the case $m >> 1/a$,

$$E_0 = \frac{\pi}{2} \sum_{q=-N}^{N} m = \frac{\pi}{2} \frac{L}{a} m \propto \frac{L}{a^2}. \tag{33}$$

So in both cases the divergence structure is the same. But what about a single mode excitation? If $m = 0$ we have

$$\omega_q = \frac{2}{a} |\sin \frac{\pi}{2N+1} q| \to \frac{2\pi q}{L}, \tag{34}$$

which takes values in the range

$$0, \frac{2\pi}{L} ... < \omega_q < \frac{2}{a}. \tag{35}$$

In the case $m >> 1/a$

$$\omega_q \to m. \tag{36}$$

These are also the highest energy modes since $m >> 1/a$. We will now summarize. The zero pint energy is infinite in the infinite volume limit. This is just a manifestation of the fact that energy is an extensive quantity so it has to be proportional to size of the system. However the ground state energy density also has an UV divergence which, in the case $m = 0$ comes from the $1/a^2$ behavior. In the continuum limit this corresponds to a divergence associated with summing up over an infinity of degrees of freedom extending over arbitrary short distance scales. In momentum space it corresponds to fluctuations of the string with arbitrary short wavelengths and therefore arbitrary large momenta and energies. In the limit $m >> 1/a$, m plays the role of a heavy mass and the energy is determined by the rest mass. Finally in the continuum limit one finds that the normal modes have relativistic dispersion relation

$$\omega_q \to \sqrt{q^2 + m^2}, \tag{37}$$

thus they can be identified with "particles". Thus the fields provide a "fabric" for the matter waves! For $am \ll 1$ the UV divergence of the zero-point energy density is related to the scale invariance of original Hamiltonian. If $m \to \infty$, in the nonrelativistic limit, we need to impose a cutoff $k < 1/a$ and then have $m > 1/a$. Thus we are finding that in order to define the theory we need to put an UV cutoff. For example to make the ground state energy density finite we need to define the Hamiltonian by,

$$H \to H(a) \to H(a) + \delta H(a). \tag{38}$$

with

$$\delta H(a) = const = - < 0|H(a)|0 > . \tag{39}$$

This is referred to as renormalization. In general in an interacting theory all masses and coupling constant will have to be made dependent on a. Physical observables are made independent of the cutoff parameter a by carefully adjusting the functional dependence of parameters on a. In the simple renormalization example given above the subtraction of the vacuum energy is irrelevant from the physical point of view since it is a constant (albeit an infinite) contribution to the Hamiltonian and cannot be measured by comparing energies of physical states.

3.1 Infrared slavery and nonperturbative methods

Using the language of renormalization we can understand how QCD can be at the same time confining (no free quarks) and asymptotically free (Q^2-independence of DIS) (cf. Fig.4).

The theory has to be regularized at some scale Λ $1/a$,which makes the QCD coupling scale-dependent, $\alpha_s \equiv g^2/4\pi = \alpha_s(\Lambda)$. This dependence has to be such that physical quantities, masses, decay widths etc. are Λ independent. QCD predicts that $\alpha(\Lambda) \propto 1/\ln(\Lambda)$ for $\Lambda \gg 1$ GeV and $\alpha(\Lambda)$ grows as $\Lambda \to 0$. When a particular observable, $\mathcal{O}$ is calculated in QCD it will depend on the coupling, and the cut-off scale, $\mathcal{O} = \mathcal{O}(\alpha_s(\Lambda), \Lambda, p_i)$. In addition it may depend on external momenta, p_i, as it is for example in the case of a scattering amplitude. The explicit Λ dependence means that the observable has to be calculated with the regulator in place. Since a physical observable should be Λ-independent one can use any value of the cut-off to calculate O ! For example in the case of parton distributions entering the DIS cross sections, the external momenta are the 4-momentum transferred to the proton, and the proton 4-momentum, or Q^2 and W. For large values of the external momenta it is possible to use perturbation theory to study how an observable depends on these momenta. This is because QCD leads to an expansion of the form

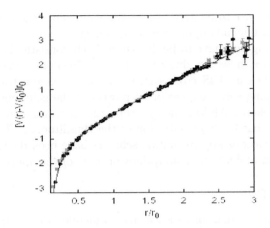

Figure 4: Potential energy, $V(r)$ as a function of the separation between static quark-antiquark sources, $r_0 \sim 0.5$ fm, computed using lattice gauge simulations[14]. At short distance the potential behaves like in electrostatics, $-\alpha/r$ with $\alpha \sim 0.2$, and it grows linearly.

$\mathcal{O} \sim \sum_n \alpha_s^n(\Lambda) c_n(p_i/\Lambda)$ with coefficients $c_n(1) \sim O(1)$. Thus if $p_i >> 1$ GeV one can choose $\Lambda \sim p_i$ and therefore have $\alpha(\Lambda) < 1$. On the other hand if there are no external large scales that and observable depends upon, which for example is the case for the matrix element

$$m_{hadron} = \frac{\langle hadron|H_{QCD}|hadron\rangle}{\langle hadron|hadron\rangle}, \tag{40}$$

determining mass of a hadron one would have to set $\Lambda \sim m_{hadron}$ and the perturbative expansion breaks down since for $\Lambda \sim m_{hadron} \leq 1GeV$, $\alpha(\Lambda)$ is predicted to be large, (even though the exact behavior of $\alpha(\Lambda)$ for small Λ is not yet fully determined). One possibility is that $alpha(\Lambda) \to 1/\Lambda^2$ for small Λ which under certain approximations can be related to a linear behavior of a confining potential between static color sources, as shown in Fig. 4.

4   Survey of nonperturbative methods

When formulated as a Hamiltonian system in a discretized space, (as any other field theory) solving QCD is equivalent to finding the spectrum of the Hamiltonian and the corresponding wave function from which other observables can be determined. The procedure gets a bit complicated since the parameters which the Hamiltonian depends upon (coupling and quark masses) are functions of the lattice spacing and have to be determined self-consistently.

In come cases, however, due to asymptotic freedom certain relations between observable *e.g.* DIS cross section at various momentum transfer, can be calculated using series expansion in powers of the coupling constant (perturbative expansion). All low energy properties, however, require in principle a full, nonperturbative diagonalization of the Hamiltonian. This is a complicated task and various approximation schemes have been developed. Here we discuss the main underlying assumptions of some of the approaches[15].

4.1 Lattice QCD

In principle lattice QCD provides a first principle (*i.e.* with no *ad hock*) assumptions tool for computing hadronic observables. In a nut shell, as we did for the simple one-dimensional scalar theory, the continuum of dynamical variables represented by quark and gluon filed operators is replaced by a denumerable set of variables defined on sites (quarks) or links between the sites (gluons) of the discretized 4-dimensional Euclidean space[16]. The passage from Minkowski to Euclidean space is done by taking $t \to i\tau$ which makes numerical computations much more efficient (eliminates oscillations, generated by $exp(iEt)$ and replaces them by exponential dumping, $exp(-E\tau)$). Observables are then calculated by representing matrix elements as path integrals, thus reducing computations to a multi-dimensional numerical integration typically done using Monte Carlo techniques.

As an example consider a simple case involving computation of the ground state energy of a simple harmonic oscillator (SHO) in one dimension using lattice path integral techniques. The relation between the quantum transition probability (in *imaginary* time) and the classical path integral is given by

$$\langle x_f \tau_f | x_i, \tau_i \rangle = \langle x_f | e^{-H(\tau_f - \tau_i)} | x_i \rangle = \int \mathcal{D}x(\tau) e^{-S[x]}. \tag{41}$$

For the SHO the Euclidean action S is given by

$$S = \int_{\tau_i}^{\tau_f} L\left(x(\tau), \frac{dx(\tau)}{d\tau}\right) = \int_{\tau_i}^{\tau_f} d\tau \frac{1}{2}\left[\left(\frac{dx(\tau)}{d\tau}\right)^2 + x^2(\tau)\right], \tag{42}$$

where we took the mass of the particle and angular frequency of the oscillator both equal to one. Note that after the replacement $t \to i\tau$ the Euclidean action is positive. Take $x_i = x_f \equiv x$ and $t_i - t_f \equiv T$. The left hand side can be evaluated by inserting a complete set of eigenstates of the Hamiltonian

$$\langle x|e^{-HT}|x\rangle = \sum_n \langle x|E_n\rangle e^{-E_n T}\langle E_n|x\rangle, \tag{43}$$

where $E_n = (n + 1/2)$. In the limit $T \to \infty$ the left hand side approaches $e^{-E_0 T}|\langle x|E_0\rangle|^2$ where E_0 is the lowest (ground state) eigenvalue of the Hamiltonian, $E_0 = 1/2$ with contributions from excited states exponentially suppressed. Thus to obtain the ground state energy one can performed the integral over x (since $\int dx|\langle x|E_0\rangle|^2 = 1$) and have,

$$E_0 = -\frac{1}{T} \lim_{T\to\infty} \log(\int dx \text{ RHS of Eq. (41) }). \tag{44}$$

The RHS is then calculated on a finite grid by splitting the time interval $[\tau_i, \tau_f]$ into N nodes,

$$\tau \to \tau_j = \tau_i + aN, j = 0, 1 \cdots N, a \equiv \frac{(\tau_f - \tau_i)}{N}. \tag{45}$$

The τ integral in the action turns into a sum over nodes

$$\int_{\tau_i}^{\tau_f} d\tau \to \sum_{j=0}^{N-1} a\left[\frac{1}{2}\left(\frac{x_{j+1} - x_j}{a}\right)^2 + \frac{1}{2}(V(x_{j+1}) + V(x_j))\right], \tag{46}$$

with $x_0 = x_N = x$ and $x_j = x(t_j)$ and the path integral is then reduced to an $N - 1$-dimensional integral over all possible values of particle's displacement at the times $\tau_i, i = 1, 2, \cdots N - 1$ (the end-points are fixed) with the integration measure given by

$$\int \mathcal{D}x(\tau) \to \int dx_1 dx_2 \cdots dx_{N-1}\left(\frac{1}{2\pi a}\right)^{\frac{N}{2}}. \tag{47}$$

In lattice filed theory the lattice spacing plays the role of UV cutoff, since distances shorter then a cannot be resolved. In the absence of other mass scales (e.g. quark masses) this is the only dimensional parameter and therefore all dimension-full quantities e.g. hadron masses will have to be given in units of the lattice spacing,

$$m_{hadron} = \frac{1}{a}f(\alpha(1/a), a). \tag{48}$$

Again, this is not a contradiction (lattice spacing is obviously unphysical) since the strength of the interactions $\alpha = \alpha(\Lambda) = \alpha(1/a)$ also depends on a. Setting physical scale can be done for example by comparing how the coupling changes with a and making sure for small a it does so in agreement with perturbative QCD. The cost of lattice QCD simulations increases with the lattice size roughly as $1/a^6$, thus it would be desirable to keep a as large as possible and at the same time keep discretization errors small. The numerical dicretization errors coming from approximating the derivatives by finite difference is a possible thing one can improve upon. The other has to do with making sure the perturbative limit is reached quickly (*i.e.* for large a), this is done by the so called tadpole improvement.

4.2 QCD sum rules

The basic goal of QCD sum rules is to extract hadronic observables by matching Green's function calculated in perturbative QCD with their hadronic representation [17]. To be more specific consider the following matrix element

$$\Pi(q)(q^\mu q^\nu - q^2 g^{\mu\nu}) \equiv \int d^4x e^{iq\cdot x} \langle 0|Tj^\mu(x), j^\nu(0)|0\rangle. \tag{49}$$

Here $j^\mu(x) = \bar{\psi}_i(x)\gamma^\mu\psi_i(0)$ is a vector current and T stands for a time-ordered product. The reason for using T-ordering is that a typical scattering amplitude is given by a matrix element of T-ordered operators which in turn follows from causality. For example the matrix element given above can be related to a cross-section for $e^+e^- \to hadrons$,

$$\sigma(e^+e^- \to hadrons) = \text{Phase space}|\bar{u}(e^+)\gamma^\mu v(e^-)\langle 0|Tj_\mu j_\nu|0\rangle|^2, \tag{50}$$

where u and v are the Dirac spinors representing the e^+ and e^- respectively. The tensor structure of the LHS in Eq. (49) follows from current conservation $\partial_\mu j^\mu = 0$.

The correlator Π can be calculated in two ways, by employing an identity

$$\sum_{hadrons} \langle 0| j^\mu(x)|hadrons\rangle\langle hadrons|j^\nu(0)0\rangle =$$

$$\sum_{quarks\&gluons} \langle 0|j^\mu(x)|quarks\&gluons\rangle\langle quarks\&gluons|j^\nu(0)0\rangle, \tag{51}$$

which simply expresses the resolution of identity in either hadronic of quark-gluon basis. But since hadrons are eigenstates of the QCD Hamiltonian and momentum operators the $x = (\mathbf{x}, t)$ dependence can be made explicit and the matrix elements in the hadronic basis becomes,

$$\Pi(q^2) = \frac{1}{\pi} \int ds \frac{\rho(s)}{s - q^2}, \tag{52}$$

where the denominator arises from the x-integral and the function ρ in the numerator describes the density of hadronic states,

$$\rho(s)(q^\mu q^\nu - s g^{\mu\nu}) \propto \sum_{hadrons} \delta(s - m_{hadron}^2)|\langle 0|j^\mu q, hadron, m_{hadron}\rangle$$

$$\langle hadron, m_{hadron} q|j^\nu|0\rangle. \tag{53}$$

For example if there is an infinitely narrow hadron (*i.e.* the sum over hadrons in the LHS in Eq. (51) is given by a single state) its contribution to ρ is given by $\rho(s) \propto \delta(s - m_h^2)$. If q^2 is much larger then masses of hadrons contributing to Eq. (49) then one can approximate $1/(q^2 - s)$ by $1/q^2$ and thus all hadrons contributing to ρ in the sum in Eq. (53) have to be retained. On the other hand if q^2 is of the order of the mass of the lowest hadron, m_0 then contributions of heavier hadrons are suppressed by m_0^2/m_i^2 with $m_i > m_0$. This suppression can be made even stronger (exponential) by performing the so called Borel transformation. As discussed earlier because of asymptotic freedom, if $|q^2|$ is larger then the typical QCD scale, 1GeV a matrix element can be evaluated using quark and gluon basis using perturbative expansion in the coupling constant (*i.e.* just like standard time-dependent perturbation theory for a scattering amplitude). Clearly it is necessary to go beyond perturbation theory if $q^2 \sim 1$ GeV^2. The inclusion of nonperturbative correction on the QCD side is typically done phenomenologically using the so called operator product expansion. By careful matching the hadronic and QCD, quark-gluon based evaluation of the correlator one can extract hadronic properties parameterizing $\rho(s)$ such as hadron masses and matrix elements $\langle 0|j^\mu|hadron\rangle$.

4.3 Self consistent methods

Dyson-Schwinger equations is a set of coupled integral equations for Green's functions [18]. These equation are "self-consistent" *i.e* in contrast to a perturbative expansion they sum-up infinite set of perturbative contributions (diagrams) to a given matrix element. For illustration consider a quark 2-point function *i.e.* the quark propagator,

$$S_F(p) = \int d^4 x e^{ipx} \langle 0|\psi(x)\bar{\psi}(0)|0\rangle. \tag{54}$$

A few lowest terms in the perturbative expansion of S_F are shown in Fig. 5a.

274

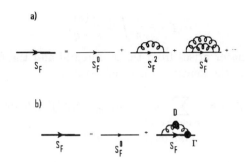

Figure 5: a) Perturbative expansion of the quark propagator. b) Self-consistent equation of the propagator in terms of full ("dressed") gluon propagator, D, and the quark-gluon vertex, Γ.

Inspecting the structure of the perturbative diagrams it is clear that the full, "dressed" propagator can be written in terms of a "dressed" quark-gluon vertex, "dressed" gluon propagator, and "dressed" quark propagator itself, as shown in Fig. 5b,

$$\Sigma(p) = \int d^4k \Gamma(p, k) D(k) S(p - k), \qquad (55)$$

where Γ is the dressed vertex, *etc.* The self-consistency is now clear : the propagator is determined by the loop integral, which itself is determined by the propagator. Similar equations can now be written for the other "dressed" quantities. The problem however is that the resulting set of coupled integral equations for the involves and infinite set of unknowns. For example unlike Eq. (55) there is an infinite set of self-consistent diagrams contributing to an equation which determines the "dressed" quark-gluon vertex. In any practical application this infinite set of equations is truncated, for example one may make a guess for Γ and/or D and solve for S. Finally, Just like in the lattice formulation calculations are typically done Euclidean space. We will not discuss these this method much further and refer the reader to an extensive literature on this subject.

4.4 The Constituent quark model

The flavor classification of hadrons proposed by Gell-Mann and Zweig, can be simply realized if the complicated quark-gluon structure of hadrons is replaced by the valence "constituent" quarks with right flavor quantum number and spin. Since the constituent quarks replace the (infinite) number of "bare" quark and gluon degrees of freedom they need to be thought of as effective quasiparticles with intrinsic structure and finite masses. In terms of this *finite* number of valence degrees of freedom one can construct effective interactions and construct hadronic wave functions. The quark model has been extremely successful in accounting for a large variety of hadronic phenomena, which quite amazing, if one considers its simplicity and rather loose connection with the underlying QCD[6,19]. As an example of a most naive nonrelativistic quark model Hamiltonian consider,

$$H = \sum_f m_f + \sum_{ff'} \frac{C}{m_{f'} m_f} \vec{\sigma}_{f'} \vec{\sigma}_f. \tag{56}$$

Here the sums run over various (quark) flavors. In this Hamiltonian the kinetic motion of the quarks has been approximated entirely by in effect constant constituent mass, and the only residual interaction left is assumed to be a chromo-magnetic interaction between spins. The mass dependence of this spin dependent interaction is taken to mimic one-gluon-exchange force and there is some evidence from lattice calculations that indeed for heavy quarks the spin-spin interaction is short-ranged just like energetic one gluon exchange. This Hamiltonian gives a good description of meson hyperfine splittings *i.e* $m_\rho - m_\pi$, $m_{K^*} - m_K$, etc if one assumes that the light, u, d *constituent* quark masses are $m = 200 - 300$ MeV Alternatively one can derive relations between masses of various hadrons like[5],

$$m_\eta - m_\pi = (4/3)(m_K - m_\pi), \tag{57}$$

for mesons and

$$3E_\Lambda - 2(E_N + E_\Xi) + E_\Sigma = 0, \tag{58}$$

for baryons, which turn out to be correct to within 10% ! The quark model however can me made much more sophisticated with complicated phenomeno-logical interactions motivated by various nonperturbative phenomena, like flux tube formation, which leads to a linear confining interaction or various relativistic effects like Thomas precession. An important consequence of the valence quark model is that it predicts that the intrinsic quantum numbers of a hadron are related to those of the valence quarks. For example for mesons, made out of a quark-antiquark pair we have

$$J = L + S, \quad P = (-1)^{L+1}, \quad C = (-1)^{L+S}, \tag{59}$$

where J, P, C are the spin, parity and charge conjugation of the mesons and L is the orbital angular momentum, S total spin of the quark-antiquark pair. One can thus give a full classification of hadrons in terms of the valence quark states.

5 QCD vs Constituent Quarks, chiral Symmetry and dynamical symmetry breaking

As mentioned earlier the quark model assumes that the bulk of hadron low energy properties comes from motion of a few quasiparticles. This is in contrast to a picture emerging from the underlying QCD where quarks enter with small masses and therefore would be expected to move highly relativistically and copiously radiate gluons which in turn would produce more quark-antiquark paris. These apparently disjoint pictures can be reconciled if one considers the property of QCD known as chiral symmetry. To illustrate who it works we will consider QCD with two light (u,d) quarks only and further assume that that quarks are massless. This is a reasonable approximation since the parameter which plays the role of the quark mass, for the u and d quarks is of the order of a few MeV *i.e.* much smaller compared to the typical light hadron mass $O(1\text{GeV})$.

In the massless quark limit QCD poses a new global symmetry associates with transformations[3]

$$\psi_f(x) \rightarrow \sum_{f'=u,d} \left[e^{-i\theta^a \frac{\tau^a}{2} \gamma_5} \right]_{ff'} \psi_{f'}(x), \tag{60}$$

where $f, f' = u, d$, τ^a, $a = 1, 2, 3$ are the three Pauli matrices and θ^a are three independent real parameters. In other words we have introduced three independent rotational angles in the two dimensional complex space spanned by the two quark flavor. These rotations form the $SU(2)$ group. On can also introduce a global phase rotation, same one for the two flavors, simply replacing τ^a by a 2×2 identity matrix. This extra phase rotation is parametrized by an additional parameter α^4 and it defines the $U(1)$ group. Here, however we will concentrate on the consequences of the $SU(2)$ rotations only. Notice the presence of the γ_5. From Noether's theorem it follows that once there is symmetry there are conserved currents. In this case three, one for each "rotation axis". These *axial* currents are given by,

$$A^{\mu,a}(x) = \sum_{f'f=u,d} \bar{\psi}_{f'}(x) \frac{\tau^a_{f'f}}{2} \gamma^\mu \gamma_5 \psi_f(x), \tag{61}$$

and the corresponding, conserved, $dQ_5^a(t)/dt = 0$, charges are given by

$$Q_5^a \equiv \int d\mathbf{x} A^{0a}(\mathbf{x}, t). \tag{62}$$

If the u and d quark masses are equal by not necessarily vanish the QCD Lagrangian is also invariant under the more familiar isospin rotations with $\gamma_5 \to 1$ in Eq. (60),

$$\psi_f \to \sum_{f'=u,d} \left[e^{-i\theta^a \frac{\tau^a}{2}} \right]_{ff'} \psi_{f'}, \tag{63}$$

from which follows conservation of the isovector-vector currents given by,

$$V^{\mu,a} = \sum_{f'f=u,d} \bar{\psi}_{f'}(x) \frac{\tau_{f'f}^a}{2} \gamma^\mu \psi_f(x), \quad Q^a \equiv \int d\mathbf{x} V^{0a}(\mathbf{x}, t). \tag{64}$$

Using Q^a and Q_5^a one can define the so called left and right charges $Q_{L,R}^a = \frac{1}{2}(Q^a \pm Q_5^a)$ and show that these from two independent sets of generators of $SU(2)$ transformations referred to as $SU(2)_L \times SU(2)_R$. This is not the case for the vector and axial charges whose commutation relations are mixed. One can also show that left and right charges are mixed under parity *i.e.* $PQ_{L,R}P^{-1} = Q_{R,L}$, wich follows from the pseudo-vector nature of the axial current $A^{\mu a}$. Assume that a given hadronic state $|hadron\rangle$, has positive parity

$$H|hadron\rangle = m_h|hadron\rangle, \quad P|hadron\rangle = +|hadron\rangle, \tag{65}$$

then since $Q_{L,R}$ are conserved generators of a symmetry the states $Q_L|hadron\rangle$ and $Q_R|hadron\rangle$ have the same energy m_h and thus so do the states,

$$|hadron\rangle_\pm \equiv \frac{1}{2}(Q_L \pm Q_R)|hadron\rangle, \tag{66}$$

but $P|hadron\rangle_\pm = \pm|hadron\rangle_\pm$ which implies that conservation of the axial-isospin charges leads to multiplets of parity degenerate hadrons! But in the physical hadronic spectrum opposite parity states have quite different masses. For example the lightest positive parity baryon (proton) has mass of 938 MeV and the lightest negative parity baryon N^* mass of 1535 MeV. One thus concludes that the symmetry has to be broken. It is straightforward to show that the quark mass term $m_f \bar{\psi}_f \psi_f$ breaks chiral symmetry, however, as mentioned earlier the $m_u/1\text{GeV} < 1\%$ and it is unlikely that light quark masses play a major role in chiral symmetry breaking. Another possibility is that the symmetry is broken *spontaneously*, *i.e.* dynamics leads to a ground state which is not symmetric under the chiral $SU(2)$ rotations. Spontaneous symmetry breaking is a familiar phenomenon. For example residual interactions between magnetic domains are rotationally invariant, but in a ferromagnet the magnetic domains point in a specific direction and such a state clearly brakes rotational invariance. Which part of the $SU(2)_L \times SU(2)_R$ is expected to be broken by the vacuum? Since isospin is a good quantum number one expects that the axial-isospin rotations are broken by the vacuum *i.e.*

$$Q_5^a |vacuum\rangle = 0, \text{ and } Q^a |vacuum\rangle = 0. \tag{67}$$

If vacuum does not respect the symmetry of the Hamiltonian then according to the Goldstone's theorem there should be a massless particle in the spectrum. In QCD these correspond to the three pions, $\pi^\pm, \pi^0$ corresponding to the three broken axial directions.

Dynamical chiral symmetry breaking leads to a complicated, non-symmetric vacuum and can also explain the success of the constituent quark model. Consider the QCD Hamiltonian,

$$H = K_q + K_g + V, \tag{68}$$

where $K's$ represent kinetic energies of free quarks and gluons and V contains all interactions. The canonical formulation of QCD requires gauge fixing, here we will assume that a physical gauge has been chosen and that the Hamiltonian depends on a set of canonical coordinates collectively denoting the quark and gluon fields. For example K_q is given by,

$$K_q = \sum_\alpha \int \frac{d\vec{k}}{(2\pi)^3} \sqrt{\vec{k}^2 + m^2} \left[b_\alpha^\dagger(\vec{k}) b_\alpha(\vec{k}) + d_\alpha^\dagger(\vec{k}) d_\alpha(\vec{k}) \right]. \tag{69}$$

Here $b^\dagger, b$ and $d^\dagger, d$ are the quark and antiquark creation and annihilation operators respectively, α denotes other quark internal quantum numbers (*i.e.* spin, flavor, color) and $m = m_u = m_d$ is the mass parameter from the QCD Lagrangian. The complete set of eigenstates of K_q and K_g is given by application of particle creation operators on the Fermi-Dirac vacuum defined by $b|0\rangle = d|0\rangle = 0$ In this basis the axial charge operator is given by,

$$Q_5^a = \sum_\alpha \int \frac{d\vec{k}}{(2\pi)^3 \sqrt{\vec{k}^2 + m^2}} \left[b_\alpha^\dagger(\vec{k}) \frac{\tau^a}{2} \vec{\sigma} \cdot \vec{k} b_\alpha(\vec{k}) - (b \to d) \right.$$
$$\left. + m \left(b_\alpha^\dagger(\vec{k}) \frac{\tau^a}{2} d_\alpha^\dagger(\vec{k}) + (b \to d) \right) \right]. \tag{70}$$

Thus in the chiral limit, $m = 0$ the ground state is axially symmetric, $Q_5^a |0\rangle = 0$. Since $V \neq 0$ the eigenstates of K are not eigenstates of the full Hamiltonian (hadrons are in the spectrum of QCD but not free quarks and gluons), but they do form a complete basis which can be used to expand the eigenstates of the full Hamiltonian. Schematically

$$|h\rangle = \sum_{N_q,N_g} |q_1, \cdots, q_{N_q}; g_1, \cdots g_{N_g}\rangle \Psi_h(q_1, \cdots g_{N_g}). \tag{71}$$

Since the ground state of QCD breaks axial symmetry it would be more advantageous, however, to use a different basis to expand a hardon state rather than the one discussed above. The new basis would build on a chiraly non-invariant, vacuum. It is possible to define such a vacuum and then introduce a complete quasiparticle basis build on top it by the same methods which lead to Cooper pairs in BCS superconductivity. In our case the paris are made out of a quark and an antiquark coupled to $J^{PC} = 0^{++}$ – vacuum quantum numbers, *i.e.* according to Eq. (59) with $L = S = 1$. Such vacuum can be obtained for example by performing the following transformation of the original chiraly invariant basis of free quarks and gluons,

$$|BCS, \phi(k = |\vec{k}|)\rangle = e^{\sum_\alpha \int \frac{d\vec{k}}{(2\pi)^3} \tan \phi(k) \left[b_\alpha^\dagger(\vec{k}) \vec{\sigma} \cdot \vec{k} d_\alpha^\dagger(-\vec{k}) + c.c \right]} |0\rangle, \tag{72}$$

where $\phi(k)$ is the Bogolubov angle which can be though off as a variational parameter to be determined by $\delta \langle BCS|H|BCS\rangle / \delta\phi = 0$. It is a straightforward to show that for $\phi(k) \neq 0$, $Q_5|BCS\rangle \neq 0$, thus the BCS angle is a measure of dynamical chiral symmetry. Even tough the BCS ground state is only an ansatz for the ground state the complete set of quasiparticle excitations build on top of this vacuum may be more efficient for diagonalizing the QCD Hamiltonian. In the quasiparticle basis the one body kinetic energy contributes of the order of a few hundred MeV to the bound state masses and the quasparticels have properties of the constituent quarks of the valence model. Thus unlike the free, partonic basis one expects that only a few quasiparticle states will contribute to a hadronic eigenstate[20,21].

We have often referred to the QCD Hamiltonian, however there is no such a thing as the QCD Hamiltonian. QCD, just like QED is formulated based on the principle of local gauge invariance, which introduces gluons as "mediators" of the strong force. Lorentz symmetry requires there to be 4 space-time components of the gluon field, but just like masless photons, gluons should have only two (not four) polarizations. In other works the theory has nonphysical degrees of freedom which have to be eliminated before a Hamiltonian can be defined. This elimination is performed by choosing a specific gauge which then defines a Hamiltonian. One choice for example is to set $A^{\mu=0,a}(\mathbf{x}, t) = 0$ the so called time-axial or Weyl gauge. The corresponding Hamiltonian has been extensively studied using lattice descreization methods and in particular in the strong coupling limit it naturally leads to the so called area law for the Wilson loop which implies that a potential between static color sources increases linearly with the distance.

Another gauge choice if the Coulomb gauge. The Coulomb gauge formulation provides a very natural starting point for building the constituent representation in accord with confinement and dynamical chiral symmetry breaking. In the Coulomb gauge the single particle spectrum contains only physical excitations, *i.e. two transverse gluon polarizations.* and it leads to a natural realization of confinement. This arises because of elimination of the non-physical degrees of freedom through the gauge choice, $\nabla \cdot \vec{A} = 0$ and results in an instantaneous interaction between color charges. This is the analog of the Coulomb potential in QED. In QCD, however, the colored Coulomb gluons can couple to transverse gluons leading to a Coulomb kernel which depends also on the dynamical gluon degrees of freed. As shown in Ref. [22] summation of the dominant IR contributions to the vacuum expectation value of the Coulomb operator results in a potential between color charges which grows linearly at large distances in an excellent agreement with lattice calculations. In a self-consistent treatment the same potential then modifies the single gluon spectral properties. In particular it leads to an effective mass for quasi-gluon excitations $O(500 - 800 \text{ GeV})$, which is also in agreement with recent lattice calculations. This could be used to justify the assumption implied by quark model that mixing between valence quarks and states with a explicit gluonic excitations is suppressed.

6   Role of Gluons

So far we have mainly discussed gluons as mediators of the strong force between color charges in analogy with photons mediating the electromagnetic interactions. Because of the strong nature of color interactions and the fact that gluons themselves carry color charge, gluons may play a much more dynamical role then photons. For example two gluons can form a bound state – glueball just like a quark and antiquark forms a mesons. Similarly gluons can be excited in the presence of quarks and lead to a whole new spectrum of excitations referred to as hybrids. There are certain hadronic reactions believed to be "gluon rich" and used to search for glueballs, *i.e.* proton-antiproton annihilation, proton-proton peripheral scattering, or radiative J/ψ decays [23]. From theoretical considerations it is expected that the lightest glueball should have spin, (J), parity (P) and charge conjugation (C) quantum numbers $J^{PC} = 0^{++}$ and mass between $1.5 - 1.7$ GeV. In this mass range, however there is a number

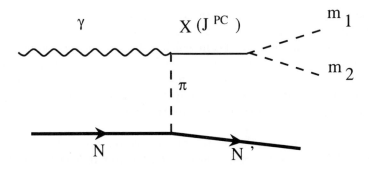

Figure 6: Schematic of a peripheral $\gamma N \to N' m_1 m_2$ two-meson photoproduction. A high energy photon scatter off e meons cloud (*e.g.* π) and produces a rapidly moving meson resonance which subsequently decays into a meson pair.

of open continuum channels, *e.g.* 2π, $K\bar{K}$, $\eta\eta$, which the glueball can decay to, making its identification complicated. The most recent analysis done by the Crystal Barrel collaboration of the $p\bar{p}$ data indicated that indeed the meson spectrum in this mass range is over-populated as compared to the number of resonances expected from the quark model. Furthermore it has been shown that the extra resonances may have a significant glueball component. The extra, gluonic degree of freedom in glueballs or hybrids lead to a possibility of states whose J^{PC} quantum number are outside of the quark model counting rules given by Eq. (59). These are refereed to as exotic and an identification of a resonance with such quantum numbers would be a "smoking gun" signature of presence of gluonic excitations in the hadronic spectrum. Only mesons can have exotic quantum numbers. For example $J^{PC} = 1^{-+}$ can be obtained by combining spin-1 quark-antiquark system with quarks in a relative s-wave ($J^{PC}_{Q\bar{Q}} = 1^{--}$) with a spin-1 gluons in a p-wave with respect to the $Q\bar{Q}$ system. In fact it is predicted that the lightest exotic states should have such quantum numbers and mass slightly below 2 GeV [24]. Because the quark pair in light exotic is in a spin-1 state it is unlikely that exotics will be strongly produced in π or K induced reactions. Indeed there some exotic meson candidates have been found in high energy πp peripheral production but the signal is only a few percent compared to production of non-exotic resonances [25]. It is expected, however that photons, being a virtual quark-antiquark state with spin-1 could be a strong source of exotic mesons in peripheral reactions, were a high energy photon scatters off the meson cloud around a proton target exciting gluonic field in the mesons [26].

The data on meson photoproduction is very limited, mainly due to technical challenges in producing high intensity, polarized photon beams. Nowadays the CEBAF electron accelerator at the Jefferson Lab has the potential to deliver very high quality electron beam which using the coherent bremsstrahlung technique can be turned into a linearly polarized photon beam. In order to perform exotic mesons photoproduction studies the energy of the accelerator has to be upgraded from the current 6 to 12 GeV, to produce $8-9$ GeV photons which is the optimal energy for covering the exotic mesons spectrum in the $1.5 - 2.5$ GeV mass range[27].

In one year of running the yields for exotic mesons production are expected to produced 250K exotic mesons which is $O(10^3)$ more then currently collected by the E852 experiment using π-induced production. With a hermetic detector with full event reconstruction capabilities even a weak exotic signal (on order of magnitude weaker then expected for photoproduction) can be well identified.

Acknowledgments

I would like to thank Jose Goity, Cynthia Keppel and Robert Willams for invitation to the school. This work was supported by the U.S. Department of Energy contract DE-FG02-87ER40365.

References

1. M. Gell-Mann, *Phys. Lett.* **8**, 214 (1964); G. Zweig, CERN preprint 8419/TH 412 (1964).
2. K. Hagiwara et al.,*Phys. Rev.* D **66**, 010001 (2002).
3. C. Itykson and J.-B. Zuber, *Quantum Field Theory* (McGarw Hill, New York, 1985).
4. L.I. Schiff, *Quantum Mechanics*, (McGraw Hill, New York, 1968).
5. T.D. Lee, *Partcle Physics and Introduction to Field Theory*, (Harwood Academic, Switerland, 1981).
6. S. Godfrey and N. Isgur, PRD **32**, 189 (1985), *and reference in there.*
7. J.I. Friedman and H.W. Kendall, ARNPS **22**, 203 (1972).
8. R.P. Feynman, *Phys. Rev. Lett.* **23**, 1415 (1969).
9. J.D. Bjorken and E.A. Paschos, *Phys. Rev.* **185**, 1975 (1969).
10. J. Ashman et al., *Phys. Lett.* B **206**, 364 (1988).
11. H. Fritzsch and M. Gell-Mann, in *Proceedings of the 16th. International Conference on High Energy Physics*, ed. J.P. Jackson and A. Roberts (National Accelerator Lab., Batavia, Illinois, 1972); H. Fritzsch, M. Gell-Mann and H. Leutwyler, *Phys. Lett.* B **47**, 365 (1973); D.J. Gross and F. Wilczek, *Phys. Rev.* D **8**, 3633 (1973).
12. O.W. Greenberg, *Phys. Rev. Lett.* **13**, 598 (1964); M.Y. Han and Y. Nambu, *Phys. Rev.* **139**, 1006 (1965).
13. T. Muta, *Foundations of Quantum Chromodynamics* (World Scientific, Singapore, 1987).
14. G.S. Bali, *QCD potentiology*, arXiv:hep-ph/0010032.
15. J.D. Walecka, *Theoretical Nuclear and particle Physics* (Oxford University Press, March 1995)
16. K.G. Wilson, *Phys. Rev.* D **10**, 2445 (1974).
17. M.A. Shifman, A.I. Vainshtein, and V.I. Zakharov, NPB **147**, 1979 (;) NPB **147**, 448 (1979).
18. C.D. Roberts and A.G. Williams, PPNP **33**, 477 (1994).
19. F.E.Close, *AnIntroduction to Quarks and Partons*, (Academic Press, London 1979).
20. A.P. Szczepaniak, E.S. Swanson, PRD **55**, 1578 (1997).
21. A.P. Szczepaniak, E.S. Swanson, PRL **87**, 072001 (2001).

22. A.P. Szczepaniak, E.S. Swanson, PRD **62**, 094027 (2000); PRD **65**, 025012 (2002); A.P. Szczepaniak, P.Krupinski, PRD **66**, 096006 (2002).
23. C. Amsler, RMP **70**, 1293 (1998); D. Barberis *et al.*, PLB **353**, 589 (1995); J.Z. Bai *et al.*, PRL **77**, 3959 (1997).
24. N. Isgur and J. Paton, PRD **31**, 2910 (1985); T. Barnes, F.E. Close, F.de Viron and J. Weyers, NPB **224**, 241 (1983).
25. D.R. Thompson *et al.*, PRL **79**, 1630 (1997); G.S. Adams *et al.*, PRL **81**, 5760 (1998); E.I. Ivanov *et al.*, PRL **86**, 3977 (2001).
26. A.P. Szczepaniak, M. Swat, PLB **516**, 72 (2001).
27. A.Dzierba *et al.*, *The Hall D Project Design Report : Searching for the QCD Exotics with a Beam of Photons* vr.4 (November 2002), http://dustbunny.physics.indiana.edu/HallD/

Student Abstracts

E00-102: TESTING THE LIMITS OF THE SINGLE PARTICLE MODEL IN ^{16}O(E,E'P)

MATTIAS ANDERSSON

Lund University, P.O. Box 118, SE-221 00 Lund, Sweden

E-mail: Mattias.Andersson@nuclear.lu.se

This experiment is an update to E89-003, which was the first physics measurement performed in Hall A at Jefferson Lab in Newport News, VA. We shall examine the momentum distribution of the protons in the ^{16}O nucleus. By doing this, we hope to determine the momentum region for which Single Particle Models such as the Relativistic Distorted Wave Impulse Approximation (RDWIA) are valid, and that where other more sophisticated models give a better description. As in E89-003, the ^{16}O(e,e'p) reaction will be studied in the quasielastic region, but this time at $Q^2 = 0.9$ (GeV/c)^2 and $\omega = 499$ MeV. The longitudinal-transverse interference response function (R_{LT}) and the longitudinal-transverse asymmetry function (A_{LT}) will be extracted from cross sections measured at a single electron angle for $E_{miss} < 60$ MeV and $p_{miss} < 515$ MeV/c. The proven self-normalizing waterfall target built by INFN and the Hall A High Resolution Spectrometers (HRS) in standard configuration will be used. The experiment is at present scheduled to run for six weeks during the fall of 2001.

STRANGENESS ELECTRO- AND PHOTO-PRODUCTION WITH THE CLAS SPECTROMETER

ROBERT BRADFORD

Department of Physics, Carnegie Mellon University, Pittsburgh PA 15213, USA
E-mail: bradford@ernest.phys.cmu.edu

The production of strange particles is one of the topics being vigorously investigated with Jefferson Lab's CLAS spectrometer. Preliminary calculations of Λ and Σ electroproduction differential cross sections demonstrate significant deviations from predictions of several isobar model calculations. In addition, a preliminary signal for the photoproduction of the Ξ^- baryon has been extracted from the G6B data set.

ELECTRON SCATTERING FROM HIGH MOMENTUM NUCLEON IN DEUTERIUM

C. Butuceanu

Department of Physics, College Of William & Mary ,
P.O. Box 8795, Williamsburg, VA, 23187-8795
E-mail: ccbutu@jlab.org

Using a 6 GeV unpolarized electron beam and the CLAS detector at TJNAF, $E94-102$ experiment measured inelastically scattered electrons in coincidence with protons emitted backwards relative to the virtual photon direction in the reaction d(e, e/p)X. Using the advantages of CLAS detector (large acceptance and out-of-plane detection of the backward proton) we will study the mentioned reaction for a large range of proton momenta $(0.25 - 0.6$ GeV) and electron kinematics $(Q^2 = 1 - 6 GeV^2, x = 0.2 - 1)$.

In a simple spectator model, the electron scatters off a forward-moving neutron inside the deuteron, and the detected backward-moving proton is an undisturbed spectator. Its measured momentum is equal and opposite to the momentum of the neutron before was struck. By measuring the semi-inclusive cross section as the function of the spectator momentum and the direction, we can study the dependence on kinematics and off-shell behaviour of the electron-nucleon cross section in the elastic, resonance, and deep inelastic regions. At the same time we gain information on the high-momentum structure of the deuteron wave function. If the virtual photon couples to a quark in a 6-quark object, the spectator picture breaks down and there will be a different dependence of the cross section on the kinematics variables (x, Q^2), and momentum of the spectator nucleon, $\vec{p}_s$. At the high relative nucleon momenta, in the kinematic region which favors short internucleon distances, the structure of the deuteron wave function will give a hint about the existence of the 6-quark objects.

I will present the features of this experiment, the physics involved and the directions of the analysis. I will review the Spectator Model and 6 Quark Bag Model, models tested in this experiment for describing the correlation nucleon-nucleon, as well as the applications of these models in DIS and Resonance regions. I will summarize the progress of the experiment and projected steps of the analysis.

References

1. Proposal for the $E94-102$ experiment, S. E. Kuhn, K. A. Griffioen, 1994.
2. C. E. Carlson, K. E. Lassila, William and Mary Report **WM-94-101**, 1994.
3. C. E. Carlson, K. E. Lassila, U. P. Sukhatme, Phys. Let. **B263**, 1991, 277.

DEEPLY VIRTUAL COMPTON SCATTERING (DVCS)

A. CAMSONNE

Hall A Jefferson Laboratory, Room A 116 Cebaf Center ,12000 Jefferson Avenue ,
Newport News 23606, USA Université Blaise Pascal/LPC IN2P3 Clermont
Ferrand, 24 Avenue des Landais, Aubière, France
E-mail: camsonne@jlab.org

The study of the electron scattering in the Deeply Inelastic Scattering
on the proton was very successful in uncovering its substructure which is
parametrized in terms of parton distributions. But the inclusive nature of
the reaction gives results averaged on all final states possible shadowing some
informations. The Virtual Comtpon Scattering on the proton is the electroproduction of photons $ep \rightarrow ep\gamma$. Since it requires a photon that can be
detected in the final state, it is an exclusive reaction which gives us an extra
degree of freedom allowing us to probe more thoroughly the proton structure.
In the Deeply Inelastic regime a new formalism has been devised called Generalized Parton Distributions (GPDs) which is able to take into account this
new parameter. These distributions are thus richer than the regular parton
distributions especially giving access to informations about correlations between partons.

Those last few years a lot of work has been put on both theoretical with the
GPDs formalism and the experimental side at many electron facilities around
the world and first results on DVCS electron helicity asymmetry from HER-
MES at the DESY electron facility (Hamburg, Germany) and from Hall B
at Jefferson Laboratory (Newport News,VA USA) have been published in
2001. Though encouraging these results showed also a high level of background events which were particularly hard to resolve from the real events
since one particle of the final state was not detected. Hence every laboratory
where a DVCS experiment is scheduled (Compass at CERN,HERMES and
HERA at DESY,Hall A and Hall B at Jefferson Laboratory) is designing or
building an additional detector to be able to detect both the proton and the
photon coming from a DVCS event to sign the exclusivity of the reaction. Every facility has specific technical difficulties to overcome but also has a specific
contribution to a better knowledge of the proton structure via the GPDs.

PRECISE DETERMINATION OF THE
PROTON CHARGE RADIUS EXPERIMENT (RPEX)

B. CLASIE

Massachusetts Institute of Technology,
Laboratory of Nuclear Science,
77 Massachusetts Ave
Cambridge MA 02139, USA
E-mail:clasie@mit.edu

At the HUGS 2002 student seminars, an introduction to RPEX (Pr. 00-02, spokespersons: H. Gao, J. Calarco) will be given. The proton rms charge radius is a fundamental quantity, and its precise determination will have significant impacts on tests of QCD and QED. It will be measured with sub 1% precision and will run at the Bates Linear Accelerator Center located in Middleton, Massachusetts. This experiment will utilize the symmetric sectors of the BLAST detector, and the super-ratio technique to extract both the proton elastic form factor ratio at low Q^2 and the proton charge radius. A laser driven polarized H/D target is being developed for use in this experiment.

SCALAR MESON TO PHOTON DECAY IN A LIGHT-FRONT CONSTITUENT QUARK MODEL

MARTIN DEWITT AND CHUENG JI

North Carolina State University,Physics Dept. Box 8202, Raleigh, NC 27695, USA
E-mail: madewitt@unity.ncsu.edu
E-mail: chueng_ji@ncsu.edu

The Standard Model predicts the existence of so-called "exotic" states such as glueballs,hybrids,and four-quark states. The presence of low lying exotics with $J^{PC} = 0^{++}$ complicates the analysis of the light scalar mesons since, given the fact that they have the same quantum numbers, they will tend to mix. Here, we present a calculation of the $f_0 \to \gamma\gamma$ transition form factor for the isoscalar states $f_0(1370)$, $f_0(1500)$, *and* $f_0(1710)$ using a light-front constituent quark model. The flavor-glue content of these states is determined using a mass-squared mixing scheme. The form factor calculation involves writing down a light-cone spin-space wavefunction for the scalar meson. The radial wavefunction is taken to be a simple harmonic oscillator wavefunction with one model parameter, which is then used as a trial wavefunction of the variational method. The spin-orbit wavefunction is determined exactly by the Melosh transformation. The full wavefunction is then used to calculate the two-photon transition form factor. In the limit as q^2 goes to zero, the form factor can be used to calculate the $f_0 \to \gamma\gamma$ decay width. Ratios of these decay widths for the different scalars can be compared with other model calculations.

PION PHOTOPRODUCTION
UP TO THE SECOND RESONANCE REGION

KURNIAWAN FOE

The George Washington University, Department of Physics,
725 Corcoran Hall Room 202
21st Street NW Washington DC 20052, USA
E-mail:kurn_foe@gwu.edu

The pion photoproduction has been extensively studied since the last four decades. This particular reaction is very useful in understanding the structure of nucleon and its resonances. We investigate this reaction up to the second resonance region. Though the differential cross section is relatively rather small (in mb units) but it implies the information about the role of nucleon resonances such as $P_{33}(1232)$, $P_{11}(1440)$, $D_{13}(1520)$, $S_{11}(1535)$, $P_{31}(1620)$. We use the πN scattering model developed by F.Gross-Y.Surya. They treated the P_{33} and D_{13} as pure spin-3/2 particles. This model describe the πN interaction very well. By extending this model to the γN and adding more higher baryon resonances, we hope there'll be a good agreement with the experimental data and some analysis (SAID and MAID). The other features of pion photoproduction; the role of P_{33} or Delta, the ratio of E_2/M_1, are also interesting to be investigated. While the experimental data for higher nucleon resonance from the γN are not produced yet, this investigation gives prominently foundation for further research.

MEASUREMENT OF THE NEUTRON ELECTRIC FORM FACTOR IN $D(\vec{E},E'\vec{N})P$ AT Q^2=0.6 AND 0.8 $(\frac{GEV}{C})^2$.

DEREK GLAZIER

Department of Physics and Astronomy, University of Glasgow, Scotland
E-mail: d.glazier@physics.gla.ac.uk

Recent double polarisation experiments at MAMI have provided accurate, model independent measurements of the Neutron Electric Form Factor, (G_E^n). Quasi-elastic, $D(\vec{e},e'\vec{n})p$ experiments, currently in progress at MAMI, will extract the polarisation components of the recoil neutron using a spin precession technique. The polarisation can then be used to calculate a value for G_E^n. This will complete the measurement of G_E^n as a function of Q^2 for $Q^2 < 1(\frac{GeV}{c})^2$. This will provide a stringent test for nucleon models, as G_E^n is particularly model sensitive in this Q^2 range.

THE DEEP VIRTUAL COMPTON SCATTERING EXPERIMENT

DAVID HAYES

Office 143 Trailer City Jefferson Lab

The DVCS experiment is scheduled to run in September, 2003. I will give a brief description of the DVCS reaction and the equipment we are building to test it. The main subject will be the recently completed DVCS test run and its relation to the experiment. I will present results from the elastic calibration runs for scintillator calibration and discuss how these are related to proton counting rates in the scintillator array.

SLAC EXPERIMENT E158: A PRECISION MEASUREMENT OF THE WEAK MIXING ANGLE IN MØLLER SCATTERING

L. J. KAUFMAN

University of Massachusetts, Department of Physics, Amherst MA 01002, USA
E-mail: ljkauf@physics.umass.edu

An overview of SLAC experiment E158 is presented. E158 makes a precision measurement of the parity-nonconserving left-right asymmetry in the scattering of longitudinally polarized electrons from the atomic electrons in a hydrogen target (Møller scattering). The asymmetry measures the effective pseudo-scalar weak neutral-current contribution to Møller scattering at an average Q^2 of 0.03 $(\text{GeV/c})^2$, and it is proportional to $(\frac{1}{4} - \sin^2 \theta_W)$, where $\sin^2 \theta_W$ is the electroweak mixing angle. The asymmetry will be measured to a precision of 7×10^{-9}, which corresponds to $\delta(\sin^2 \theta_W) \sim 0.0007$. A comparison of this measurement with precision asymmetry measurements of the Z^0 resonance provides the first statistically significant measurement of $\sin^2 \hat{\theta}_W(M_Z^2) - \sin^2 \hat{\theta}_W(0)$, testing the electroweak theory at the quantum loop level away from the Z^0 resonance. The experimental setup suited to the challenges of this precision measurement is described.

References

1. SLAC Proposal E158 (K. S. Kumar, spokesperson, E. W. Hughes and P. A. Souder, deputy spokespersons) (1997).

RADIATIVE VECTOR MESON DECAYS AT RADPHI EXPERIMENT

M. KORNICER

University of Connecticut
E-mail: kornicer@phys.uconn.edu

The existence of isoscalar $f_0(980)$ and isovector $a_0(980)$ hadronic resonances is well established but their quark structure is not yet known. Their
masses and widths are to low to fit in the $q\bar{q}$ scalar nonet picture, and it
has been suggested that they might be meson molecules or 4-quark states.
The Radphi experiment at Jlab is designed to study rare radiative decays of
$\phi(1020)$ meson, particularly $\phi \to a_0\gamma$ and $\phi \to f_0\gamma$ branching ratios. By measuring the ratio of these branching ratios, which is sensitive to the a_0 and f_0
quark content, it is expected to reveal the substructure of these mesons. The
ϕ meson is produced in flight by diffractive photoproduction in the $\gamma p \to \phi p$
reaction, using tagged photon beam with energy between 4-5.6 GeV. The lead
glass calorimeter together with the three-level trigger system is used to select
all-neutral final states. The known radiative decays of vector mesons $\omega \to \pi\gamma$
and $\phi \to \eta\gamma$ are used to calibrate the lead glass detector (LGD). The LGD calibration is prerequisite for answering the question posed by the rare radiative
ϕ decays.

THE $G0^0$ EXPERIMENT AT JLAB

J. LENOBLE

RM 12/C124 Mail Stop 12H2, Jefferson Laboratory, 12000 Jefferson Avenue,
Newport News VA 23606 , USA
E-mail: lenoble@jlab.org

The $G0^0$ experiment aims at measuring the parity violation asymmetries in elastic electron-proton scattering. The structure of the nucleon is not very well understood in the frame of QCD, i.e. in terms of the degrees of freedom that appear in the QCD Lagrangian. As the asymmetries are sensitive to the interference of the neutral weak and eletromagnetic amplitudes, the vector neutral weak form factors of the nucleons, G_E^Z and G_M^Z, two quantities precisely defined in the context of QCD, can be extracted from them. Assuming isospin symmetry that gives a relationship between the structure of the proton and the neutron, and by measuring assymetries at both forward and backward scattering angles, it is possible to determine the contributions of the quarks u, d and s to the form factors.

The $G0^0$ experiment will also extract separately the effective axial current of the nucleon, G_A^e by measuring quasi-elastic assymetries using a deuterium target in the backward angle. The axial current of the $N\Delta$ transition will also be measured by detecting in addition the inelastic electrons in the backward scattering experiment.

In the so-called "forward angle geometry", polarized electrons of $3 GeV$ are sent onto a $20 cm$ liquid hydrogen target and recoil protons are detected at angles around $70^o \pm 10^o$. This will allow to cover a Q^2 region of $.12 - 1 (GeV/c)^2$. The $40 \mu A$ beam delivered by the CEBAF facility at JLab is pulsed at $31.25 MHz$. The magnetic analysis of the recoil protons is done by a superconducting toroidal magnet having 8 sectors. The apparatus is completed by a series of pairs of plastic scintillators arranged as 16 independent detectors per sector and a specially designed counting electronics for acquiring directly TOF spectra that will allow to spatially separate elastic protons from background.

In the backward measurement, the $G0^0$ spectrometer will be turned and electrons scattered on liquid hydrogen and liquid helium targets will be detected. Cerenkov detectors will be added to the setup to discriminate electrons from pions.

The experiment will be carried out in Hall C at JLab. It is a collaboration of approximately 90 phycists from Caltech, Carnegie-Mellon, Connecticut, Illinois, IPN-Orsay, ISN-Grenoble, Jefferson Lab, Kentucky, Louisiana Tech, Manitoba, Massachussets, Maryland, New Mexico State, Norfolk State, Northern British Columbia, TRIUMF, Virginia Polutech, William & Mary, and Yerevan.

INVESTIGATION OF THE ^{3}HE(E,E'PN) AND ^{16}O(E,E'PN) REACTIONS

D. MIDDLETON

Department of Physics and Astronomy, University of Glasgow, University Avenue
Glasgow, G12 8QQ, Scotland
E-mail: d.middleton@physics.gla.ac.uk

A measurement of the (e,e'pn) reaction has been carried out in the 3 spectrometer facility of the A1-collaboration at the Mainz 855 MeV electron microtron MAMI. The targets studied were ^{3}He and ^{16}O and some data were also taken with a ^{2}H target for calibration purposes. A magnetic spectrometer was used to detect the scattered electrons, a large plastic scintillator hodoscope was used to detect the protons and a large time-of-flight scintillator array was used for detection of the neutrons. The ^{3}He data were taken at an energy transfer of 220 MeV and momentum transfer of 375 MeV/c. The ^{16}O data were taken at super-parallel kinematics with an energy transfer of 215 MeV and a momentum transfer of 316 MeV/c. For the kinematics studied the reaction cross section is sensitive to nucleon-nucleon correlations and two-body mechanisms like Δ-excitation and meson exchange currents.

PHASE STRUCTURE OF THE NON-LINEAR σ-MODEL
WITH OSCILLATOR REPRESENTATION METHOD

Yuriy Mishchenko

Department of Physics, North Carolina State University
Raleigh, North Carolina 27695-8202 USA

RS: Chueng-R. Ji

Department of Physics, North Carolina State University
Raleigh, North Carolina 27695-8202 USA

Non-Linear σ-model plays an important role in many areas of theoretical physics. Been initially uintended as a simple model for chiral symmetry breaking, this model exhibits such nontrivial effects as spontaneous symmetry breaking, asymptotic freedom and sometimes is considered as an effective field theory for QCD. Besides, non-linear σ-model can be related to the strong-coupling limit of $O(N)$ ϕ^4-theory, continuous limit of N-dim. system of quantum spins, fermion gas and many others and takes important place in undertanding of how symmetries are realized in quantum field theories. Because of this variety of connections, theoretical study of the critical properties of σ-model is interesting and important.

Oscillator representation method is a theoretical tool for studying the phase structure of simple QFT models. It is formulated in the framework of the canonical quantization and is based on the view of the unitary non-equivalent representations as possible phases of a QFT model. Successfull application of the ORM to ϕ^4 and ϕ^6 theories in 1+1 and 2+1 dimensions motivates its study in more complicated models such as non-linear σ-model.

In our talk we introduce ORM, establish its connections with variational approach in QFT. We then present results of ORM in non-linear σ-model and try to interprete them from the variational point of view. Finally, we point out possible directions for further research in this area.

DEEPLY VIRTUAL MESON PRODUCTION: THEORETICAL INTERPRETATION AND EXPERIMENTAL CONSIDERATIONS

LUDYVINE MORAND

CEA Saclay

The process under study here is the exclusive electroproduction of mesons on the nucleon, in the Bjorken regime (Q^2, ν large and x_B finite). The concept of Generalized Parton Distributions (GPDs) which appeared recently to interpretate such exclusive reactions in this regime is reviewed. I emphasize on the richness of GPDs and especially the fact that they provide new information on the nucleon spin structure, which may shed the light on the "spin puzzle". Furthermore, the upcoming experimental program at JLab aiming at measuring exclusive electroproduction of the vector mesons ρ, ω and ϕ in the Bjorken regime with the Cebaf Large Acceptance Spectrometer (CLAS) is presented. I discuss the relevant considerations for exclusive physics and what first step in the broad program of determining GPDs we hope to be able to make.

NUCLEAR MEDIUM EFFECTS ON ANALYZING POWER INVESTIGATED WITH A PROTON KNOCKOUT REACTION

R. NEVELING

Department of Physics, University of Stellenbosch, Stellenbosch, 7600, South Africa
E-mail: retief@physics.sun.ac.za

Exclusive measurement of quasi-free nucleon scattering provides a direct mechanism to study modifications of the free nucleon-nucleon (NN) interaction in nuclear medium. It is known that due to the density dependence of the NN interaction, quasi-free proton scattering from light targets (A$\leq$40) at medium energies ($\geq$400 MeV) yields analyzing powers that are substantially reduced with regards to Distorted Wave Impulse Approximation (DWIA) calculations that utilize the free NN interaction. However, at lower energies ($\leq$200 MeV) the free NN interaction appears to be adequate for the description of proton knockout from light targets. In order to extend the current data-set to knockout from a heavy target, the 200 MeV polarized proton beam of the National Accelerator Centre (Faure, South Africa) was utilized to study the $^{208}Pb(p, 2p)^{207}$Tl knockout reaction. The measured analyzing powers display the so-called 'quenching' effect with regards to standard DWIA calculations, as found for quasi-free scattering at higher energies. Agreement between theory and experiment is shown to improve only marginally when density dependent NN interactions are incorporated within the DWIA.

SEARCH FOR THE $\eta(1295), \eta(1400)$ **AND** $\eta(1480)$ **IN THE REACTION** $\mathrm{p\bar{p}} \to 2\pi^+2\pi^-\eta$ **AT REST**

Jörg Reinnarth

Helmholtz-Institut für Strahlen- und Kernphysik,
Nussallee 14-16,
D-53115 bonn, Germany
E-mail: reinnart@iskp.uni-bonn.de

The Standardmodel predicts, beyond pure $\bar{q}q$-states, exotic particles like glueballs and hybrids, which carry gluonic degrees of freedom. In the mass region of about 1250 MeV-1500 MeV, three $J^{PC} = 0^{-+}$ states are known, the $\eta(1295), \eta(1400)$ and $\eta(1480)$. All theses states cannot be pure $\bar{q}q$-states, thus the $\eta(1400)$ is discussed to be a glueball candidate.

Data was taken with the Crystal Barrel detector at LEAR at CERN. 6026 events of the reaction $\mathrm{p\bar{p}} \to 2\pi^+2\pi^-\eta$ were fitted and the branching ratio of the reaction $\mathrm{p\bar{p}} \to \pi^+\pi^-\eta(1400)$ determined to $(1.64\pm0.46)\cdot10^{-3}$. The partial wave analysis shows clear evidence for a $\eta(1400)$ and also hints for the $\eta(1480)$. The $\eta(1295)$ is not needed to describe the data. The results will be presented and a new interpretation of the eta spectrum will be given.

INVESTIGATION OF THE E-MESON IN THE REACTION
P$\bar{\text{P}}$ → $\pi^+\pi^-\pi^+\pi^-\eta$

JÖRG REINNARTH

Helmholtz Institut für Strahlen- und Kernphysik,
Nussalle 14-16, 53115 Bonn, Germany

In this paper we provide an overview on the theory and technique used to perform a partial wave analysis of the reaction p$\bar{\text{p}}$ → $\pi^+\pi^-\pi^+\pi^-\eta$. First, we will show why this channel is so interesting, and what the main questions in this meson sector are. Afterwards, we will present the main techniques of the partial wave analysis and at the end we will review the results and give a short interpretation.

1. Motivation for a partial wave analysis of the reaction p$\bar{\text{p}}$ → $\pi^+\pi^-\pi^+\pi^-\eta$

Gluons carry color charges of QCD and are members of an SU(3)-color octet. They can bind to color singlets, so called glueballs, or to q$\bar{\text{q}}$g-states, often called hybrids. A large number of experiments has been carried out to find such exotic particles. Exotic particles can carry the same quantum numbers as pure q$\bar{\text{q}}$-states. It is then difficult to decide, whether a state is pure q$\bar{\text{q}}$, exotic or mixed.

The $\eta(1440)$, first called ι (greek: the first) because it was believed to be the first glueball, is produced strongly in J/ψ decays where an enhancement of gluonic degrees of freedom is expected. This interpretation was supported by the similarity of the $\eta(1295)$ and $\pi(1300)$ masses suggesting ideal mixing. The corresponding s$\bar{\text{s}}$-state should carry a mass of about 1550 MeV. For this interpretation the $\eta(1440)$ is too light and an exotic one was favored. A further hint against the s$\bar{\text{s}}$ interpretation of the $\eta(1440)$ was the fact that the $\eta(1440)$ was produced in π^-p but not in K^-p scattering.

After some years it turned out that there is a mass splitting of the $\eta(1440)$ into the $\eta(1400)$ and $\eta(1480)$. Today three states close in mass are known, the $\eta(1295)$, $\eta(1400)$ and $\eta(1480)$, which cannot be all pure q$\bar{\text{q}}$-states. The questions arises which of these particles is exotic or whether there is another explanation. To answer these questions a partial wave analysis on $p\bar{p} \to 4\pi\eta$ was carried out.

2. $p\bar{p}$ annihilation at rest

The most significant data on $p\bar{p}$-annihilation come from the Crystal Barrel experiment at LEAR (Low Energy Antiproton Ring) at CERN [11]. The experiment collected large statistics of data on $p\bar{p}$ annihilation both at rest and in flight. A comprehensive review can be found in [3]. We concentrate on $p\bar{p}$ annihilation into $4\pi\eta$.

2.1. *Quantum numbers of the $p\bar{p}$ system*

Protonium is a system of two spin-$\frac{1}{2}$ particles, of a proton and an antiproton, and carries the same quantum numbers as mesons. The spins of the quarks can combine to either total spin $S = 0$ or total spin S = 1. In addition to the total spin, we can have orbital angular momentum l between the $q\bar{q}$ pair. Hence, l and S can combine to total angular momentem $J = l \otimes S$, where $J = |l - S|, |l - S + 1|, ..., |l + S|$. We can use l, S and J to construct the $I^G J^{PC}$ quantum numbers of the mesons, which follow :

$$
\begin{array}{ll}
\text{finalspin} & J = l + S \\
\text{parity} & P = (-1)^{l+1} \\
\text{C} - \text{parity} & C = (-1)^{l+S} \\
\text{G} - \text{parity} & G = (-1)^{l+S+I}
\end{array}
$$

Proton and antiproton are isospin $\frac{1}{2}$-particles, so they can couple to isospin 0 and 1. The allowed antiproton-proton quantum numbers are given in Table 1:

$^{2S+1}L_J$	$I^G \left(J^{PC} \right)$	
$^1S_0$	$1^-(0^{-+})$	$\mathbf{0^+(0^{-+})}$
$^3S_1$	$1^+(1^{--})^\star$	$0^-(1^{--})$
$^1P_1$	$\mathbf{1^+(1^{+-})}$	$0^-(1^{+-})$
$^3P_0$	$1^-(0^{++})$	$\mathbf{0^+(0^{++})}$
$^3P_1$	$1^-(1^{++})$	$\mathbf{0^+(1^{++})}$
$^3P_2$	$1^-(2^{++})$	$\mathbf{0^+(2^{++})}$

Table 1.: Quantum numbers of Protonium. Because of G-parity only the bold marked initial states decay to $4\pi\eta$.

The four pions carry negative G-parity and the η has positiv G-parity; the initial proton-antiproton system is restricted to positive G-parity.

Since G-parity is conserved in strong interaction, $p\bar{p}$ annihilation into $2\pi^+ 2\pi^- \eta$ is only allowed for these $p\bar{p}$ initial states having positive G-parity. To extract intermediate states, a partial wave analysis has to be carried out.

2.2. Annihilation into $\pi^+\pi^-\pi^+\pi^-\eta$

The data have been taken at the Low-Energy-Antiproton-Ring at LEAR at CERN. Low energy antiprotons (200 MeV/c) are extracted and brought to rest in a liquid hydrogen target. The annihilation products are then detected in the Crystal Barrel detector, which has been described in [11]. The target is surrounded by a pair of cylindrical multi-wire proportional chambers (PWC's) and a 23-layer cylindrical drift chamber (JDC). The JDC ist surrounded by a 1380 crystal CsI(Tl) barrel calorimeter. The crystals point towards the target center and the calorimeter covers polar angles between $12°$ and $168°$ degrees and 2π in azimuth. The useful acceptance for shower detection is 95% of 4π.

2.2.1. Event reconstruction

The data of the current analysis have been collected using a four-prong trigger. The trigger selects events with four tracks reaching the outermost layers of the JDC. The data are then reconstructed and those events satisfying the following criteria retain:

- Less than 15 photons in the Crystal Barrel calorimeter.
- Exactly four long charged tracks in the JDC. - For each electromagnetic shower, the energy deposited in the central crystal should exceed 14 MeV. This removes spurious hits due to shower fluctuations.
- Events containing photons centered in the crystals adjacted to the beam pipe are rejected due to possible shower leakage.
- Exactly two photons.

Data surviving these cuts are then submitted to a 4-c constraint kinematic fit to the hypothesis $p\bar{p} \to \pi^+\pi^-\pi^+\pi^-\gamma\gamma$. In a second step, a series of higher constraint kinematic fits are performed in which the $\gamma\gamma$ pairs are constrained to be either π^0 or η. From the 7.3 million four-prong triggers events, 6 023 survive these cuts, with a confidence level larger then 10%.

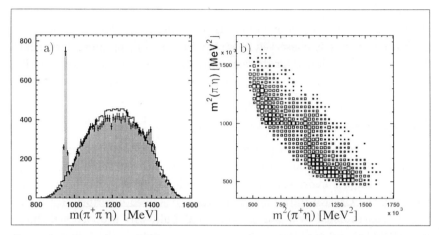

Figure 1.: *The $\pi^+\pi^-\eta$ invariant mass combination (a). The Dalitz plot of the decay $\eta(1440) \to \pi^+\pi^-\eta$.*

In Fig. 1.a) we show the $\pi^+\pi^-\eta$ invariant mass combinations with the phase-distribution. The most striking feature is the η'-peak, while the $\eta(1400)$ is also visible. Figure 1.b) shows the Dalitz plot of the reaction $\eta(1400) \to \pi^+\pi^-\eta$. The decay $\eta(1440) \to a_0(980)\pi$ is visible in the structure along the $\pi\eta$-axes. A partial wave analysis extracts all other contributions to the Dalitz plot, which are not clearly visible

In the following section the main ideas of the techniques of the partial wave analysis will be explained.

3. The partial wave analysis

Partial wave analysis is a technique which attempts to fit the production and subsequent decay of a meson by examining the mass and angular distribution of the system. A theoretical model of the decay of mesons is used to form the decay amplitude. Different parametrisations are fitted to the data by varying parameters like mass, width and complex production strength of the introduced intermediate resonances. Comparing the likelihoods of the results, the best data description will be found.

The description of sequential two-body decays takes place within the framework of the isobar model. Each subsequent two-body decay lies on a straight line in the center of mass system of the resonance. For $p\bar{p} \to 2\pi^+2\pi^-\eta$, a specific decay chain looks like:

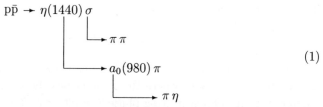

$$\tag{1}$$

Since the flight direction of the resonance is preferred, the angular distribution of the decay particles to this axis can be determined.

We now want to build the transition amplitude for such a decay chain. The amplitude has to take into account the dynamical part of the decay (phase space, mass and width of the particles), the angular distribution, which depends on the quantum numbers of the particles, the dynamical part of the daughter particles, and Clebsch Gordan coefficients taking into account the different particle combinations.

3.1. *The dynamical amplitude*

By fitting Breit-Wigner amplitudes to the data, we can in principle determine the masses and the widths of a state. In case of more than one resonance which strongly overlap, the Breit-Wigner functions can no longer be added because the unitarity condition is broken. The simple Breit-Wigner amplitudes do not even provide a natural mechanism to include thresholds. Therefore, the K-Matrix formalism is used, which stems from scattering theory and we write down the transition from some initial state to a final state. For a clear overview see [5].

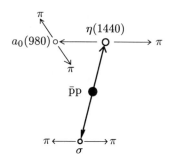

Figure 2.: One possible decay chain of the decay $p\bar{p} \rightarrow \eta(1440)\sigma$

The scattering matrix S takes an initial state to a final state

$$S_{fi} = < f|S|i >$$ (2)

This matrix is unitary $(SS^* = 1)$ and we can rewrite S in terms of the T-matrix as $S = I + 2iT$, where the T describes the transition and can be written in terms of a scattering phase $T = e^{i\delta} \sin \delta$. T can be expressed by the K-matrix, where $K = K^*$

$$K^{-1} = T^{-1} + iT$$ (3)

This can be inverted to yield:

$$T = K(I - iK)^{-1} = (I - iK)^{-1}K$$ (4)

Because of $T = e^{i\delta} \sin \delta$ we get the familiar form for a resonance:

$$K = \tan \delta$$ (5)

The F-Vector, the production and decay vector for a resonance, which decays into two open channels is given by:

$$F_A(m) = \frac{m_0 \sqrt{\Gamma_0}}{m_0^2 - m^2 - im_0 \left(\frac{\rho_1(m)}{\rho_1(m_0)} \Gamma_1 B_{L_1}^2 + \frac{\rho_2(m)}{\rho_2(m_0)} \Gamma_2 B_{L_2}^2 \right)} \left(\begin{array}{c} \sqrt{\frac{\Gamma_1}{\rho_1(m_0)}} B_{L_1} \\ \sqrt{\frac{\Gamma_2}{\rho_2(m_0)}} B_{L_2} \end{array} \right)$$ (6)

ρ denotes the phase space factor, which results from the masses of the initial and final particles in the reaction $A \to 1\ 2$:

$$\rho(m) = \sqrt{\left[1 - \left(\frac{m_1 + m_2}{m_A} \right)^2 \right] \left[1 - \left(\frac{m_1 - m_2}{m_A} \right)^2 \right]}$$ (7)

The barrier-factor $B_{A_i}^l(q, q_0)$ is the ratio of the centrifugal barrier factors with decay-momentum q and q_0 at the position of the resonance maximum.

$$B_{A_i}^L(q, q_0) = \frac{F_L(q)}{F_L(q_0)}$$ (8)

The decay-momentum q_i for a special decay-channel i can be calculated with the help of phase space factors ρ:

$$q_i(m) = m \frac{\rho_i(m)}{2}$$ (9)

The angular momentum barrier factors are computed as a function of $z = (q/q_R)^2$, $(Q_R = 197 \text{ MeV}/c)$, and are given as follows:

$$F_0(q) = 1 \tag{10}$$

$$F_1(q) = \sqrt{\frac{2z}{z+1}} \tag{11}$$

$$F_2(q) = \sqrt{\frac{13z^2}{(z-3)^2 + 9z}} \tag{12}$$

$$F_3(q) = \sqrt{\frac{277z^3}{z(z-15)^2 + 9(2z-5)^2}} \tag{13}$$

The T-Matrix is used to describe the decay of the daughter particles:

$$T_1(m) = \frac{m_0 \frac{\rho_i(m)}{\rho_i^0} \cdot \Gamma_i \cdot B_{L_i}^2}{m_0^2 - m^2 - im_0 \left(\frac{\rho_1(m)}{\rho_1(m_0)} \Gamma_1 B_{L_1}^2 + \frac{\rho_2(m)}{\rho_2(m_0)} \Gamma_2 B_{L_2}^2 \right)} \tag{14}$$

By fitting these amplitudes to the data, we can determine the masses and widths of a state. To determine the J^{PC} of a state, we have to look at the angular distributions of the decay products with respect to some initial state, which is included in the final amplitude through the helicity amplitude H.

3.2. *The helicity amplitudes*

The helicity is defined as the projection of the total spin, $\vec{J} = \vec{l} + \vec{s}$, into the direction of motion of the particle $\vec{p}$. Because $\vec{l}$ and $\vec{p}$ are orthogonal to each other, we arrive at:

$$\lambda = \frac{\vec{J} \cdot \vec{p}}{|\vec{p}|} = \frac{\vec{l} \cdot \vec{p}}{|\vec{p}|} + m_S = m_S \qquad \text{with} \quad -|\vec{s}| \le m_S \le |\vec{s}|, \tag{15}$$

where m_s is the projection of the particle's spin along the z axis. The helicity operator is invariant under rotations and boosts along $\vec{p}$. We generalize particle 2 in the decay $A \to 1\,2$ being emitted in some arbitrary direction, (θ, ϕ). We can go from this frame to the frame where particle 2 is moving along the z axis via a rotation in θ and ϕ. The helicity states refer to a system Σ_3, which can be reached through two different rotations from the initial system Σ_1.

$$R(\theta, \phi) = R_{y_2}(\theta) R_{z_1}(\phi) \tag{16}$$

Figure 3 shows these rotations. The first rotation transforms system Σ_1 into system Σ_2 by a rotation about z_1 by an angel ϕ. In the next step the new system is transformed about z_2 by θ into system Σ_3. A last rotation about z_3 does not change the direction of z_3 so that the last angle ψ is chosen to be zero.

The eigenvalue $|jm\rangle$ is irreducible under this transformation, which yields the following:

Doing this classical rotations, we use three Euler angles to accomplish the rotation of the reference frame. The full rotation $R(\theta, \phi)$ in spin space is

$$R(\theta, \phi) = e^{\hat{n} \cdot \vec{J}} = e^{i\phi J_z} \cdot e^{iJ_y \theta}. \tag{17}$$

$$R(\theta, \phi, \psi = 0) \, |jm\rangle = \sum_{m'=-j}^{j} D_{m'm}^{j}(\theta, \phi, \psi = 0) \, |jm'\rangle \tag{18}$$

$$R(\theta, \phi) = D_{mm'}^{J}(\theta, \phi) = e^{im'\phi} d_{mm'}^{J}(\theta) \tag{19}$$

We are now able to rotate our helicity state from system 3 where particle 2 is moving along the z axis to system 1 where it is moving in some arbitrary direction (θ, ϕ)

$$|p, \theta, \phi, \lambda_1, \lambda_2, M\rangle_1 = D_{M\lambda}^{J}(-\theta, -\phi) \, |p, \lambda_1, \lambda_2\rangle_3 \quad \text{with} \quad \lambda = \lambda_1 - \lambda_2 \tag{20}$$

The transition of state A into 1 and 2 with helicities λ_1 and λ_2 is

$$f_{\lambda_1 \lambda_2, M}(\theta, \phi) = \langle \, |\vec{p}|, \lambda_1, \lambda_2, M \, | \, T \, | \, M' \rangle = D^{*J}_{M\lambda}(-\theta, -\phi) \langle \lambda_1 \lambda_2 \, | \, T \, | \, M' \rangle$$
$$= D_{\lambda M}^{J}(\theta, \phi) \, T_{\lambda_1 \lambda_2} = e^{iM\phi} d_{\lambda M}^{J}(\theta) \, T_{\lambda_1 \lambda_2} \tag{21}$$

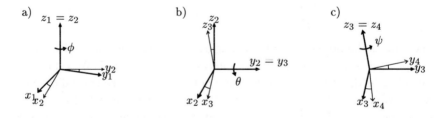

Figure 3.: a) Rotation about the z_1-axis by an angel ϕ. b) Rotation about the y_2-axis by an angel θ. c) Rotation about the y_2-axis by an angel ψ.

This transition-amplitude is a matrix with $(2s_1 + 1) \cdot (2s_2 + 1)$ rows and $(2J+1)$ columns. The angular distribution of the decay particles is given by $D^J_{M\lambda}(\theta, \phi)$, while the decay dynamics is placed into the $(2s_1 + 1) \times (2s_2 + 1)$-dimensional matrix $T_{\lambda_1 \lambda_2}$ summing over all possible l and s values with an unknown complex coefficient, α_{ls} for each of these elements.

$$T_{\lambda_1 \lambda_2} = \sum_{ls} \alpha_{ls} \langle J\lambda \,|\, l\,s\,0\,\lambda \rangle \langle s\,\lambda \,|\, s_1\,s_2\,\lambda_1 - \lambda_2 \rangle \qquad (22)$$

In this case, we are interested in a sequential decay into 5 particles so that the formalism has to be extended. Let us consider the decay

$$A \to [1 \to 1_a\, 1_b][2 \to 2_a\, 2_b] \qquad (23)$$

We can calculate a transition matrix for each of the three decays, $f(A \to 1\, 2)$, $f(1 \to 1_a\, 1_b)$ and $f(2 \to 2_a\, 2_b)$. The final transition matrix is given by the vector product:

$$f_T = [f(2) \times f(1)] \times f(A) = \sum_{\lambda(2), \lambda(1)} (f_{\lambda(1_a)\lambda(1_b),\lambda_1} \times f_{\lambda(2_a)\lambda(2_b),\lambda_2}) \cdot f_{\lambda(1)\lambda(2),\lambda(A)} \qquad (24)$$

Let us now take a look at a simple example

$$f_0(1370) \to \rho(\to \pi\pi)\rho(\to \pi\pi) \qquad (25)$$

In the case of the decay $\rho \to \pi\pi$ ($1^- \to 0^+0^+$), the angular momentum l is 1 as the total spin $J = 1$. The spin of both pions is $s = s_1 = s_2 = 0$, so that the helicity states are $\lambda = \lambda_1 = \lambda_2 = 0$. The T-matrix is given by:

$$T_{\lambda_1=0,\lambda_2=0} = \sum_{ls} \alpha_{ls} \langle J\lambda \,|\, l\,s\,0\,\lambda \rangle \langle s\,\lambda \,|\, s_1\,s_2\,\lambda_1 - \lambda_2 \rangle \qquad (26)$$

$$T_{0,0} = \langle 10|1000\rangle\langle 00|0000\rangle = 1 \qquad (27)$$

The transition amplitude f - with one row $(= (2s_\pi) + 1 \cdot (2s_\pi + 1))$ and three columns $(= 2J + 1)$ - arises to:

$$f_{00,M} = \left(D^1_{01}(\theta_\rho, \phi_\rho)\ D^1_{00}(\theta_\rho, \phi_\rho)\ D^1_{0-1}(\theta_\rho, \phi_\rho) \right) = \left(e^{i\phi}\tfrac{\sin\theta}{\sqrt{2}}\ \cos\theta\ -e^{-i\phi}\tfrac{\sin\theta}{\sqrt{2}} \right) \qquad (28)$$

The tensor product $f_{Tensor} = f(\rho_1 \to \pi\pi) \times f(\rho_2 \to \pi\pi)$ consists of nine columns:

$$f_{Tensor} = \begin{pmatrix} e^{i(\phi_1+\phi_2)}\frac{\sin\theta_1\sin\theta_2}{2} & e^{i\phi_1}\frac{\sin\theta_1\cos\theta_2}{\sqrt{2}} & -e^{i(\phi_1-\phi_2)}\frac{\sin\theta_1\sin\theta_2}{\sqrt{2}} \\ e^{i\phi_2}\frac{\sin\theta_2\cos\theta_1}{\sqrt{2}} & \cos\theta_1\cos\theta_2 & -e^{-i\phi_2}\frac{\cos\theta_1\sin\theta_2}{\sqrt{2}} \\ -e^{-i(\phi_1-\phi_2)}\frac{\sin\theta_1\sin\theta_2}{2} & -e^{-i\phi_1}\frac{\sin\theta_1\cos\theta_2}{\sqrt{2}} & e^{-i(\phi_1+\phi_2)}\frac{\sin\theta_1\sin\theta_2}{2} \end{pmatrix} \qquad (29)$$

The decay $f_0(1370) \to \rho\rho$

In this case $0^+ \to 1^- 1^-$ the total angular momentum is $J = 0$ and both spins of the ρ's are $s_1 = s_2 = 1$. The angular momentum l is 0, so that the final spin is $s = 0$. λ_1 and λ_2 can carry the values 1, 0 and -1, so that nine possible combination for $\lambda = \lambda_1 - \lambda_2$ exist. The T-matrix is given by:

$$T_{\lambda_1=0,\lambda_2=0} = \sum_{ls} \alpha_{ls} \langle J\lambda | l\, s\, 0\, \lambda \rangle \langle s\lambda | s_1\, s_2\, \lambda_1 - \lambda_2 \rangle \qquad (30)$$

$$T_{\lambda_1,\lambda_2} = \langle 0\lambda | 0\, 0\, 0\, \lambda \rangle \langle 0\lambda | 1\, 1\, \lambda_1 - \lambda_2 \rangle \qquad (31)$$

Calculating all possible combination we achieve

λ_1	λ_2	$\lambda = \lambda_1 - \lambda_2$	$\langle 0\lambda\|0\,0\,0\,\lambda\rangle$	$\langle 0\lambda\|s_1\,s_2\,\lambda_1 - \lambda_2\rangle$		T_{λ_1,λ_2}	
1	1	0	1	$\sqrt{\frac{1}{3}}$	$T_{1,1}$	$=$	$\sqrt{\frac{1}{3}}$
1	0	1	0	0	$T_{1,0}$	$=$	0
1	-1	2	0	0	$T_{1,-1}$	$=$	0
0	1	-1	0	0	$T_{0,1}$	$=$	0
0	0	0	1	$-\sqrt{\frac{1}{3}}$	$T_{0,0}$	$=$	$-\sqrt{\frac{1}{3}}$
0	-1	1	0	0	$T_{0,-1}$	$=$	0
-1	1	-2	0	0	$T_{-1,1}$	$=$	0
-1	0	-1	0	0	$T_{-1,0}$	$=$	0
-1	-1	0	1	$\sqrt{\frac{1}{3}}$	$T_{-1,-1}$	$=$	$\sqrt{\frac{1}{3}}$

The D-function is 1 if $J = \lambda = M = 0$, and we find for $f_{\lambda_1,\lambda_2,M}(\theta,\phi) = D^J_{\lambda M}(\theta,\phi) \cdot T_{\lambda_1 \lambda_2}$:

$$f(f_0 \to \rho\rho) = (\frac{1}{3}, 0, 0, 0, -\frac{1}{3}, 0, 0, 0, \frac{1}{3})^T \qquad (32)$$

We now calculate the final product f_T by

$$f_T = \left[f(\rho_1 \to \pi^+\pi^-) \otimes f(\rho_2 \to \pi^+\pi^-) \right] \cdot f(f_0(1370) \to \rho_1\rho_2) \qquad (33)$$

$$f_T = \sqrt{\frac{1}{3}} \cdot e^{i(\phi_1+\phi_2)} \frac{\sin\theta_1 \sin\theta_2}{2} + \left(-\sqrt{\frac{1}{3}} \right) \cdot \cos\theta_1 \cos\theta_2 \qquad (34)$$

$$+ \sqrt{\frac{1}{3}} \cdot e^{-i(\phi_1+\phi_2)} \frac{\sin\theta_1 \sin\theta_2}{2} \qquad (35)$$

$$= \frac{1}{\sqrt{3}} [\cos(\phi_1 + \phi_2) \cdot \sin\theta_1 \sin\theta_2 - \cos\theta_1 \cos\theta_2] \qquad (36)$$

In this analysis, the helicity amplitude and the amplitude for the dynamics form the final amplitude. To understand the acceptance of the detector system, good detector resolution is crucial.

3.3. *Final amplitude*

The final amplitude which is used for data description is given by the sum over all decay channels and the sum over all different combinations for each decay channel :

$$A_k = \sum_{j=1}^{decay\ channel} \sum_{i=1}^{\#combinations} CG_i \cdot F_{A_{ij}} \cdot T_{B_{ij}} \cdot T_{C_{ij}} \cdot H_j \cdot B_L \qquad (37)$$

CG_i Clebsch-Gordan-Coefficents, $F_{A_{ij}}$ F-Vector which describes the production and decay of the daughter resonances, T decay amplitude for the daughter particles H_j B_L

3.4. *The technique of the partial wave analysis*

We are now able to produce a theoretical distribution of the reaction $p\bar{p} \to 4\pi\eta$. Using the conservation of quantum numbers due to strong interaction, we calculate all possible decay chains and calculate the helicity, dynamical and final amplitudes for each decay chain. These amplitudes are added up coherently for each initial state and summed up incoherently over the initial states. Using a loglikelihood fit, we calculate for a set of parameters (mass, width, amplitude, phases) the value, called FCN, which is an indicator for the quality of the description. The parameters are now varied, to get the best description of the data. We use MINUIT, which automatically varies prenamed parameters and optimizes the FCN.

In the first step of a partial wave analysis, different combinations of resonances are tested. We try to observe if a special decay channel increases the FCN value significantly, which means that it is clearly needed to describe the data. Afterwards the masses and widths of a single resonance are varied freely and MINUIT tries to find the optimum of these parameters while all other parameters (amplitudes and production strengths) are varied freely.

In the last step, we use scans, where we change masses, using fixed widths, in steps and take a look at the change of the FCN. The FCN should carry an optimum at the mass of the resonance and all other masses should give a worse FCN value. Seeing a maximum in such a scan gives hints for resonances in the decay chains.

4. The partial wave analysis

The $\pi^+\pi^-\pi^+\pi^-\eta$ data were analyzed using a maximum likelihood fit. In p$\bar{\text{p}}$-annihilation, 90% of all decays stem from $^1S_0$ or $^3S_1$ initial states. So only these two states and the $^3P_0$-state are included in these analyses. We start with amplitudes for the following reaction chains:

- p$\bar{\text{p}}$ $(^1S_0)\to\eta f_0(1370)$; $f_0(1370)\to\sigma\sigma$
  $$f_0(1370)\to\rho\rho$$
- p$\bar{\text{p}}$ $(^1S_0)\to\sigma\eta(1410)$; $\eta(1410)\to a_0(980)\pi$
  $$\eta(1410)\to\sigma\eta$$
- p$\bar{\text{p}}$ $(^1S_0)\to\pi a_0(1450)$; $a_0(1450)\to a_0(980)\sigma$
- p$\bar{\text{p}}$ $(^3S_1)\to\rho f_1(1285)$; $f_1(1285)\to a_0(980)\pi$
  $$f_1(1285)\to\sigma\eta$$
- p$\bar{\text{p}}$ $(^3S_1)\to\eta b_1(1235)$; $b_1(1235)\to\rho\sigma$
- p$\bar{\text{p}}$ $(^3P_0)\to\eta f_1(1285)$; $f_1(1285)\to\rho\rho$

We refer to this compilation as our reference fit. Its comparison with data is demonstrated in Figures 4.

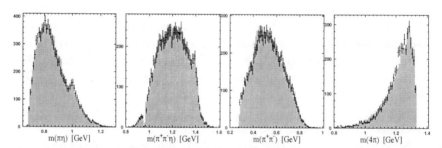

Figure 4.: *Result of our reference fit: invariant mass distributions $m(\pi\eta)$, $m(\pi^+\pi^-\eta)$, $m(\pi^+\pi^-)$ and $m(4\pi)$. Data: error bars, Fit: shaded region.*

4.1. *The E-meson*

Starting from our reference fit, we search for a optimum mass of a pseudoscalar resonance with a width of 55 MeV. Fig. 5.(a) shows the log likelihood as a function of the assumed mass. The need for a pseudoscalar state at 1400 MeV is obvious. Also a small bump at a mass about 1240 MeV is visible, which is shown more clearly in Fig. 6 (b). The question arises if this bump can be ascribed to the $\eta(1295)$, which will be explored in the next section. The local maximum of the $\eta(1400)$ is explored by plotting the likelihood as a function of mass for fixed widths or as a function of width (Fig. 5.a) for fixed masses (Fig. 5.b). Likelihood changes less than 9 correspond to 3σ intervals. We deduce

$$M = (1405 \pm 7) \text{ MeV} \qquad \text{and} \qquad \Gamma = (55 \pm 5) \text{ MeV} \qquad (38)$$

which agrees very well with the PDG averages of (1405 ± 5) and (56 ± 7) MeV, respectively. The branching ratio for the decay chain $p\bar{p} \rightarrow \pi^+\pi^- E; E \rightarrow \pi^+\pi^-\eta$, is determined to $(1.64 \pm 0.46) \cdot 10^{-3}$.

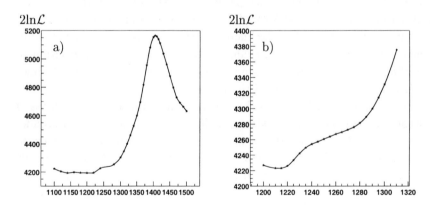

Figure 5.: *(a): Scan for a 0^+0^{-+} resonance with a fixed width of 55 MeV, (b) more detailed scan in the mass region of the $\eta(1295)$*

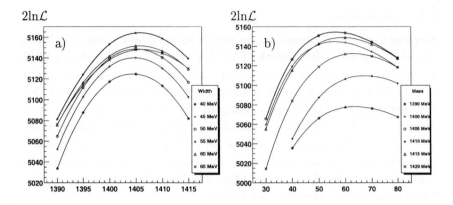

Figure 6.: *(a): Scans for a 0^+0^{-+} resonance with fixed widths between (40-65) MeV, (b) scan for the E-meson width with fixed masses between (1390-1420) MeV.*

4.2. *Search for further η resonances*

The $\eta(1295)$ has precisely-known parameters. Mass and width are known with an accuracy of a few MeV. Hence we tried to introduce this meson with its PDG values. The likelihood $(2\ln\mathcal{L})$ increases by 22 for 4 additional parameters, a marginal change only. Its fractional contribution to the $\pi^+\pi^-\pi^+\pi^-\eta$ final state is found to be less than $3\cdot10^{-5}$, about 50 times smaller than the contribution of the E-meson.

For the nominal $\eta(1295)$ width the likelihood as a function of the mass of a second η-resonance shows a strange behavior with 2 peaks and a broader structure (see Fig. 7.a). At low masses a peak at 1180 MeV of insignificant strength appears. Scanning the width at the fixed PDG mass of 1296 MeV (see Fig. 7.b), no maximum is found, the likelihood continues to increase with width. These observations lead us to conclude that we have no evidence for the $\eta(1295)$.

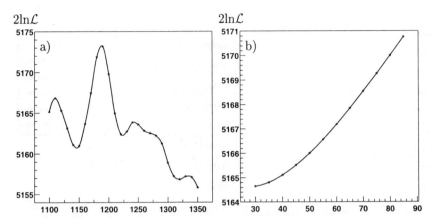

Figure 7.: *(a): Scan for an additional 0^+0^{-+} resonance with fixed width of 65 MeV,(b) with fixed mass of 1295 MeV.*

4.2.1. *The $\eta(1480)$*

The E(1440) is possibly split into a low-mass $\eta(1400)$ and a high mass $\eta(1480)$ [1]. We searched for an additional high-mass state and find indeed some evidence for it. For a width of 56 MeV, the scan optimizes for a mass of (1480-1500) MeV. The likelihood change is 28. Scanning the width including several masses, a maximum at a width of (30-50) MeV shows up, depending on the mass of the resonance. But also in this scan hints for the $\eta(1480)$ are visible. If an $\eta(1480)$ exists, it contributes to the $4\pi\eta$ channel by $(2.6 \pm 0.5) \cdot 10^{-4}$.

5. Summary and discussion

The $E \rightarrow \pi^+\pi^-\eta$ decay is clearly established in our data. Mass and width of the E-meson are fully compatible with previous determinations of these quantities (Table 5) and compare quite well with our previous finding [4] that the E-meson decays with approximately equal rates via $a_0(980)\pi$ and $\eta\sigma$ into $\eta\pi\pi$. We searched for the $\eta(1295)$ but did not find evidence for its production in $p\bar{p}$ annihilation. Even if the $f_1(1285)$ is excluded from the fit, there is no $\eta(1295)$. There is evidence that the E could be split into a $\eta(1410)$ and a $\eta(1480)$. We find hints for an extra 0^{-+}-state at a mass of 1480 MeV. The contribution of this state is 5 times smaller than the contribution of the $\eta(1400)$.

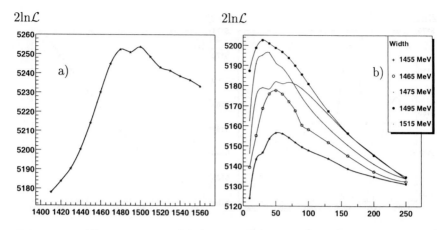

Figure 8.: *(a): Scan for an additional 0^+0^{-+} resonance with fixed width of 55 MeV,(b) with fixed masses of 1480 MeV*

$\eta(1440)$				
	Mass (MeV)	Width	$\Gamma(a_0(980))/\Gamma(\eta\sigma)$	$\Gamma(a_0(980))/\Gamma(\eta\pi\pi)$
This analysis	1405±7	55±5	0.62±0.14	0.38±0.09
PDG	1405±5	56±7	—	—
Amsler [8]	1413	49±8	0.4±0.2	0.29±0.10
Alde [6]	1424±6	85±18	—	0.19±0.04
Amsler [4]	1409±3	86±10	1.28±0.2	0.56±0.04±0.03
OBELIX [?]	1385±7	—	0.70±0.12±0.20	—

$\eta(1480)$				
	Mass (MeV)	Width	$\Gamma(a_0(980))/\Gamma(\eta\sigma)$	$\Gamma(a_0(980))/\Gamma(\eta\pi\pi)$
This analysis	1490±20	55±25	—	—
PDG	1475±5	81±11	—	—

Table 2.: Comparison of $\eta(1400)$ and $\eta(1480)$ decay modes as determined in different reactions.

5.1. *Possible Explanation*

Until today the explanation of the nature of the three particles, $\eta(1295)$, $\eta(1400)$ and $\eta(1480)$, given in the PDG identifies the $\eta(1295)$ as the first radial excitation of the η (2^1S_0), the $\eta(1480)$ as the first radial excitation of the η' (2^1S_0), which leaves the $\eta(1400)$ to be an exotic candidate.

Calculation of J. Suh within the $^3P_0$-model give room for another interpretation [10]. In Figre 9.a the Breit-Wigner resonance shape of a decaying particle with a mass of 1440 MeV/c^2 is shown. This is the transition amplitude with a maximum at the resonance mass. He calculated the decay amplitude for a decay of an $\eta(1440)$ into $a_0\pi$. As visible in Figure 9.b the decay amplitude has a pole in the wave function at the resonance mass. The final wave function of the particle, which is seen in nature, is the product of the transition amplitude and the decay amplitude and is shown in Figure 9.c. Because of the pole of the decay function at the resonance mass, the final amplitude gets two maxima and an $\eta(1440)$ decaying into $a_0\pi$ is visible in two different structures, the $\eta(1400)$ and the $\eta(1480)$. The position of these maxima in calculation depends on the parameters, so that in Figure 9.c the maxima do not lie at 1400 MeV/c^2 and 1480 MeV/c^2. The $\eta(1400)$ and $\eta(1480)$ could stem from one particle with a mass about 1440 MeV/c^2.

5.2. *New interpretation*

As shown, the $\eta(1400)$ and $\eta(1480)$ could correspond to one η with a mass about 1440 MeV/c^2. So just the $\eta(1295)$ and the $\eta(1440)$ have to be placed in the nonetts. In this analysis we do not find a hint on the $\eta(1295)$, which also has not been seen in $\gamma\gamma$-production at LEP. If the $\eta(1295)$ really was the radial excitation of the η, the branching ratio of p$\bar{\text{p}} \rightarrow \eta(1295)\pi\pi$ should be in the order of magnitude of p$\bar{\text{p}} \rightarrow \pi(1300)\pi\pi$ ans so be clearly visible. But it is not. The branching ratio of p$\bar{\text{p}} \rightarrow \eta(1400)\pi\pi$ is indeed of the same order of magnitude, so that we interpret the $\eta(1400)$ as the radial excitation of the η. The $\eta(1295)$ gets an exotic interpretation whereby the existence of the $\eta(1295)$ is questionable.

18

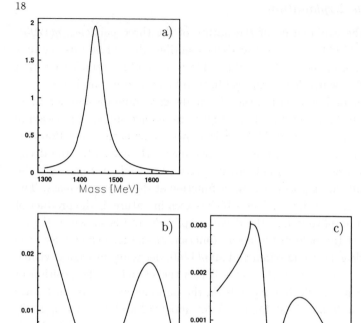

Figure 9.: The final amplitude c) is the product of the transition amplitude a) and the decay amplitude b)

References

1. K. Hagiwara *et al.*, Phys. Rev. **D 66** (2002)
2. E. Aker *et al.*, Nucl. Instrum. Methods **A 321** 69 (1992)
3. C. Amsler, Rev. Mod. Phys. **70** 1293 (1998)
4. C. Amsler *et al.*, Phys. Lett. **B 358** 389 (1995)
5. S. U. Chung, J.Brose, R. Hackmann, E. Klempt, S. Spanier and C. Strassburger, Ann. Phys. (Leipzig) **4** 4004 (1995)
6. D. Alde *et al.* Phys. Atom. Nucl. **60** 386 (1997)
7. J. Z. Bai *et al.* Phys. Lett. **B446** 356
8. A. Abele *et al.* Nucl. Phys. **B514** 45
9. E. Klempt, Meson Spectroscopy: Glueballs, Hybrids and qq̄ Mesons, Summary of the lessons for the summer school in Zuoz.
10. J. Suh, PhD Thesis, Untersuchung des E-Mesons in der Proton-Antiproton-Vernichtung in Ruhe
11. Crystal Barrel Collaboration, E. Aker *et al.*, Nucl. Instrum. Meth. **A 321** (1992) 69

REAL COMPTON SCATTERING AT JEFFERSON LAB

M. ROEDELBRONN AND A. DANAGOULIAN

Nuclear Physics Lab, Loomis Laboratory of Physics, 1110 West Green Street,
Urbana IL 61801, USA
E-mail: roedelbr@uiuc.edu, aregjan@jlab.org

An experiment has been approved to measure the exclusive, unpolarized cross section for Real Compton Scattering in the energy range 3-6 GeV and over a wide angular range (65-130° scattering angle in the CM frame), as well as to measure the longitudinal and transverse components of the polarization transfer to the recoil proton at a single kinematic point. Together, these measurements will test models of the reaction mechanism and possibly determine new structure functions of the proton that are related to the same nonforward parton densities that determine the elastic electron scattering scattering form factors and the parton densities. The experiment utilizes an untagged bremsstrahlung photon beam and the standard Hall A cryogenic liquid hydrogen targets. The scattered photon is detected in a photon spectrometer, currently undergoing construction and testing. The coincident recoil proton is detected in one of the Hall A magnetic High Resolution Spectrometers (HRS), and its polarization components are measured in the existing Focal Plane Polarimeter. At present this experiment is scheduled to run in January 2002.

HIGHER TWIST CORRECTIONS TO PARITY VIOLATING DEEP INELASTIC ASYMMETRY

GIAN FRANCO SACCO

California Institute of Technology , 391 S. Holliston Ave., Pasadena CA 91125, USA

E-mail: giangi@krl.caltech.edu

One of the best tool to investigate parity violating phenomena is to consider the asymmetry $\frac{\sigma_R - \sigma_L}{\sigma_R + \sigma_L}$ where $\sigma_{R(L)}$ indicates the cross section of a right handed (left handed) polarized electron beam scattering on an unpolarized nuclear target. Such asymmetry assumes a relatively easy form in the parton model, and if sea quark and mass corrections are neglected, then it does not depend on the nuclear structure [1]. Our goal is to study the importance of such corrections in different kinematics regions. In the next years such asymmetry might be used to extract a precise value of the Weinberg angle so it is necessary to know and take into account all the possible corrections to the naive parton model.

By using a code provided by the CTEQ group we were able to study the behaviour of the asymmetry as a function of Q^2 and the Bjorken variable x . The result was that our prediction differ utmost one percent with respect the one in [1]. Besides sea quark and mass corrections, another type of correction could be important in the analysis of the asymmetry: those are the so called higher twist corrections which are usually suppressed by $\frac{1}{Q^2}$, where $Q^2 = -q^2$ with q equal to the momentum transferred. Such corrections have to be taken into account , eventually, when the momentum transferred is of the order of a *Gev*. Unfortunately higher twists are given by matrix elements of quark currents between hadronic states, and due to the lack of knowledge of the hadronic wave functions, we have to rely on some model. In the future we plan to calculate some of these higher twist effects by using the bag model, to get an estimate of the order of magnitude in which they might enter the asymmetry.

References

1. R.N. Cahn, F.J. Gilman *Phys. Rev.* D **17**, 1313 (1978).

G0: A STRANGE EXPERIMENT

J. A. Secrest

*Department of Physics and Astronomy, College of William and Mary, Williamsburg
VA 23187 USA
E-mail: secrest@jlab.org*

G0 is a parity violation experiment with a dedicated large acceptance spectrometer. This experiment will measure parity violating asymmetries in electron-proton scattering to probe the $s\bar{s}$ sea of the nucleon between $0.1 \leq Q^2 \leq 1.0$ GeV^2 at a beam energy of 3.0 GeV. The strange quark contributions to the magnetic and charge densities of the proton will be determined by measuring the strange form factors $G_M^s(Q^2)$ and $G_E^s(Q^2)$. I will discuss what makes G0 different from other strange experiments, the experimental set-up with an emphasis on the detectors, and what remains to be done before the experiment begins in 2002.

MEASUREMENT OF THE NEUTRON (^{3}HE) SPIN STRUCTURE FUNCTIONS AND THE GENERALIZED GDH SUM AT JEFFERSON LAB

K. Slifer for the Jefferson Lab E94010 Collaboration
Temple University, Philadelphia PA 19122, USA
E-mail: kslifer@temple.edu

In order to investigate the Q^2 dependence of the generalized Gerasimov-Drell-Hearn (GDH) sum for the neutron, we have measured the spin dependent longitudinal and transverse $^3\vec{\text{He}}(\vec{e}, e')$ cross sections for momentum transfers ranging from 0.1 to 1 GeV^2 and excitation energies across the quasielastic, resonance and deep inelastic region. The longitudinally polarized ($P_{\text{beam}} \simeq 70\%$) electron beam of energy ranging from 0.86 to 5.1 GeV of the Thomas Jefferson National Accelerator Facility was scattered at a fixed angle, off a high pressure polarized ($P_{\text{targ}} \simeq 35\%$) ^{3}He target in Hall A. The direction of the target polarization was maintained either parallel or perpendicular to the incident electron beam. This measurement allows the extraction of the polarized neutron spin structure functions g_1 and g_2 and the evaluation of the extended GDH sum on both ^{3}He and the neutron. This sum, when compared to theoretical models, will help us to understand the transition from the perturbative to the non-perturbative regime of QCD.

FOCAL-PLANE POLARIMETER

STEFFEN STRAUCH

Rutgers, The State University of New Jersey, Piscataway, New Jersey, 08854

The preferential spin orientation of a particle in the final state of a nuclear reaction is an observable which carries information about the underlying reaction mechanism. In this introductory lecture I will discuss the basic principles of polarimetry of spin $\frac{1}{2}$ particles, focal-plane proton polarimeter and the extraction of the proton recoil polarization.

1 Introduction

The polarization of a particle is defined as its preferential spin orientation. Compared to cross section measurements polarization degrees of freedom make a much richer variety of observables accessible and thus provide a powerful tool to study nuclear reactions. I will present two examples for proton recoil polarization that demonstrate this: deuteron photodisintegration $d(\gamma, \vec{p})$ and electron scattering $A(\vec{e}, e'\vec{p})$.

Deuteron photodisintegration is governed by 12 independent complex helicity amplitudes $F_{i\pm}$, which are T-matrix elements between the initial and final states.[1] The unpolarized differential cross section is proportional to the sum of the squares of these amplitudes $d\sigma/d\Omega \propto \sum_{i=1}^{6} \sum_{\epsilon=\pm} |F_{i\epsilon}|^2$ and does not contain phase information. The induced polarization p_y of the recoiling proton, however, is proportional to another combination of amplitudes with interferences between the various amplitudes and reveals phase information: $p_y \propto \text{Im} \sum_{i=1}^{3} \left[F_{i+}^* F_{(i+3)-} + F_{i-}^* F_{(i+3)+} \right]$. In coincidence electron scattering off an unpolarized target the $(\vec{e}, e'\vec{N})$ physical observables are expressed in terms of 18 response functions, which are bilinear combinations of the nuclear electromagnetic current operator[2]. Thirteen of these response functions are exclusively accessible by recoil polarization measurements of the ejected nucleon. It is not the total number of accessible response functions which makes the polarization measurement so attractive, it is the option to pick specific responses which are most sensitive to certain aspects under study.

Recoil polarization experiments at electron scattering facilities only became possible with the advent of high luminosity electron beams. Table 1 gives an overview of experiments done at the Focal Plane Polarimeter (FPP) in Jefferson Lab Hall A[3]. Experiments have also been done at Bates[4] and MAMI[5].

Table 1. Proton recoil polarization experiments on the nucleon and on few and many body nuclei at Jefferson Lab Hall A. Reference to the experiments is given in the right column.

	Nucleon	
$p(\vec{e}, e'\vec{p})$	Proton electromagnetic form factors (G_E/G_M)	6
$p(\vec{e}, e'\vec{p})\pi^0$	Nucleon structure in N to Δ transition	7
$p(\vec{\gamma}, \vec{p})\pi^0$	Test of asymptotic scaling	8
	Few and many body nuclear systems	
$d(\vec{\gamma}, \vec{p})n$	Test of asymptotic scaling	9
$d(\vec{e}, e'\vec{p})$	Nuclear Response	10
$^4\text{He}(\vec{e}, e'\vec{p})$	Nuclear Response, medium effects	11
$^{16}\text{O}(\vec{e}, e'\vec{p})$	Nuclear Response, medium effects	12

2 Polarimetry

Proton polarimeters take advantage of the asymmetry in the scattering of polarized protons off spin-zero nuclei due to the spin-orbit ($\vec{L} \cdot \vec{S}$) dependent part of the nuclear force. At intermediate energies (~ 100 MeV to a few GeV) the polarization analyzer has been almost exclusively carbon. The scattering process is illustrated in Fig. 1. Panel (a) shows two trajectories of protons with equal momenta $\vec{p}_1$ and $\vec{p}_2$ assuming an on average attractive nucleon-nucleus potential. With the same transverse proton spin orientation, $\vec{S}$ pointing 'up'

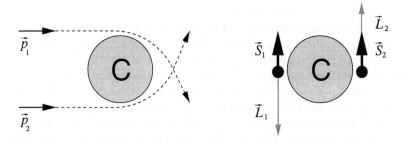

Figure 1. Scattering of polarized protons off a spin zero carbon nucleus. Trajectories in the scattering plane (a). Proton spin and orbital angular momentum (b).

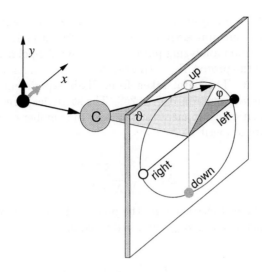

Figure 2. Schematic polarimeter with a polarized proton scattering off a carbon nucleus leading to an angular distribution which reveals the transverse polarization components of the proton.

in panel (b), the scalar product $\vec{L} \cdot \vec{S}$ is of opposite sign for both protons because of the different sign in the orbital angular momentum ($\vec{L} = \vec{r} \times \vec{p}$). This gives rise to a more attractive potential for one proton and to a less attractive potential for the other proton and finally to an asymmetry in the number of particles scattered to the left, N_L, and to the right, N_R. This left-right asymmetry ϵ is proportional to the polarization of the proton P and to the carbon analyzing power A_C which is an important quantity characterizing a polarimeter.

$$\epsilon = \frac{N_L - N_R}{N_L + N_R} = A_C P \tag{1}$$

As shown in more general in Fig. 2 the polarization components of the proton transverse to the proton's momentum (P_x and P_y) result in a specific angular distribution[13]

$$\sigma(\vartheta, \varphi) = \sigma_0(\vartheta) \cdot [1 + A_C(\vartheta, T_{CC})(P_y \cos \varphi - P_x \sin \varphi)] \tag{2}$$

of the proton scattering cross section σ off the carbon nucleus with polar angle ϑ and azimuthal angle φ. In the orientation of Fig. 2 the P_x polarization component leads to an up-down asymmetry and P_y to a left-right asymmetry. False or instrumental asymmetries are neglected in Eq. (2) and in the following. The carbon analyzing power depends on the polar angle ϑ and the kinetic energy of the proton T_{CC}. An example of such an angular distribution is shown in Fig. 3. The data, taken from JLab experiment 93-049[11], were integrated over a ϑ range from $10°$ to $26°$. The statistical precision ΔP of the extracted polarization components depends on the number of incident protons N_0 and the figure of merit F,[14]

$$\Delta P = \sqrt{\frac{2}{N_0 F^2}}. \tag{3}$$

The figure of merit characterizes the polarimeter performance in the scattering angle domain $(\vartheta_{min}, \vartheta_{max})$ and is defined as

$$F^2 = \int_{\vartheta_{min}}^{\vartheta_{max}} \eta(\vartheta) A_C^2(\vartheta) d\vartheta, \tag{4}$$

where the polarimeter efficiency is the fraction of scatterers $\eta(\vartheta) = N(\vartheta)/N_0$ with polar angles ϑ. Small angle scattering with ϑ of the order of $3°$ or less is dominated by multiple coulomb scattering and this reaction has no analyzing power. The figure of merit of a polarimeter can be maximized

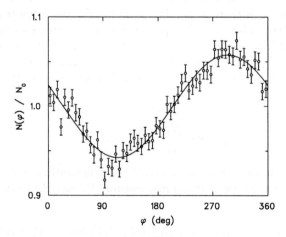

Figure 3. Example of an angular distribution of polarized protons scattered off a carbon analyzer. Data are taken from $^1H(\vec{e}, e'\vec{p})$. The solid line is a fit of Eq. (2) to the data.

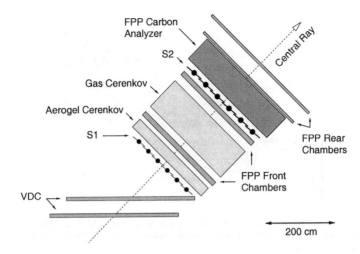

FPP Carbon
Analyzer

Central Ray

S2

Gas Cerenkov

Aerogel Cerenkov

S1

FPP Rear
Chambers

FPP Front
Chambers

VDC

200 cm

Figure 4. Detector package for one of the High Resolution Spectrometers in Jefferson Lab Hall A with vertical drift chambers (VDC), scintillators (S1 and S2), Čerenkov detectors and the FPP (including front and rear chambers and carbon analyzer). The scale is approximate.

by choosing an optimal thickness for the analyzer. An increase in carbon thickness increases the probability of useful pC interactions. At the same time, however, there is also an increase in absorptive interactions which can reduce the efficiency and an increase in multiple Coulomb scattering which reduces the average analyzing power at small angles and the tracking resolution. An increased analyzer thickness also reduces the average kinetic energy of the protons. Since the carbon analyzing power peaks at about $T_{CC} = 200$ MeV this energy reduction increases the average analyzing power for energies larger than 200 MeV.

Examples of parameterizations of carbon analyzing powers and figure of merits in different regimes of kinetic energy and carbon thicknesses are those given in Ref. [13,14,15].

3   Focal Plane Polarimeter

Focal plane polarimeters are typically setup as part of the detector system of spectrometers. Figure 4 shows the detector package for one of the High Resolution Spectrometers in Jefferson Lab Hall A together with the polarimeter[3]. The polarimeter consists of two front and two rear straw chambers to de-

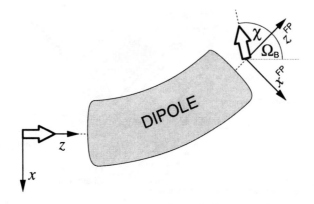

Figure 5. Schematic view of spin precession in the magnetic field of a dipole. The proton spin is indicated an open arrow. The precession angle χ is the difference between the spin and momentum rotation.

termine the incident and scattered proton track. In between is a segmented carbon analyzer with a maximum of thickness of 51 cm. Similar devices were built at Bates[4] and MAMI[5].

Since the FPP is in the focal plane of the spectrometer, one must connect the polarization measured in the focal plane (FP) to the polarizations of interest at the target. This does not only include a change in coordinate system, but also the precession of the proton spin in the magnetic fields of the spectrometer. The relationship is given by a rotation. In case of a pure dipole field the rotation matrix has the simple form

$$
\begin{pmatrix} P_x^{FP} \\ P_y^{FP} \\ P_z^{FP} \end{pmatrix} = \underbrace{\begin{pmatrix} \cos\chi & 0 & -\sin\chi \\ 0 & 1 & 0 \\ \sin\chi & 0 & \cos\chi \end{pmatrix}}_{S_{ij}} \cdot \begin{pmatrix} P_x + hP_x' \\ P_y + hP_y' \\ P_z + hP_z' \end{pmatrix} \tag{5}
$$

where the net polarization of the recoiling proton is the sum $\vec{P} + h\vec{P}'$, whith the induced polarization $\vec{P}$, the polarization transfer coefficient $\vec{P}'$ and the longitudinal beam polarization h. The spin precession angle χ is defined in Fig. 5. It is proportional to the bending angle Ω_B of the trajectory and the energy of the proton E_p (see Ref. [16])

$$
\chi = (\mu_p - 1)\frac{E_p}{m_p}\Omega_B. \tag{6}
$$

Here, m_p and μ_p are the proton mass and magnetic moment in units of the nuclear magneton. For the general case the spin transport coefficients S_{ij} can be calculated with a computer code, like the differential-algebra-based code COSY[17], together with a model of the spectrometer magnetic fields.

4 Extraction of Polarization Observables

The induced and transferred polarization components are the quantities to be determined. This can be done by means of the maximum likelihood technique (see e.g. Ref. [18]). The likelihood function L is defined as a product of the probability densities for each of the N events

$$L = \prod_{i=1}^{N} f(\varphi_i; \vec{P}, \vec{P}'). \tag{7}$$

Only those events are included in the product where the full cone of $\varphi_i \in [0°, 360°)$ would have been observed with the same probability (cone test). As mentioned earlier we also neglect possible instrumental asymmetries. The probability density, as expressed by the asymmetries ϵ_x and ϵ_y, is then given by

$$f(\varphi) = \frac{1}{2\pi} \left(1 + \epsilon_x \sin\varphi + \epsilon_y \cos\varphi \right) \tag{8}$$

The asymmetries are linear functions of, in the most general case, the induced and transferred polarization components.

$$\epsilon_x = -A_C \cdot \left[S_{xx}(P_x + hP'_x) + S_{xy}(P_y + hP'_y) + S_{xz}(P_z + hP'_z) \right] \tag{9}$$

$$\epsilon_y = A_C \cdot \left[S_{yx}(P_x + hP'_x) + S_{yy}(P_y + hP'_y) + S_{yz}(P_z + hP'_z) \right] \tag{10}$$

The analyzing power A_C and the spin transport coefficients S_{ij} which take into account the change of coordinate system and the proton spin precession in the spectrometer's magnetic fields are known or calculated on a event by event basis. The electron beam polarization h is either positive or negative depending on the helicity state of the beam at the time the particular event was taken. The likelihood function can be rewritten as

$$L = \prod_{i=1}^{N} \frac{1}{2\pi} \left(1 + \sum_k \lambda_{k,i} P_k + \sum_k \lambda'_{k,i} P'_k \right) \tag{11}$$

where $k = x, y, z$ and the coefficients λ are given for each event individually by

$$\begin{aligned}
\lambda_x &= A_C(S_{yx}\cos\varphi - S_{xx}\sin\varphi), & \lambda'_x &= h\lambda_x, \\
\lambda_y &= A_C(S_{yy}\cos\varphi - S_{xy}\sin\varphi), & \lambda'_y &= h\lambda_y, \\
\lambda_z &= A_C(S_{yz}\cos\varphi - S_{xz}\sin\varphi), & \lambda'_z &= h\lambda_z.
\end{aligned} \tag{12}$$

The maximum likelihood estimations $\hat{P}_k$, $\hat{P}'_k$ are those which maximize the likelihood function L and are given by the solutions of the coupled nonlinear equations $\partial \ln L/\partial P_k = 0$ and $\partial \ln L/\partial P'_k = 0$. These solutions are not easily obtained. A linearization of these, whose justification I can not discuss here, simplifies the problem to the linear equation system

$$\underbrace{\begin{pmatrix} \sum \lambda_x \\ \sum \lambda_y \\ \vdots \\ \sum \lambda'_z \end{pmatrix}}_{B} = \underbrace{\begin{pmatrix} \sum \lambda_x\lambda_x & \sum \lambda_x\lambda_y & \cdots & \sum \lambda_x\lambda'_z \\ \sum \lambda_y\lambda_x & \sum \lambda_y\lambda_y & \cdots & \sum \lambda_y\lambda'_z \\ \vdots & \vdots & \ddots & \vdots \\ \sum \lambda'_z\lambda_x & \sum \lambda'_z\lambda_y & \cdots & \sum \lambda'_z\lambda'_z \end{pmatrix}}_{F} \begin{pmatrix} \hat{P}_x \\ \hat{P}_y \\ \vdots \\ \hat{P}'_z \end{pmatrix}. \tag{13}$$

The sums are carried out over all events. The polarization estimates are thus given by

$$(\hat{P}_x, \hat{P}_y, \ldots, \hat{P}'_z)^T = F^{-1}B \tag{14}$$

with covariance matrix F^{-1}.

Acknowledgements

This work was supported by U.S. National Science Foundation grant PHY 9803860. Southeastern Universities Research Association (SURA) manages the Thomas Jefferson National Accelerator under DOE contract DE-AC05-84ER40150.

References

1. V.P. Barannik et al., Nucl. Phys. **A451**, 751 (1986).
2. A. Picklesimer, J.W. Van Orden, Phys. Rev. **C 35**, 266 (1987).
3. Jefferson Lab HRS Focal Plane Polarimeter Page, http://hallaweb.jlab.org/equipment/dctectors/fpp.html.
4. B.D. Milbrath et al., Phys. Rev. Lett. **80**, 452 (1998).
5. Th. Pospischil, doctoral thesis, Mainz (2000) and Th. Pospischil et al., submitted to Nucl. Instr. Methods **A**.

6. M.K. Jones *et al.*, Phys. Rev. Lett. **84**, 1398 (2000).

7. S. Frullani, J.J. Kelly and A. Sarty (spokespeople), Jefferson Lab Experiment 91-011.

8. R. Ent and P. Ulmer (spokespeople), Jefferson Lab Experiment 93-049.

9. K. Wijesooriya *et al.*, Phys. Rev. Lett. **86**, 2975 (2001).

10. J. Finn and P. Ulmer (spokespeople), Jefferson Lab Experiment 89-019.

11. R. Ent and P. Ulmer (spokespeople), Jefferson Lab Experiment 93-049.

12. S. Malov *et al.*, Phys. Rev. C **62**, 057302 (2000).

13. E. Aprile-Giboni *et al.*, Nucl. Instr. Meth. **215**, 147 (1983).

14. B. Bonin *et al.*, Nucl. Instr. Meth. **A288**, 379 (1990).

15. R.D. Ransome *et al.*, Nucl. Instr. Meth. **201**, 315 (1982).

16. V. Bargmann, L. Michel and V.L. Telegdi, Phys. Rev. Lett. **2**, 435 (1959).

17. M. Bertz, *COSY Infinity User's Manual*, MSU, unpublished, 1996.

18. D. Besset *et al.*, Nucl. Instr. Meth. **166**, 515 (1979).

NEUTRINO MASS
MSW OSCILLATIONS
&
SEESAW MECHANISM

HASAN YÜKSEL

University of Wisconsin, Department of Physics,
1150 University Avenue
Madison WI 53706, USA
E-mail:yuksel@nucth.physics.wisc.edu

In the Hugs 2002 student seminar sessions, I treat a number of topical issues in neutrino physics: the phenomenology of the MSW mechanism; a brief discussion of global analysis of atmospheric and solar neutrinos; Dirac and Majorana neutrino masses and the Seesaw mechanism as an explanation for the smallness of the neutrino mass. MSW model assumes that neutrinos are created in flavor eigenstates without any definite mass but they evolve in mass eigenstates with definite mass. Latest experiments confirms the neutrino oscillations. Considering MSW oscillations in calculations provides an acceptable solution for measured neutrino flux deficiency in solar and atmospheric neutrinos. Neutrinos do not have mass in the standard model. Experimentally, it is known that neutrino mass is exceedingly small, if not zero. The seesaw mechanism predicts left handed light neutrinos with right handed heavy partners.

CONSTITUENT COUNTING RULE

LINGYAN ZHU

Lab. for Nuclear Science and Dept. of Physics, MIT
77 Massachusetts Avenue, Cambridge MA 02139, USA
E-mail: lyzhu@mit.edu

The constituent counting rule was showed from different exclusive experiments, including the photodisintegration and photoproduction experiments in Jlab. Some theories, including simple dimensional counting, Perturbative Chromodynamics(pQCD), Quark-Gluon Strings model(QGS) and Skewed Parton Distribution(SPD), were presented towards this behavior. But the early onset and oscillation around the scaling, as well as the breaking of Hadron Helicity Conservation(HHC), need further developments.